INEQUALITIES: A JOURNEY INTO LINEAR ANALYSIS

Contains a wealth of inequalities used in linear analysis, and explains in detail how they are used. The book begins with Cauchy's inequality and ends with Grothendieck's inequality, in between one finds the Loomis–Whitney inequality, maximal inequalities, inequalities of Hardy and of Hilbert, hypercontractive and logarithmic Sobolev inequalities, Beckner's inequality, and many, many more. The inequalities are used to obtain properties of function spaces, linear operators between them, and of special classes of operators such as absolutely summing operators.

This textbook complements and fills out standard treatments, providing many diverse applications: for example, the Lebesgue decomposition theorem and the Lebesgue density theorem, the Hilbert transform and other singular integral operators, the martingale convergence theorem, eigenvalue distributions, Lidskii's trace formula, Mercer's theorem and Littlewood's 4/3 theorem.

It will broaden the knowledge of postgraduate and research students, and should also appeal to their teachers, and all who work in linear analysis.

D. J. H. Garling is Emeritus Reader in Mathematical Analysis at the University of Cambridge and a Fellow of St John's College, Cambridge.

INEQUALITIES: A JOURNEY INTO LINEAR ANALYSIS

D. J. H. GARLING

CAMBRIDGE
UNIVERSITY PRESS

CAMBRIDGE
UNIVERSITY PRESS

University Printing House, Cambridge CB2 8BS, United Kingdom

Published in the United States of America by Cambridge University Press, New York

Cambridge University Press is part of the University of Cambridge.

It furthers the University's mission by disseminating knowledge in the pursuit of education, learning and research at the highest international levels of excellence.

www.cambridge.org
Information on this title: www.cambridge.org/9780521699730

First published 2007

A catalogue record for this publication is available from the British Library

Library of Congress Cataloguing in Publication data

Garling, D. J. H.
    Inequalities: a journey into linear analysis / D. J. H. Garling.
        p. cm.
    Includes bibliographical references and index.
    ISBN 978-0-521-69973-0 (pbk.: alk. paper) – ISBN 978-0-521-87624-7 (hardback: alk. paper) 1. Inequalities (Mathematics) 2. Functional analysis. I. Title.
    QA295.G319 2007
    515'.26–dc22

2007015539

ISBN 978-0-521-87624-7 Hardback
ISBN 978-0-521-69973-0 Paperback

Contents

Introduction

Inequalities lie at the heart of a great deal of mathematics. G.H. Hardy reported Harald Bohr as saying 'all analysts spend half their time hunting through the literature for inequalities which they want to use but cannot prove'. Inequalities provide control, to enable results to be proved. They also impose constraints; for example, Gromov's theorem on the symplectic embedding of a sphere in a cylinder establishes an inequality that says that the radius of the cylinder cannot be too small. Similar inequalities occur elsewhere, for example in theoretical physics, where the uncertainty principle (which is an inequality) and Bell's inequality impose constraints, and, more classically, in thermodynamics, where the second law provides a fundamental inequality concerning entropy.

Thus there are very many important inequalities. This book is not intended to be a compendium of these; instead, it provides an introduction to a selection of inequalities, not including any of those mentioned above. The inequalities that we consider have a common theme; they relate to problems in real analysis, and more particularly to problems in linear analysis. Incidentally, they include many of the inequalities considered in the fascinating and ground-breaking book *Inequalities*, by Hardy, Littlewood and Pólya [HaLP 52], originally published in 1934.

The first intention of this book, then, is to establish fundamental inequalities in this area. But more importantly, its purpose is to put them in context, and to show how useful they are. Although the book is very largely self-contained, it should therefore principally be of interest to analysts, and to those who use analysis seriously.

The book requires little background knowledge, but some such knowledge is very desirable. For a great many inequalities, we begin by considering sums of a finite number of terms, and the arguments that are used here lie at the heart of the matter. But to be of real use, the results must be extended to infinite sequences and infinite sums, and also to functions and integrals. In order to be really useful, we need a theory of measure and integration which includes suitable limit theorems. In a preliminary chapter, we give a brief account of what we need to know; the details will not be needed, at least in the early chapters, but a familiarity with the ideas and results of the theory is a great advantage.

Secondly, it turns out that the sequences and functions that we consider are members of an appropriate vector space, and that their 'size', which is involved in the inequalities that we prove, is described by a norm. We establish basic properties of normed spaces in Chapter 4. Normed spaces are the subject of linear analysis, and, although our account is largely self-contained, it is undoubtedly helpful to have some familiarity with the ideas and results of this subject (such as are developed in books such as *Linear analysis* by Béla Bollobás [Bol 90] or *Introduction to functional analysis* by Taylor and Lay [TaL 80]. In many ways, this book provides a parallel text in linear analysis.

Looked at from this point of view, the book falls naturally into two unequal parts. In Chapters 2 to 13, the main concern is to establish inequalities between sequences and functions lying in appropriate normed spaces. The inequalities frequently reveal themselves in terms of the continuity of certain linear operators, or the size of certain sublinear operators. In linear analysis, however, there is interest in the structure and properties of linear operators themselves, and in particular in their spectral properties, and in the last four chapters we establish some fundamental inequalities for linear operators.

This book journeys into the foothills of linear analysis, and provides a view of high peaks ahead. Important fundamental results are established, but I hope that the reader will find him- or herself hungry for more. There are brief Notes and Remarks at the end of each chapter, which include suggestions for further reading: a partial list, consisting of books and papers that I have enjoyed reading. A more comprehensive guide is given in the monumental *Handbook of the geometry of Banach spaces* [JoL 01,03] which gives an impressive overview of much of modern linear analysis.

The Notes and Remarks also contain a collection of exercises, of a varied nature: some are five-finger exercises, but some establish results that are needed later. Do them!

Linear analysis lies at the heart of many areas of mathematics, including for example partial differential equations, harmonic analysis, complex analysis and probability theory. Each of them is touched on, but only to a small extent; for example, in Chapter 9 we use results from complex analysis to prove the Riesz-Thorin interpolation theorem, but otherwise we seldom use the powerful tools of complex analysis. Each of these areas has its own collection of important and fascinating inequalities, but in each case it would be too big a task to do them justice here.

I have worked hard to remove errors, but undoubtedly some remain. Corrections and further comments can be found on a web-page on my personal home page at `www.dpmms.cam.ac.uk`

1
Measure and integral

1.1 Measure

Many of the inequalities that we shall establish originally concern finite sequences and finite sums. We then extend them to infinite sequences and infinite sums, and to functions and integrals, and it is these more general results that are useful in applications.

Although the applications can be useful in simple settings – concerning the Riemann integral of a continuous function, for example – the extensions are usually made by a limiting process. For this reason we need to work in the more general setting of measure theory, where appropriate limit theorems hold. We give a brief account of what we need to know; the details of the theory will not be needed, although it is hoped that the results that we eventually establish will encourage the reader to master them. If you are not familiar with measure theory, read through this chapter quickly, and then come back to it when you find that the need arises.

Suppose that Ω is a set. A *measure* ascribes a size to some of the subsets of Ω. It turns out that we usually cannot do this in a sensible way for all the subsets of Ω, and have to restrict attention to the *measurable* subsets of Ω. These are the 'good' subsets of Ω, and include all the sets that we meet in practice. The collection of measurable sets has a rich enough structure that we can carry out countable limiting operations.

A *σ-field* Σ is a collection of subsets of a set Ω which satisfies

(i) if (A_i) is a sequence in Σ then $\cup_{i=1}^{\infty} A_i \in \Sigma$, and

(ii) if $A \in \Sigma$ then the complement $\Omega \setminus A \in \Sigma$.

Thus

(iii) if (A_i) is a sequence in Σ then $\cap_{i=1}^{\infty} A_i \in \Sigma$.

The sets in Σ are called Σ-*measurable* sets; if it is clear what Σ is, they are simply called the *measurable sets*.

Here are two constructions that we shall need, which illustrate how the conditions are used. If (A_i) is a sequence in Σ then we define the *upper limit* $\overline{\lim} A_i$ and the *lower limit* $\underline{\lim} A_i$:

$$\overline{\lim} A_i = \cap_{i=1}^{\infty} \left(\cup_{j=i}^{\infty} A_j \right) \quad \text{and} \quad \underline{\lim} A_i = \cup_{i=1}^{\infty} \left(\cap_{j=i}^{\infty} A_j \right).$$

Then $\overline{\lim} A_i$ and $\underline{\lim} A_i$ are in Σ. You should verify that $x \in \overline{\lim} A_i$ if and only if $x \in A_i$ for infinitely many indices i, and that $x \in \underline{\lim} A_i$ if and only if there exists an index i_0 such that $x \in A_i$ for all $i \geq i_0$.

If Ω is the set $\mathbf{N}$ of natural numbers, or the set $\mathbf{Z}$ of integers, or indeed any countable set, then we take Σ to be the collection $P(\Omega)$ of all subsets of Ω. Otherwise, Σ will be a proper subset of $P(\Omega)$. For example, if $\Omega = \mathbf{R}^d$ (where $\mathbf{R}$ denotes the set of real numbers), we consider the collection of *Borel sets*; the sets in the smallest σ-field that contains all the open sets. This includes all the sets that we meet in practice, such as the closed sets, the G_δ sets (countable intersections of open sets), the F_σ sets (countable unions of closed sets), and so on. The Borel σ-field has the fundamental disadvantage that we cannot give a straightforward definition of what a Borel set looks like – this has the consequence that proofs must be indirect, and this gives measure theory its own particular flavour.

Similarly, if (X, d) is a metric space, then the Borel sets of X are sets in the smallest σ-field that contains all the open sets. [Complications can arise unless (X, d) is separable (that is, there is a countable set which is dense in X), and so we shall generally restrict attention to separable metric spaces.]

We now give a size (non-negative, but possibly infinite or zero) to each of the sets in Σ. A *measure* on a σ-field Σ is a mapping μ from Σ into $[0, \infty]$ satisfying

(i) $\mu(\emptyset) = 0$, and

(ii) if (A_i) is a sequence of disjoint sets in Σ then $\mu(\cup_{i=1}^{\infty} A_i) = \sum_{i=1}^{\infty} \mu(A_i)$: μ is *countably additive*.

The most important example that we shall consider is the following. There exists a measure λ *(Borel measure)* on the Borel sets of $\mathbf{R}^d$ with the property that if A is the rectangular parallelopiped $\prod_{i=1}^{d}(a_i, b_i)$ then $\lambda(A)$ is the product $\prod_{i=1}^{d}(b_i - a_i)$ of the length of its sides; thus λ gives familiar geometric objects their natural measure. As a second example, if Ω is a countable set, we can define $\#(A)$, or $|A|$, to be the number of points, finite or infinite, in A; $\#$ is *counting measure*. These two examples are radically different: for counting measure, the one-point sets $\{x\}$ are *atoms*; each has positive measure, and any subset of it has either the same measure or zero measure. Borel measure on $\mathbf{R}^d$ is *atom-free*; no subset is an atom. This is equivalent to requiring that if A is a set of non-zero measure A, and if $0 < \beta < \mu(A)$ then there is a measurable subset B of A with $\mu(B) = \beta$.

Countable additivity implies the following important continuity properties:

(iii) if (A_i) is an increasing sequence in Σ then

$$\mu(\cup_{i=1}^\infty A_i) = \lim_{i\to\infty} \mu(A_i).$$

[Here and elsewhere, we use 'increasing' in the weak sense: if $i < j$ then $A_i \subseteq A_j$. If $A_i \subset A_j$ for $i < j$, then we say that (A_i) is 'strictly increasing'. Similarly for 'decreasing'.]

(iv) if (A_i) is a decreasing sequence in Σ and $\mu(A_1) < \infty$ then

$$\mu(\cap_{i=1}^\infty A_i) = \lim_{i\to\infty} \mu(A_i).$$

The finiteness condition here is necessary and important; for example, if $A_i = [i, \infty) \subseteq \mathbf{R}$, then $\lambda(A_i) = \infty$ for all i, but $\cap_{i=1}^\infty A_i = \emptyset$, so that $\lambda(\cap_{i=1}^\infty A_i) = 0$.

We also have the following consequences:

(v) if $A \subseteq B$ then $\mu(A) \leq \mu(B)$;

(iv) if (A_i) is any sequence in Σ then $\mu(\cup_{i=1}^\infty A_i) \leq \sum_{i=1}^\infty \mu(A_i)$.

There are many circumstances where $\mu(\Omega) < \infty$, so that μ only takes finite values, and many where $\mu(\Omega) = 1$. In this latter case, we can consider μ as a probability, and frequently denote it by $\mathbf{P}$. We then use probabilistic language, and call the elements of Σ 'events'.

A *measure space* is then a triple (Ω, Σ, μ), where Ω is a set, Σ is a σ-field of subsets of Ω (the *measurable sets*) and μ is a measure defined on Σ. In order to avoid tedious complications, we shall restrict our attention to σ-*finite* measure spaces: we shall suppose that there is an increasing sequence (C_k) of measurable sets of finite measure whose union is Ω. For example, if λ is Borel measure then we can take $C_k = \{x : |x| \leq k\}$.

Here is a useful result, which we shall need from time to time.

Proposition 1.1.1 (The first Borel–Cantelli lemma) *If (A_i) is a sequence of measurable sets and $\sum_{i=1}^\infty \mu(A_i) < \infty$ then $\mu(\overline{\lim} A_i) = 0$.*

Proof For each i, $\mu(\overline{\lim} A_i) \leq \mu(\cup_{j=i}^\infty A_j)$, and $\mu(\cup_{j=i}^\infty A_j) \leq \sum_{j=i}^\infty \mu(A_j) \to 0$ as $i \to \infty$. □

If $\mu(A) = 0$, A is called a *null set*. We shall frequently consider properties which hold except on a null set: if so, we say that the property holds *almost everywhere*, or, in a probabilistic setting, *almost surely*.

1.2 Measurable functions

We next consider functions defined on a measure space (Ω, Σ, μ). A real-valued function f is *Σ-measurable*, or more simply *measurable*, if for each real α the set $(f > \alpha) = \{x \colon f(x) > \alpha\}$ is in Σ. A complex-valued function is *measurable* if its real and imaginary parts are. (When $\mathbf{P}$ is a probability measure and we are thinking probabilistically, a measurable function is called a *random variable*.) In either case, this is equivalent to the set $(f \in U) = \{x \colon f(x) \in U\}$ being in Σ for each open set U. Thus if Σ is the Borel σ-field of a metric space, then the continuous functions are measurable. If f and g are measurable then so are $f + g$ and fg; the measurable functions form an algebra $\mathcal{M} = \mathcal{M}(\Omega, \Sigma, \mu)$. If f is measurable then so is $|f|$. Thus in the real case $\mathcal{M}$ is a lattice: if f and g are measurable, then so are $f \vee g = \max(f, g)$ and $f \wedge g = \min(f, g)$.

We can also consider the Borel σ-field of a compact Hausdorff space (X, τ): but it is frequently more convenient to work with the *Baire σ-field*: this is the smallest σ-field containing the closed G_δ sets, and is the smallest σ-field for which all the continuous real-valued functions are measurable. When (X, τ) is metrizable, the Borel σ-field and the Baire σ-field are the same.

A measurable function f is a *null function* if $\mu(f \neq 0) = 0$. The set $\mathcal{N}$ of null functions is an ideal in $\mathcal{M}$. In practice, we identify functions which are equal almost everywhere: that is, we consider elements of the quotient space $M = \mathcal{M}/\mathcal{N}$. Although these elements are equivalence classes of functions, we shall tacitly work with representatives, and treat the elements of M as if they were functions.

What about the convergence of measurable functions? A fundamental problem that we shall frequently consider is 'When does a sequence of measurable functions converge almost everywhere?' The first Borel–Cantelli lemma provides us with the following useful criterion.

Proposition 1.2.1 *Suppose that (f_n) is a decreasing sequence of non-negative measurable functions. Then $f_n \to 0$ almost everywhere if and only if $\mu((f_n > \epsilon) \cap C_k) \to 0$ as $n \to \infty$ for each k and each $\epsilon > 0$.*

Proof Suppose that (f_n) converges almost everywhere, and that $\epsilon > 0$. Then $((f_n > \epsilon) \cap C_k)$ is a decreasing sequence of sets of finite measure, and if $x \in \cap_n (f_n > \epsilon) \cap C_k$ then $(f_n(x))$ does not converge to 0. Thus, by condition (iv) above, $\mu((f_n > \epsilon) \cap C_k) \to 0$ as $n \to \infty$.

For the converse, we use the first Borel–Cantelli lemma. Suppose that the condition is satisfied. For each n there exists N_n such that $\mu((f_{N_n} > 1/n) \cap C_n) < 1/2^n$. Then since $\sum_{n=1}^{\infty} \mu((f_{N_n} > 1/n) \cap C_n) < \infty$,

$\mu(\overline{\lim}((f_{N_n} > 1/n) \cap C_n)) = 0$. But if $x \notin \overline{\lim}((f_{N_n} > 1/n) \cap C_n)$ then $f_n \to 0$.   □

Corollary 1.2.1 *A sequence (f_n) of measurable functions converges almost everywhere if and only if*

$$\mu\left((\sup_{m,n \geq N} |f_m - f_n| > \epsilon) \cap C_k \right) \to 0 \quad as \ N \to \infty$$

for each k and each $\epsilon > 0$.

It is a straightforward but worthwhile exercise to show that if $f(x) = \lim_{n \to \infty} f_n(x)$ when the limit exists, and $f(x) = 0$ otherwise, then f is measurable.

Convergence almost everywhere cannot in general be characterized in terms of a topology. There is however a closely related form of convergence which can. We say that $f_n \to f$ *locally in measure* (or *in probability*) if $\mu((|f_n - f| > \epsilon) \cap C_k) \to 0$ as $n \to \infty$ for each k and each $\epsilon > 0$; similarly we say that (f_n) is locally *Cauchy in measure* if $\mu((|f_m - f_n| > \epsilon) \cap C_k) \to 0$ as $m, n \to \infty$ for each k and each $\epsilon > 0$. The preceding proposition, and another use of the first Borel–Cantelli lemma, establish the following relations between these ideas.

Proposition 1.2.2 *(i) If (f_n) converges almost everywhere to f, then (f_n) converges locally in measure.*

(ii) If (f_n) is locally Cauchy in measure then there is a subsequence which converges almost everywhere to a measurable function f, and $f_n \to f$ locally in measure.

Proof (i) This follows directly from Corollary 1.2.1.

(ii) For each k there exists N_k such that $\mu((|f_m - f_n| > 1/2^k) \cap C_k) < 1/2^k$ for $m, n > N_k$. We can suppose that the sequence (N_k) is strictly increasing. Let $g_k = f_{N_k}$. Then $\mu((|g_{k+1} - g_k| < 1/2^k) \cap C_k) < 1/2^k$. Thus, by the First Borel–Cantelli Lemma, $\mu(\overline{\lim}((|g_{k+1} - g_k| > 1/2^k) \cap C_k)) = 0$. But $\overline{\lim}(|g_{k+1} - g_k| > 1/2^k) \cap C_k) = \overline{\lim}(|g_{k+1} - g_k| > 1/2^k)$. If $x \notin \overline{\lim}(|g_{k+1} - g_k| > 1/2^k)$ then $\sum_{k=1}^{\infty} |g_{k+1}(x) - g_k(x)| < \infty$, so that $(g_k(x))$ is a Cauchy sequence, and is therefore convergent.

Let $f(x) = \lim g_k(x)$, when this exists, and let $f(x) = 0$ otherwise. Then (g_k) converges to f almost everywhere, and locally in measure. Since $(|f_n - f| > \epsilon) \subseteq (|f_n - g_k| > \epsilon/2) \cup (|g_k - f| > \epsilon/2)$, it follows easily that $f_n \to f$ locally in measure.   □

In fact, there is a complete metric on M under which the Cauchy sequences are the sequences which are locally Cauchy in measure, and the convergent sequences are the sequences which are locally convergent in measure. This completeness result is at the heart of very many completeness results for spaces of functions.

If A is a measurable set, its *indicator function* I_A, defined by setting $I_A(x) = 1$ if $x \in A$ and $I_A(x) = 0$ otherwise, is measurable. A *simple function* is a measurable function which takes only finitely many values, and which vanishes outside a set of finite measure: it can be written as $\sum_{i=1}^{n} \alpha_i I_{A_i}$, where $A_1, \ldots, A_n$ are measurable sets of finite measure (which we may suppose to be disjoint).

Proposition 1.2.3 *A non-negative measurable function f is the pointwise limit of an increasing sequence of simple functions.*

Proof Let $A_{j,n} = (f > j/2^n)$, and let $f_n = \frac{1}{2^n} \sum_{j=1}^{4^n} I_{A_{j,n} \cap C_n}$. Then (f_n) is an increasing sequence of simple functions, which converges pointwise to f. □

This result is extremely important; we shall frequently establish inequalities for simple functions, using arguments that only involve finite sums, and then extend them to a larger class of functions by a suitable limiting argument. This is the case when we consider integration, to which we now turn.

1.3 Integration

Suppose first that $f = \sum_{i=1}^{n} \alpha_i I_{A_i}$ is a non-negative simple function. It is then natural to define the integral as $\sum_{i=1}^{n} \alpha_i \mu(A_i)$. It is easy but tedious to check that this is independent of the representation of f. Next suppose that f is a non-negative measurable function. We then define

$$\int_\Omega f \, d\mu = \sup\{\int g \, d\mu : g \text{ simple}, 0 \le g \le f\}.$$

A word about notation: we write $\int_\Omega f \, d\mu$ or $\int f \, d\mu$ for brevity, and $\int_\Omega f(x) \, d\mu(x)$ if we want to bring attention to the variable (for example, when f is a function of more than one variable). When integrating with respect to Borel measure on $\mathbf{R}^d$, we shall frequently write $\int_{\mathbf{R}^d} f(x) \, dx$, and use familiar conventions such as $\int_a^b f(x) \, dx$. When $\mathbf{P}$ is a probability measure, we write $\mathbf{E}(f)$ for $\int f \, d\mathbf{P}$, and call $\mathbf{E}(f)$ the *expectation* of f.

We now have the following fundamental continuity result:

Proposition 1.3.1 (The monotone convergence theorem) *If (f_n) is an increasing sequence of non-negative measurable functions which converges pointwise to f, then $(\int f_n \, d\mu)$ is an increasing sequence and $\int f \, d\mu = \lim_{n \to \infty} \int f_n \, d\mu$.*

Corollary 1.3.1 (Fatou's lemma) *If (f_n) is a sequence of non-negative measurable functions then $\int (\liminf f_n) \, d\mu \leq \liminf \int f_n \, d\mu$. In particular, if f_n converges almost everywhere to f then $\int f \, d\mu \leq \liminf \int f_n \, d\mu$.*

We now turn to functions which are not necessarily non-negative. A measurable function f is *integrable* if $\int f^+ \, d\mu < \infty$ and $\int f^- \, d\mu < \infty$, and in this case we set $\int f \, d\mu = \int f^+ \, d\mu - \int f^- \, d\mu$. Clearly f is integrable if and only if $\int |f| \, d\mu < \infty$, and then $|\int f \, d\mu| \leq \int |f| \, d\mu$. Thus the integral is an absolute integral; fortuitous cancellation is not allowed, so that for example the function $\sin x / x$ is not integrable on $\mathbf{R}$. Incidentally, integration with respect to Borel measure extends proper Riemann integration: if f is Riemann integrable on $[a, b]$ then f is equal almost everywhere to a Borel measurable and integrable function, and the Riemann integral and the Borel integral are equal.

The next result is very important.

Proposition 1.3.2 (The dominated convergence theorem) *If (f_n) is a sequence of measurable functions which converges pointwise to f, and if there is a measurable non-negative function g with $\int g \, d\mu$ such that $|f_n| \leq g$ for all n, then $\int f_n \, d\mu \to \int f \, d\mu$ as $n \to \infty$.*

This is a precursor of results which will come later; provided we have some control (in this case provided by the function g) then we have a good convergence result. Compare this with Fatou's lemma, where we have no controlling function, and a weaker conclusion.

Two integrable functions f and g are equal almost everywhere if and only if $\int |f - g| \, d\mu = 0$, so we again identify integrable functions which are equal almost everywhere. We denote the resulting space by $L^1 = L^1(\Omega, \Sigma, \mu)$; as we shall see in Chapter 4, it is a vector space under the usual operations.

Finally, we consider repeated integrals. If (X, Σ, μ) and (Y, T, ν) are measure spaces, we can consider the σ-field $\sigma(\Sigma \times T)$, which is the smallest σ-field containing $A \times B$ for all $A \in \Sigma$, $B \in T$, and can construct the product measure $\mu \times \nu$ on $\sigma(\Sigma \times T)$, with the property that $(\mu \times \nu)(A \times B) = \mu(A)\nu(B)$. Then the fundamental result, usually referred to as *Fubini's theorem*, is that everything works very well if $f \geq 0$ or if $f \in L^1(X \times Y)$:

$$\int_{X \times Y} f \, d(\mu \times \nu) = \int_X \left(\int_Y f \, d\nu \right) d\mu = \int_Y \left(\int_X f \, d\mu \right) d\nu.$$

In fact the full statement is more complicated than this, as we need to discuss measurability, but these matters need not concern us here.

This enables us to interpret the integral as 'the area under the curve'. Suppose that f is a non-negative measurable function on (Ω, Σ, μ). Let $A_f = \{(\omega, x) : 0 \leq x < f(\omega)\} \subseteq \Omega \times \mathbf{R}^+$. Then

$$(\mu \times \lambda)(A_f) = \int_\Omega \left(\int_{\mathbf{R}^+} I_{A_f} \, d\lambda \right) d\mu$$

$$= \int_\Omega \left(\int_0^{f(\omega)} d\lambda \right) d\mu(\omega) = \int_\Omega f \, d\mu.$$

The same argument works for the set $S_f = \{(\omega, x) : 0 \leq x < f(\omega)\}$.

This gives us another way to approach the integral. Suppose that f is a non-negative measurable function. Its *distribution function* λ_f is defined as $\lambda_f(t) = \mu(f > t)$, for $t \geq 0$.

Proposition 1.3.3 *The distribution function λ_f is a decreasing right-continuous function on $(0, \infty)$, taking values in $[0, \infty]$. Suppose that (f_n) is an increasing sequence of non-negative functions, which converges pointwise to $f \in M$. Then $\lambda_{f_n}(u) \nearrow \lambda_f(u)$ for each $0 < u < \infty$.*

Proof Since $(|f| > u) \subseteq (|f| > v)$ if $u > v$, and since $(|f| > u_n) \nearrow (|f| > v)$ if $u_n \searrow v$, it follows that λ_f is a decreasing right-continuous function on $(0, \infty)$.

Since $(f_n > u) \nearrow (f > u)$, $\lambda_{f_n}(u) \nearrow \lambda_f(u)$ for each $0 < u < \infty$.   □

Proposition 1.3.4 *Suppose that f is a non-negative measurable function on (Ω, Σ, μ), that ϕ is a non-negative measurable function on $[0, \infty)$, and that $\Phi(t) = \int_0^t \phi(s) \, ds$. Then*

$$\int_\Omega \Phi(f) \, d\mu = \int_0^\infty \phi(t) \lambda_f(t) \, dt.$$

Proof We use Fubini's theorem. Let $A_f = \{(\omega, x) : 0 \le x < f(\omega)\} \subseteq \Omega \times \mathbf{R}^+$. Then

$$\int_\Omega \Phi(f)\, d\mu = \int_\Omega \left(\int_0^{f(\omega)} \phi(t)\, dt \right) d\mu(\omega)$$

$$= \int_{\Omega \times \mathbf{R}^+} I_{A_f}(\omega, t) \phi(t)\, (d\mu(\omega) \times d\lambda(t))$$

$$= \int_0^\infty \left(\int_\Omega I_{A_f}(\omega, t) \phi(t)\, d\mu(\omega) \right) dt$$

$$= \int_0^\infty \phi(t) \lambda_f(t)\, dt.$$

$\square$

Taking $\phi(t) = 1$, we obtain the following.

Corollary 1.3.2 *Suppose that f is a non-negative measurable function on (Ω, Σ, μ). Then*

$$\int_\Omega f\, d\mu = \int_0^\infty \lambda_f(t)\, dt.$$

Since λ_f is a decreasing function, the integral on the right-hand side of this equation can be considered as an improper Riemann integral. Thus the equation can be taken as the definition of $\int_\Omega f\, d\mu$. This provides an interesting alternative approach to the integral.

1.4 Notes and remarks

This brief account is adequate for most of our needs. We shall introduce further ideas when we need them. For example, we shall consider vector-valued functions in Chapter 4. We shall also prove further measure theoretical results, such as the Lebesgue decomposition theorem (Theorem 5.2.1) and a theorem on the differentiability of integrals (Theorem 8.8.1) in due course, as applications of the theory that we shall develop.

There are many excellent textbooks which give an account of measure theory; among them let us mention [Bar 95], [Bil 95], [Dud 02], [Hal 50], [Rud 79] and [Wil 91]. Note that a large number of these include probability theory as well. This is very natural, since in the 1920s Kolmogoroff explained how measure theory can provide a firm foundation for probability theory. Probability theory is an essential tool for analysis, and we shall use ideas from probability in the later chapters.

2

The Cauchy–Schwarz inequality

2.1 Cauchy's inequality

In 1821, Volume I of Cauchy's *Cours d'analyse de l'École Royale Polytechnique* [Cau 21] was published, putting his course into writing 'for the greatest utility of the students'. At the end there were nine notes, the second of which was about the notion of inequality. In this note, Cauchy proved the following.

Theorem 2.1.1 (Cauchy's inequality) *If $a_1, \ldots, a_n$ and $b_1, \ldots, b_n$ are real numbers, then*

$$\sum_{i=1}^{n} a_i b_i \leq \left(\sum_{i=1}^{n} a_i^2 \right)^{1/2} \left(\sum_{i=1}^{n} b_i^2 \right)^{1/2} .$$

Equality holds if and only if $a_i b_j = a_j b_i$ for $1 \leq i, j \leq n$.

Proof Cauchy used Lagrange's identity:

$$\left(\sum_{1=1}^{n} a_i b_i \right)^2 + \sum_{\{(i,j):i<j\}} (a_i b_j - a_j b_i)^2 = \left(\sum_{i=1}^{n} a_i^2 \right) \left(\sum_{i=1}^{n} b_i^2 \right) .$$

This clearly establishes the inequality, and also shows that equality holds if and only if $a_i b_j = a_j b_i$, for all i, j. □

Cauchy then used this to give a new proof of the Arithmetic Mean–Geometric Mean inequality, as we shall see in the next chapter, but gave no other applications. In 1854, Buniakowski extended Cauchy's inequality to integrals, approximating the integrals by sums, but his work remained little-known.

2.2 Inner-product spaces

In 1885, Schwarz [Schw 85] gave another proof of Cauchy's inequality, this time for two-dimensional integrals. Schwarz's proof is quite different from Cauchy's, and extends to a more general and more abstract setting, which we now describe.

Suppose that V is a real vector space. An *inner product* on V is a real-valued function $(x, y) \to \langle x, y \rangle$ on $V \times V$ which satisfies the following:

(i) (bilinearity)

$$\langle \alpha_1 x_1 + \alpha_2 x_2, y \rangle = \alpha_1 \langle x_1, y \rangle + \alpha_2 \langle x_2, y \rangle,$$
$$\langle x, \beta_1 y_1 + \beta_2 y_2 \rangle = \beta_1 \langle x, y_1 \rangle + \beta_2 \langle x, y_2 \rangle,$$

for all x, x_1, x_2, y, y_1, y_2 in V and all real $\alpha_1, \alpha_2, \beta_1, \beta_2$;

(ii) (symmetry)

$$\langle y, x \rangle = \langle x, y \rangle \quad \text{for all } x, y \text{ in } V;$$

(iii) (positive definiteness)

$$\langle x, x \rangle > 0 \quad \text{for all non-zero } x \text{ in } V.$$

For example, if $V = \mathbf{R}^d$, we define the *usual* inner product, by setting $\langle z, w \rangle = \sum_{i=1}^d z_i w_i$ for $z = (z_i), w = (w_i)$.

Similarly, an *inner product* on a complex vector space V is a function $(x, y) \to \langle x, y \rangle$ from $V \times V$ to the complex numbers $\mathbf{C}$ which satisfies the following:

(i) (sesquilinearity)

$$\langle \alpha_1 x_1 + \alpha_2 x_2, y \rangle = \alpha_1 \langle x_1, y \rangle + \alpha_2 \langle x_2, y \rangle,$$
$$\langle x, \beta_1 y_1 + \beta_2 y_2 \rangle = \overline{\beta_1} \langle x, y_1 \rangle + \overline{\beta_2} \langle x, y_2 \rangle,$$

for all x, x_1, x_2, y, y_1, y_2 in V and all complex $\alpha_1, \alpha_2, \beta_1, \beta_2$;

(ii) (the Hermitian condition)

$$\langle y, x \rangle = \overline{\langle y, x \rangle} \quad \text{for all } x, y \text{ in } V;$$

(iii) (positive definiteness)

$$\langle x, x \rangle > 0 \quad \text{for all non-zero } x \text{ in } V.$$

For example, if $V = \mathbf{C}^d$, we define the *usual* inner product, by setting $\langle z, w \rangle = \sum_{i=1}^d z_i \overline{w_i}$ for $z = (z_i), w = (w_i)$.

A (real or) complex vector space V equipped with an inner product is called an *inner-product* space. If x is a vector in V, we set $\|x\| = \langle x, x \rangle^{1/2}$. Note that we have the following *parallelogram law*:

$$\|x + y\|^2 + \|x - y\|^2 = (\langle x, x \rangle + \langle x, y \rangle + \langle y, x \rangle + \langle y, y \rangle)$$
$$+ (\langle x, x \rangle - \langle x, y \rangle - \langle y, x \rangle + \langle y, y \rangle)$$
$$= 2\|x\|^2 + 2\|y\|^2.$$

2.3 The Cauchy–Schwarz inequality

In what follows, we shall consider the complex case: the real case is easier.

Proposition 2.3.1 (The Cauchy–Schwarz inequality) *If x and y are vectors in an inner product space V, then*

$$|\langle x, y \rangle| \leq \|x\| \cdot \|y\|,$$

with equality if and only if x and y are linearly dependent.

Proof This depends upon the quadratic nature of the inner product. If $y = 0$ then $\langle x, y \rangle = 0$ and $\|y\| = 0$, so that the inequality is trivially true.

Otherwise, let $\langle x, y \rangle = re^{i\theta}$, where $r = |\langle x, y \rangle|$. If λ is real then

$$\left\| x + \lambda e^{i\theta} y \right\|^2 = \langle x, x \rangle + \left\langle \lambda e^{i\theta} y, x \right\rangle + \left\langle x, \lambda e^{i\theta} y \right\rangle + \left\langle \lambda e^{i\theta} y, \lambda e^{i\theta} y \right\rangle$$
$$= \|x\|^2 + 2\lambda |\langle x, y \rangle| + \lambda^2 \|y\|^2.$$

Thus $\|x\|^2 + 2\lambda |\langle x, y \rangle| + \lambda^2 \|y\|^2 \geq 0$. If we take $\lambda = -\|x\| / \|y\|$, we obtain the desired inequality.

If equality holds, then $\|x + \lambda e^{i\theta} y\| = 0$, so that $x + \lambda e^{i\theta} y = 0$, and x and y are linearly dependent. Conversely, if x and y are linearly dependent, then $x = \alpha y$, and $|\langle x, y \rangle| = |\alpha| \|y\|^2 = \|x\| \|y\|$. $\square$

Note that we obtain Cauchy's inequality by considering $\mathbf{R}^d$, with its usual inner product.

Corollary 2.3.1 $\|x + y\| \leq \|x\| + \|y\|$, *with equality if and only if either $y = 0$ or $x = \alpha y$, with $\alpha \geq 0$.*

Proof We have

$$\|x + y\|^2 = \|x\|^2 + \langle x, y \rangle + \langle y, x \rangle + \|y\|^2$$
$$\leq \|x\|^2 + 2\|x\| \cdot \|y\| + \|y\|^2$$
$$= (\|x\| + \|y\|)^2.$$

Equality holds if and only if $\Re \langle x, y \rangle = \|x\| \cdot \|y\|$, which is equivalent to the condition stated.   □

Since $\|\lambda x\| = |\lambda| \|x\|$, and since $\|x\| = 0$ if and only if $x = 0$, this corollary says that the function $x \to \|x\|$ is a *norm* on V. We shall consider norms in Chapter 4.

As our second example of inner product spaces, we consider spaces of functions. Suppose that (Ω, Σ, μ) is a measure space. Let $\mathcal{L}^2 = \mathcal{L}^2(\Omega, \Sigma, \mu)$ denote the set of complex-valued measurable functions on Ω for which

$$\int_\Omega |f|^2 \, d\mu < \infty.$$

It follows from the parallelogram law for scalars that if f and g are in $\mathcal{L}^2$ then

$$\int_\Omega |f + g|^2 \, d\mu + \int_\Omega |f - g|^2 \, d\mu = \int_\Omega |f|^2 \, d\mu + \int_\Omega |g|^2 \, d\mu,$$

so that $f + g$ and $f - g$ are in $\mathcal{L}^2$. Since λf is in $\mathcal{L}^2$ if f is, this means that $\mathcal{L}^2$ is a vector space.

Similarly, since

$$|f(x)|^2 + |g(x)|^2 - 2|f(x)\overline{g(x)}| = (|f(x)| - |g(x)|)^2 \geq 0,$$

it follows that

$$2 \int_\Omega |f\bar{g}| \, d\mu \leq \int_\Omega |f|^2 \, d\mu + \int_\Omega |g|^2 \, d\mu,$$

with equality if and only if $|f| = |g|$ almost everywhere, so that $f\bar{g}$ is integrable. We set

$$\langle f, g \rangle = \int_\Omega f\bar{g} \, d\mu.$$

This function is sesquilinear, Hermitian and positive semi-definite. Further, $\langle f, f \rangle = 0$ if and only if $f = 0$ almost everywhere. We therefore identify functions which are equal almost everywhere, and denote the resulting quotient space by $L^2 = L^2(\Omega, \Sigma, \mu)$. L^2 is again a vector space, and the value of the integral $\int_\Omega f\bar{g} \, d\mu$ is unaltered if we replace f and g by equivalent functions. We can therefore define $\langle f, g \rangle$ on $L^2 \times L^2$: this is now an *inner product*. Consequently, we have the following result.

Theorem 2.3.1 (Schwarz' inequality) *If $f, g \in L^2(\Omega, \Sigma, \mu)$, then*

$$\left| \int_\Omega f\bar{g} \, d\mu \right| \leq \left(\int_\Omega |f|^2 \, d\mu \right)^{1/2} \left(\int_\Omega |g|^2 \, d\mu \right)^{1/2},$$

with equality if and only if f and g are linearly dependent.

More particularly, when $\Omega = \mathbf{N}$, and μ is counting measure, we write

$$l_2 = \left\{ x = (x_i) : \sum_{i=1}^\infty |x_i|^2 < \infty \right\}.$$

Then if x and y are in l_2 the sum $\sum_{i=1}^\infty x_i y_i$ is absolutely convergent and

$$\left| \sum_{i=1}^\infty x_i \bar{y}_i \right| \leq \sum_{i=1}^\infty |x_i||y_i| \leq \left(\sum_{i=1}^\infty |y_i|^2 \right)^{1/2} \left(\sum_{i=1}^\infty |x_i|^2 \right)^{1/2}.$$

We shall follow modern custom, and refer to both Cauchy's inequality and Schwarz' inequality as the Cauchy–Schwarz inequality.

2.4 Notes and remarks

Seen from this distance, it now seems strange that Cauchy's inequality did not appear in print until 1821, and stranger still that Schwarz did not establish the result for integrals until more than sixty years later. Nowadays, inner-product spaces and Hilbert spaces have their place in undergraduate courses, where the principal difficulty that occurs is teaching the correct pronunciation of Cauchy and the correct spelling of Schwarz.

We shall not spend any longer on the Cauchy–Schwarz inequality, but it is worth noting how many of the results that follow can be seen as extensions or generalizations of it.

An entertaining account of the Cauchy–Schwarz inequality and related results is given in [Ste 04].

Exercises

2.1 Suppose that $\mu(\Omega) < \infty$ and that $f \in L^2(\mu)$. Show that

$$\int_\Omega |f| \, d\mu \leq (\mu(\Omega))^{1/2} \left(\int_\Omega |f|^2 \, d\mu \right)^{1/2}.$$

The next two inequalities are useful in the theory of hypercontractive semigroups.

2.2 Suppose that $r > 1$. Using Exercise 2.1, applied to the function $f(x) = 1/\sqrt{x}$ on $[1, r^2]$, show that $2(r - 1) \le (r + 1) \log r$.

2.3 Suppose that $0 < s < t$ and that $q > 1$. Using Exercise 2.1, applied to the function $f(x) = x^{q-1}$ on $[s, t]$, show that

$$(t^q - s^q)^2 \le \frac{q^2}{2q - 1}(t^{2q-1} - s^{2q-1})(t - s).$$

2.4 Suppose that $\mathbf{P}$ is a Borel probability measure on $\mathbf{R}$. The *characteristic function* $f_{\mathbf{P}}(u)$ is defined (for real u) as

$$f_{\mathbf{P}}(u) = \int_{\mathbf{R}} e^{ixu}\, d\mathbf{P}(x).$$

(i) Prove the *incremental inequality*

$$|f_{\mathbf{P}}(u + h) - f_{\mathbf{P}}(u)|^2 \le 4(1 - \Re f_{\mathbf{P}}(h)).$$

(ii) Prove the *Harker–Kasper inequality*

$$2(\Re f_{\mathbf{P}}(u))^2 \le 1 + \Re f_{\mathbf{P}}(2u).$$

This inequality, proved in 1948, led to a substantial breakthrough in determining the structure of crystals.

2.5 Suppose that g is a positive measurable function on Ω and that $\int_{\Omega} g\, d\mu = 1$. Show that if $f \in L^1(\mu)$ then

$$\int_{\Omega} f\, d\mu \le \left(\int_{\Omega} (f^2/g)\, d\mu \right)^{1/2}.$$

3

The arithmetic mean–geometric mean inequality

3.1 The arithmetic mean–geometric mean inequality

The arithmetic mean–geometric mean inequality is perhaps the most famous of all inequalities. It is beloved by problem setters.

Theorem 3.1.1 (The arithmetic mean–geometric mean inequality)
Suppose that $a_1, \ldots, a_n$ are positive numbers. Then

$$(a_1 \ldots a_n)^{1/n} \le \frac{a_1 + \cdots + a_n}{n},$$

with equality if and only if $a_1 = \cdots = a_n$.

The quantity $g = (a_1 \ldots a_n)^{1/n}$ is the *geometric mean* of $a_1, \ldots, a_n$, and the quantity $a = (a_1 + \cdots + a_n)/n$ is the *arithmetic mean*.

Proof We give three proofs here, and shall give another one later.

First we give Cauchy's proof [Cau 21]. We begin by proving the result when $n = 2^k$, proving the result by induction on k. Since

$$(a_1 + a_2)^2 - 4a_1a_2 = (a_1 - a_2)^2 \ge 0,$$

the result holds for $k = 1$, with equality if and only if $a_1 = a_2$.

Suppose that the result holds when $n = 2^{k-1}$. Then

$$a_1 \ldots a_{2^{k-1}} \le \left(\frac{a_1 + \cdots + a_{2^{k-1}}}{2^{k-1}} \right)^{2^{k-1}}$$

and

$$a_{2^{k-1}+1} \ldots a_{2^k} \le \left(\frac{a_{2^{k-1}+1} + \cdots + a_{2^k}}{2^{k-1}} \right)^{2^{k-1}},$$

so that

$$a_1 \ldots a_{2^k} \leq \left(\left(\frac{a_1 + \cdots + a_{2^{k-1}}}{2^{k-1}} \right) \left(\frac{a_{2^{k-1}+1} + \cdots + a_{2^k}}{2^{k-1}} \right) \right)^{2^{k-1}}$$

But

$$(a_1 + \cdots + a_{2^{k-1}})(a_{2^{k-1}+1} + \cdots + a_{2^k}) \leq \tfrac{1}{4}(a_1 + \cdots + a_{2^k})^2,$$

by the case $k = 1$. Combining these two inequalities, we obtain the required inequality. Further, equality holds if and only if equality holds in each of the inequalities we have established, and this happens if and only if

$$a_1 = \cdots = a_{2^{k-1}} \quad \text{and} \quad a_{2^{k-1}+1} = \cdots = a_{2^k},$$

and

$$a_1 + \cdots + a_{2^{k-1}} = a_{2^{k-1}+1} + \cdots + a_{2^k},$$

which in turn happens if and only if $a_1 = \cdots = a_{2^k}$.

We now prove the result for general n. Choose k such that $2^k > n$, and set a_j equal to the arithmetic mean a for $n < j \leq 2^k$. Then, applying the result for 2^k, we obtain

$$a_1 \ldots a_n . a^{2^k - n} \leq a^{2^k}.$$

Multiplying by $a^{n - 2^k}$, we obtain the inequality required. Equality holds if and only if $a_i = a$ for all i.

The second proof involves the *method of transfer*. We prove the result by induction on the number d of terms a_j which are different from the arithmetic mean a. The result is trivially true, with equality, if $d = 0$. It is not possible for d to be equal to 1. Suppose that the result is true for all values less than d, and that d terms of $a_1, \ldots, a_n$ are different from a. There must then be two indices i and j for which $a_i > a > a_j$. We now transfer some of a_i to a_j; we define a new sequence of positive numbers by setting $a'_i = a, a'_j = a_i + a_j - a$, and $a'_k = a_k$ for $k \neq i, j$. Then $a'_1, \ldots, a'_j$ has the same arithmetic mean a as $a_1, \ldots, a_n$, and has less than d terms different from a. Thus by the inductive hypothesis, the geometric mean g' is less than or equal to a. But

$$a'_i a'_j - a_i a_j = aa_i + aa_j - a^2 - a_i a_j = (a_i - a)(a - a_j) > 0,$$

so that $g < g'$. This establishes the inequality, and also shows that equality can only hold when all the terms are equal.

The third proof requires results from analysis. Let

$$\Delta = \{x = (x_1, \ldots, x_n) \in \mathcal{R}^n \colon x_i \geq 0 \text{ for } 1 \leq i \leq n, x_1 + \cdots + x_n = na\}.$$

Δ is the set of n-tuples $(x_1, \ldots, x_n)$ of non-negative numbers with arithmetic mean a. It is a closed bounded subset of $\mathcal{R}^n$. The function $\pi(x) = x_1 \cdots x_n$ is continuous on Δ, and so it attains a maximum value at some point $c = (c_1, \ldots, c_n)$. [This basic result from analysis is fundamental to the proof; early versions of the proof were therefore defective at this point.] Since $\pi(a, \ldots, a) = a^n > 0$, $p(c) > 0$, and so each c_i is positive. Now consider any two distinct indices i and j. Let p and q be points of $\mathcal{R}^n$, defined by

$$p_i = 0, \quad p_j = c_i + c_j, \quad p_k = c_k \quad \text{for } k \neq i, j,$$
$$q_i = c_i + c_j, \quad q_j = 0, \quad q_k = c_k \quad \text{for } k \neq i, j.$$

Then p and q are points on the boundary of Δ, and the line segment $[p, q]$ is contained in Δ. Let $f(t) = (1-t)p + tq$, for $0 \leq t \leq 1$, so that f maps $[0, 1]$ onto the line segment $[p, q]$. $f(c_i/(c_i + c_j)) = c$, so that c is an interior point of $[p, q]$. Thus the function $g(t) = \pi(f(t))$ has a maximum at $c_i/(c_i + c_j)$. Now

$$g(t) = t(1-t)(c_i + c_j)^2 \prod_{k \neq i, j} c_k, \quad \text{so that} \quad \frac{dg}{dt} = (1 - 2t)(c_i + c_j)^2 \prod_{k \neq i, j} c_k,$$

and

$$\frac{dg}{dt}\left(\frac{c_i}{c_i + c_j}\right) = (c_j - c_i)(c_i + c_j)^2 \prod_{k \neq i, j} c_k = 0.$$

Thus $c_i = c_j$. Since this holds for all pairs of indices i, j, the maximum is attained at $(a, \ldots, a)$, and at no other point. $\qquad \square$

We shall refer to the arithmetic mean–geometric mean inequality as the AM–GM inequality.

3.2 Applications

We give two applications of the AM–GM inequality. In elementary analysis, it can be used to provide polynomial approximations to the exponential function.

Proposition 3.2.1 *(i) If $nt > -1$, then $(1-t)^n \geq 1 - nt$.*
(ii) If $-x < n < m$ then $(1 + x/n)^n \leq (1 + x/m)^m$.
(iii) If $x > 0$ and $\alpha > 1$ then $(1 - x/n^\alpha)^n \to 1$.
(iv) $(1 + x/n)^n$ converges as $n \to \infty$, for all real x.

Proof (i) Take $a_1 = 1 - nt$ and $a_2 = \cdots = a_n = 1$.

(ii)   Let $a_1 = \cdots = a_n = 1 + x/n$, and $a_{n+1} = \cdots = a_m = 1$. Then

$$(1 + x/n)^{n/m} = (a_1 \ldots a_m)^{1/m} \le (a_1 + \cdots + a_m)/m = 1 + x/m.$$

(iii)   Put $t = x/n^\alpha$. Then if $n^\alpha > x$, $1 - x/n^{\alpha-1} \le (1 - x/n^\alpha)^n < 1$, by (i), and the result follows since $1 - x/n^{\alpha-1} \to 1$ as $n \to \infty$.

If $x < 0$ then, for $n > -x$, $((1 + x/n)^n)$ is an increasing sequence which is bounded above by 1, and so it converges, to $e(x)$ say. If $x > 0$, then

$$(1 + x/n)^n (1 - x/n)^n = (1 - x^2/n^2)^n \to 1,$$

so that $(1 + x/n)^n$ converges, to $e(x)$ say, where $e(x) = e(-x)^{-1}$.   □

We set $e = e(1) = \lim_{n\to\infty} (1 + 1/n)^n$.

Carleman [Car 23] established an important inequality used in the study of quasi-analytic functions (the *Denjoy–Carleman theorem*: see for example [Hör 90], Theorem 1.3.8). In 1926, Pólya [Pól 26] gave the following elegant proof, which uses the AM–GM inequality.

Theorem 3.2.1 (Carleman's inequality) *Suppose that (a_j) is a sequence of positive numbers for which $\sum_{j=1}^{\infty} a_j < \infty$. Then*

$$\sum_{n=1}^{\infty} (a_1 \ldots a_n)^{1/n} < e \sum_{j=1}^{\infty} a_j.$$

Proof Let $m_n = n(1 + 1/n)^n$, so that $m_1 \cdots m_n = (n + 1)^n$, and let $b_n = m_n a_n$. Then

$$(n + 1)(a_1 \ldots a_n)^{1/n} = (b_1 \ldots b_n)^{1/n} \le (b_1 + \cdots + b_n)/n,$$

so that

$$\sum_{n=1}^{\infty} (a_1 \ldots a_n)^{1/n} \le \sum_{n=1}^{\infty} \frac{1}{n(n+1)} \left(\sum_{j=1}^{n} b_j \right)$$

$$= \sum_{j=1}^{\infty} b_j \left(\sum_{n=j}^{\infty} \frac{1}{n(n+1)} \right)$$

$$= \sum_{j=1}^{\infty} \frac{b_j}{j} = \sum_{j=1}^{\infty} \left(1 + \frac{1}{j} \right)^j a_j < e \sum_{j=1}^{\infty} a_j.$$

□

3.3 Notes and remarks

The AM–GM inequality has been around for a long time, and there are many proofs of it: 52 are given in [BuMV 87]. The first two proofs that we have given are truly elementary, using only the algebraic properties of an ordered field. The idea behind the second proof is called the *method of transfer*: it will recur later, in the proof of Theorem 7.7.1. It was introduced by Muirhead [Mui 03] to prove Theorem 7.9.2, which provides a far-reaching generalization of the AM–GM inequality.

The salient feature of the AM–GM inequality is that it relates additive and multiplicative averages: the logarithmic and exponential functions provide a link between addition and multiplication, and we shall use these to generalize the AM–GM inequality, in the next chapter.

Exercises

3.1 The *harmonic mean* h of n positive numbers $a_1, \ldots, a_n$ is defined as $(\sum_{j=1}^{n}(1/a_j)/n)^{-1}$. Show that the harmonic mean is less than or equal to the geometric mean. When does equality occur?

3.2 Show that a d-dimensional rectangular parallelopiped of fixed volume has least surface area when all the sides have equal length. Show that solving this problem is equivalent to establishing the AM–GM inequality.

3.3 Suppose that $a_1, \ldots, a_n$ are n positive numbers. Show that if $1 < k < n$ then

$$(a_1 \ldots a_n)^{1/n} \leq \binom{n}{k}^{-1} \sum_{i_1 < \cdots < i_k} (a_{i_1} \ldots a_{i_k})^{1/k} \leq \frac{a_1 + \cdots + a_n}{n}.$$

3.4 With the terminology of Proposition 3.2.1, show that $e(x)e(y) = e(x+y)$, that $e = e(1) = \sum_{j=0}^{\infty} 1/j!$ and that $e(x) = \sum_{j=0}^{\infty} x^j/j!$

3.5 Let $t_n = n^n/n!$. By considering the ratios t_{n+1}/t_n, show that $n^n < e^n n!$

3.6 Suppose that (a_n) and (f_n) are sequences of positive numbers such that $\sum_{n=1}^{\infty} a_n = \infty$ and $f_n \to f > 0$ as $n \to \infty$. Show that

$$\left(\sum_{n=1}^{N} f_n a_n \right) \bigg/ \left(\sum_{n=1}^{N} a_n \right) \to f \quad \text{as } N \to \infty.$$

3.7 Show that the constant e in Carleman's inequality is best possible. [Consider finite sums in the proof, and strive for equality.]

4

Convexity, and Jensen's inequality

4.1 Convex sets and convex functions

Many important inequalities depend upon convexity. In this chapter, we shall establish *Jensen's inequality*, the most fundamental of these inequalities, in various forms.

A subset C of a real or complex vector space E is *convex* if whenever x and y are in C and $0 \leq \theta \leq 1$ then $(1 - \theta)x + \theta y \in C$. This says that the real line segment $[x, y]$ is contained in C. Convexity is a real property: in the complex case, we are restricting attention to the underlying real space. Convexity is an affine property, but we shall restrict our attention to vector spaces rather than to affine spaces.

Proposition 4.1.1 *A subset C of a vector space E is convex if and only if whenever $x_1, \ldots, x_n \in C$ and $p_1, \ldots, p_n$ are positive numbers with $p_1 + \cdots + p_n = 1$ then $p_1 x_1 + \cdots + p_n x_n \in C$.*

Proof The condition is certainly sufficient. We prove necessity by induction on n. The result is trivially true when $n = 1$, and is true for $n = 2$, as this reduces to the definition of convexity. Suppose that the result is true for $n - 1$, and that $x_1, \ldots, x_n$ and $p_1, \ldots, p_n$ are as above. Let

$$y = \frac{p_{n-1}}{p_{n-1} + p_n} x_{n-1} + \frac{p_n}{p_{n-1} + p_n} x_n.$$

Then $y \in C$ by convexity, and

$$p_1 x_1 + \cdots + p_n x_n = p_1 x_1 + \cdots + p_{n-2} x_{n-2} + (p_{n-1} + p_n) y \in C,$$

by the inductive hypothesis. □

A real-valued function f defined on a convex subset C of a vector space E is *convex* if the set

$$U_f = \{(x, \lambda) \colon x \in C, \lambda \geq f(x)\} \subseteq E \times \mathbf{R}$$

of points on and above the graph of f is convex. That is to say, if $x, y \in C$ and $0 \leq \theta \leq 1$ then

$$f((1 - \theta)x + \theta y) \leq (1 - \theta)f(x) + \theta f(y).$$

f is *strictly convex* if

$$f((1 - \lambda)x + \lambda y) < (1 - \lambda)f(x) + \lambda f(y)$$

whenever x and y are distinct points of C and $0 < \lambda < 1$. f is *concave* (*strictly concave*) if $-f$ is convex (strictly convex).

We now use Proposition 4.1.1 to prove the simplest version of Jensen's inequality.

Proposition 4.1.2 (Jensen's inequality: I) *If f is a convex function on a convex set C, and $p_1, \ldots, p_n$ are positive numbers with $p_1 + \cdots + p_n = 1$, then*

$$f(p_1 x_1 + \cdots + p_n x_n) \leq p_1 f(x_1) + \cdots + p_n f(x_n).$$

If f is strictly convex, then equality holds if and only if $x_1 = \cdots = x_n$.

Proof The first statement follows by applying Proposition 4.1.1 to U_f. Suppose that f is strictly convex, and that $x_1, \ldots, x_n$ are not all equal. By relabelling if necessary, we can suppose that $x_{n-1} \neq x_n$. Let

$$y = \frac{p_{n-1}}{p_{n-1} + p_n} x_{n-1} + \frac{p_n}{p_{n-1} + p_n} x_n,$$

as above. Then

$$f(y) < \frac{p_{n-1}}{p_{n-1} + p_n} f(x_{n-1}) + \frac{p_n}{p_{n-1} + p_n} f(x_n),$$

so that

$$\begin{aligned}
f(p_1 x_1 + \cdots + p_n x_n) &= f(p_1 x_1 + \cdots + p_{n-2} x_{n-2} + (p_{n-1} + p_n)y) \\
&\leq p_1 f(x_1) + \cdots + p_{n-2} f(x_{n-2}) + (p_{n-1} + p_n)f(y) \\
&< p_1 f(x_1) + \cdots + p_n f(x_n).
\end{aligned}$$

$\square$

Although this is very simple, it is also very powerful. Here for example is an immediate improvement of the AM–GM inequality.

Proposition 4.1.3 *Suppose that $a_1, \ldots, a_n$ are positive, that $p_1, \ldots, p_n$ are positive and that $p_1 + \cdots + p_n = 1$. Then*

$$a_1^{p_1} \ldots a_n^{p_n} \leq p_1 a_1 + \cdots + p_n a_n,$$

with equality if and only if $a_1 = \cdots = a_n$.

Proof The function e^x is strictly convex (see Proposition 4.2.1), and so

$$e^{p_1 x_1} \ldots e^{p_n x_n} = e^{p_1 x_1 + \cdots + p_n x_n} \leq p_1 e^{x_1} + \cdots + p_n e^{x_n}$$

for any real $x_1, \ldots, x_n$, with equality if and only if $x_1 = \cdots = x_n$. The result follows by making the substitution $x_i = \log a_i$. □

We can think of Proposition 4.1.2 in the following way. We place masses $p_1, \ldots, p_n$ at the points $(x_1, f(x_1)), \ldots, (x_n, p(x_n))$ on the graph of f. This defines a measure on $E \times \mathbf{R}$. Then the centre of mass, or *barycentre*, of these masses is at the point

$$(p_1 x_1 + \cdots + p_n x_n, p_1 f(x_1) + \cdots + p_n f(x_n)),$$

and this lies above the graph, because f is convex. For a more sophisticated version, we replace the measure defined by the point masses by a more general measure. In order to obtain the corresponding version of Jensen's inequality, we need to study convex functions in some detail, and also need to define the notion of a barycentre with some care.

4.2 Convex functions on an interval

Let us consider the case when E is the real line $\mathbf{R}$. In this case the convex subsets are simply the intervals in $\mathbf{R}$. First let us consider differentiable functions.

Proposition 4.2.1 *Suppose that f is a differentiable real-valued function on an open interval I of the real line $\mathbf{R}$. Then f is convex if and only if its derivative f' is an increasing function. It is strictly convex if and only if f' is strictly increasing.*

Proof First suppose that f is convex. Suppose that $a < b < c$ are points in I. Then by convexity,

$$f(b) \leq \frac{c-b}{c-a} f(a) + \frac{b-a}{c-a} f(c).$$

Rearranging this, we find that

$$\frac{f(b) - f(a)}{b - a} \leq \frac{f(c) - f(a)}{c - a} \leq \frac{f(c) - f(b)}{c - b}.$$

Thus if $a < b \leq c < d$ are points in I,

$$\frac{f(b) - f(a)}{b - a} \leq \frac{f(d) - f(c)}{d - c}.$$

It follows from this that f' is increasing.

Conversely, suppose that f' is increasing. Suppose that $x_0 < x_1$ are points in I and that $0 < \theta < 1$: let $x_\theta = (1 - \theta)x_0 + \theta x_1$. Applying the mean-value theorem, there exist points $x_0 < c < x_\theta < d < x_1$ such that

$$f(x_\theta) - f(x_0) = (x_\theta - x_0)f'(c) = \theta(x_1 - x_0)f'(c),$$
$$f(x_1) - f(x_\theta) = (x_1 - x_0)f'(d) = (1 - \theta)(x_1 - x_0)f'(d).$$

Multiplying the first equation by $-(1 - \theta)$ and the second by θ, and adding, we find that

$$(1 - \theta)f(x_0) + \theta f(x_1) - f(x_\theta) = (1 - \theta)\theta(x_1 - x_0)(f'(d) - f'(c)) \geq 0.$$

If f' is strictly increasing then this inequality is strict, so that f is strictly convex. If it is not strictly increasing, so that there exist $y_0 < y_1$ in I with $f'(x) = f'(y_0)$ for $y_0 \leq x \leq y_1$, then $f(x) = f(y_0) + (x - y_0)f'(y_0)$ for $y_0 \leq x \leq y_1$, and f is not strictly convex.   □

We now drop the requirement that f is differentiable. Suppose that f is a convex function on an open interval I, and that $x \in I$. Suppose that $x + t$ and $x - t$ are in I, and that $0 < \theta < 1$. Then (considering the cases where $t > 0$ and $t < 0$ separately) it follows easily from the inequalities above that

$$\theta(f(x) - f(x - t)) \leq f(x + \theta t) - f(x) \leq \theta(f(x + t) - f(x)),$$

so that

$$|f(x + \theta t) - f(x)| \leq \theta \max(|f(x + t) - f(x)|, |f(x) - f(x - t)|),$$

and f is Lipschitz continuous at x. (A function f from a metric space (X, d) to a metric space (Y, ρ) is *Lipschitz continuous at* x_0 if there is a constant C such that $\rho(f(x), f(x_0)) \leq Cd(x, x_0)$ for all $x \in X$. f is a *Lipschitz function* if there is a constant C such that $\rho(f(x), f(z)) \leq Cd(x, z)$ for all $x, z \in X$.)

We can go further. If $t > 0$, it follows from the inequalities above, and the corresponding ones for $f(x - \theta t)$, that

$$\frac{f(x) - f(x - t)}{t} \leq \frac{f(x) - f(x - \theta t)}{\theta t}$$
$$\leq \frac{f(x + \theta t) - f(x)}{\theta t} \leq \frac{f(x + t) - f(x)}{t},$$

so that the right and left derivatives

$$D^+ f(x) = \lim_{h \searrow 0} \frac{f(x + h) - f(x)}{h} \quad \text{and} \quad D^- f(x) = \lim_{h \searrow 0} \frac{f(x - h) - f(x)}{-h}$$

both exist, and $D^+ f(x) \geq D^- f(x)$. Similar arguments show that $D^+ f$ and $D^- f$ are increasing functions, that $D^+ f$ is right-continuous and $D^- f$ left-continuous, and that $D^- f(x) \geq D^+ f(y)$ if $x > y$. Consequently, if $D^+ f(x) \neq D^- f(x)$ then $D^+ f$ and $D^- f$ have jump discontinuities at x. Since an increasing function on an interval has only countably many discontinuities, it follows that $D^+ f(x)$ and $D^- f(x)$ are equal and continuous, except at a countable set of points. Thus f is differentiable, except at this countable set of points.

Proposition 4.2.2 *Suppose that f is a convex function on an open interval I of $\mathbf{R}$, and that $x \in I$. Then there is an affine function a on $\mathbf{R}$ such that $a(x) = f(x)$ and $a(y) \leq f(y)$ for $y \in I$.*

Proof Choose λ so that $D^- f(x) \leq \lambda \leq D^+ f(x)$. Let $a(y) = f(x) + \lambda(y - x)$. Then a is an affine function on $\mathbf{R}$, $a(x) = f(x)$ and $a(y) \leq f(y)$ for $y \in I$. □

Thus f is the supremum of the affine functions which it dominates.

We now return to Jensen's inequality. Suppose that μ is a probability measure on the Borel sets of a (possibly unbounded) open interval $I = (a, b)$. In analogy with the discrete case, we wish to define the barycentre $\bar{\mu}$ to be $\int_I x \, d\mu(x)$. There is no problem if I is bounded; if I is unbounded, we require that the identity function $i(x) = x$ is in $L^1(\mu)$: that is, $\int_I |x| \, d\mu(x) < \infty$. If so, we define $\bar{\mu}$ as $\int_I x \, d\mu(x)$. Note that $\bar{\mu} \in I$.

Theorem 4.2.1 (Jensen's inequality: II) *Suppose that μ is a probability measure on the Borel sets of an open interval I of $\mathbf{R}$, and that μ has a barycentre $\bar{\mu}$. If f is a convex function on I with $\int_I f^- \, d\mu < \infty$ then $f(\bar{\mu}) \leq \int_I f \, d\mu$. If f is strictly convex then equality holds if and only if $\mu(\{\bar{\mu}\}) = 1$.*

A probability measure μ whose mass is concentrated at just one point x, so that $\mu(\{x\}) = 1$ and $\mu(\Omega \setminus \{x\}) = 0$, is called a *Dirac measure*, and is denoted by δ_x.

Proof The condition on f ensures that $\int_I f \, d\mu$ exists, taking a value in $(-\infty, \infty]$. By Proposition 4.2.2, there exists an affine function a on $\mathbf{R}$ with $a(\bar{\mu}) = f(\bar{\mu})$ and $a(y) \leq f(y)$ for all $y \in I$. Then

$$f(\bar{\mu}) = a(\bar{\mu}) = \int_I a \, d\mu \leq \int_I f \, d\mu.$$

If f is strictly convex then $f(y) - a(y) > 0$ for $y \neq \bar{\mu}$, so that equality holds if and only if $\mu(I \setminus \{\bar{\mu}\}) = 0$. $\qquad\square$

An important special case of Theorem 4.2.1 arises in the following way. Suppose that p is a non-negative measurable function on an open interval I, and that $\int_I p \, d\lambda = 1$. Then we can define a probability measure $p \, d\lambda$ by setting

$$p \, d\lambda(B) = \int_B p \, d\lambda = \int_I p I_B \, d\lambda,$$

for each Borel set B. If $\int_I |x| p(x) \, d\lambda(x) < \infty$, then $p \, d\lambda$ has barycentre $\int_I x p(x) \, d\lambda(x)$. We therefore have the following corollary.

Corollary 4.2.1 *Suppose that p is a non-negative measurable function on an open interval I, that $\int_I p \, d\lambda = 1$ and that $\int_I |x| p(x) \, d\lambda(x) < \infty$. If f is a convex function on I with $\int_I p(x) f^-(x) \, d\lambda(x) < \infty$ then*

$$f\left(\int_I x p(x) \, d\lambda(x)\right) \leq \int_I f(x) p(x) \, d\lambda.$$

If f is strictly convex then equality cannot hold.

4.3 Directional derivatives and sublinear functionals

We now return to the case where E is a vector space. We consider a *radially open* convex subset C of a vector space E: a subset C of E is *radially open* if whenever $x \in C$ and $y \in E$ then there exists $\lambda_0 = \lambda_0(x, y) > 0$ such that $x + \lambda y \in C$ for $0 < \lambda \leq \lambda_0$. Suppose that f is a convex function on C, that $x \in C$ and that $y \in E$. Then arguing as in the real case, the function $(f(x + \lambda y) - f(x))/\lambda$ is an increasing function of λ on $(0, \lambda_0(x, y))$ which is bounded below, and so we can define the *directional derivative*

$$D_y(f)(x) = \lim_{\lambda \searrow 0} \frac{f(x + \lambda y) - f(x)}{\lambda}.$$

This has important properties that we shall meet again elsewhere. A real-valued function p on a real or complex vector space E is *positive homogeneous* if $p(\alpha x) = \alpha p(x)$ when α is real and positive and $x \in E$; it is *subadditive* if $p(x + y) \leq p(x) + p(y)$, for $x, y \in E$, and it is *sublinear* or a *sublinear functional* if it is both positive homogeneous and subadditive.

Proposition 4.3.1 *Suppose that f is a convex function on a radially open convex subset C of a vector space E, and that $x \in C$. Then the directional derivative $D_y(f)(x)$ at x is a sublinear function of y, and $f(x+y) \geq f(x) + D_y(f)(x)$ for $x, x + y \in C$.*

Proof Positive homogeneity follows from the definition of the directional derivative. Suppose that $y_1, y_2 \in E$. There exists λ_0 such that $x + \lambda y_1$ and $x + \lambda y_2$ are in C for $0 < \lambda < \lambda_0$. Then by convexity $x + \lambda(y_1 + y_2) \in C$ for $0 < \lambda < \lambda_0/2$ and

$$f(x + \lambda(y_1 + y_2)) \leq \tfrac{1}{2}f(x + 2\lambda y_1) + \tfrac{1}{2}f(x + 2\lambda y_1),$$

so that

$$D_{y_1+y_2}(f)(x) \leq \tfrac{1}{2}D_{2y_1}(f)(x) + \tfrac{1}{2}D_{2y_2}(f)(x) = D_{y_1}(f)(x) + D_{y_2}(f)(x).$$

The final statement follows from the fact that $(f(x + \lambda y) - f(x))/\lambda$ is an increasing function of λ. $\qquad\square$

Radially open convex sets and sublinear functionals are closely related.

Proposition 4.3.2 *Suppose that V is a radially open convex subset of a real vector space E and that $0 \in V$. Let $p_V(x) = \inf\{\lambda > 0 : x \in \lambda V\}$. Then p_V is a non-negative sublinear functional on E and $V = \{x : p_V(x) < 1\}$.*

Conversely, if p is a sublinear functional on E then $U = \{x : p(x) < 1\}$ is a radially open convex subset of E, $0 \in U$, and $p_U(x) = \max(p(x), 0)$ for each $x \in E$.

The function p_U is called the *gauge* of U.

Proof Since V is radially open, $p_V(x) < \infty$ for each $x \in E$. p_V is positive homogeneous and, since V is convex and radially open, $x \in \lambda V$ for $\lambda > p_V(x)$, so that $\{\lambda > 0 : x \in \lambda V\} = (p_V(x), \infty)$. Suppose that $\lambda > p_V(x)$ and $\mu > p_V(y)$. Then $x/\lambda \in V$ and $y/\mu \in V$, and so, by convexity,

$$\frac{x + y}{\lambda + \mu} = \frac{\lambda}{(\lambda + \mu)} \frac{x}{\lambda} + \frac{\mu}{(\lambda + \mu)} \frac{y}{\mu} \in V,$$

so that $x + y \in (\lambda + \mu)V$, and $p_V(x + y) < \lambda + \mu$. Consequently p_V is subadditive. If $p_V(x) < 1$ then $x \in V$. On the other hand, if $x \in V$ then since V is radially open $(1 + \lambda)x = x + \lambda x \in V$ for some $\lambda > 0$, so that $p_V(x) \leq 1/(1 + \lambda) < 1$.

For the converse, if $x, y \in U$ and $0 \leq \lambda \leq 1$ then

$$p((1 - \lambda)x + \lambda y) \leq (1 - \lambda)p(x) + \lambda p(y) < 1,$$

so that $(1 - \lambda)x + \lambda y \in U$: U is convex. Since $p(0) = 0$, $0 \in U$. If $x \in U$, $y \in E$ and $\lambda > 0$ then $p(x + \lambda y) \leq p(x) + \lambda p(y)$, so that if $0 < \lambda < (1 - p(x))/(1 + p(y))$ then $x + \lambda y \in U$, and so U is radially open. If $p(x) > 0$ then $p(x/p(x)) = 1$, so that $x \in \lambda U$ if and only if $\lambda > p(x)$; thus $p_U(x) = p(x)$. If $p(x) \leq 0$, then $p(\lambda x) \leq 0 < 1$ for all $\lambda > 0$. Thus $x \in \lambda U$ for all $\lambda > 0$, and $p_U(x) = 0$. □

4.4 The Hahn–Banach theorem

Does an analogue of Proposition 4.2.2 hold for an arbitrary vector space E? The answer to this question is given by the celebrated Hahn–Banach theorem. We shall spend some time proving this, and considering some of its consequences, and shall return to Jensen's inequality later.

Recall that a linear functional on a vector space is a linear mapping of the space into its field of scalars.

Theorem 4.4.1 (The Hahn–Banach theorem) *Suppose that p is a sub-linear functional on a real vector space E, that F is a linear subspace of E and that f is a linear functional on F satisfying $f(x) \leq p(x)$ for all $x \in F$. Then there is a linear functional h on E such that*

$$h(x) = f(x) \ \text{for } x \in F \quad \text{and} \quad h(y) \leq p(y) \ \text{for } y \in E.$$

Thus h extends f, and still respects the inequality.

Proof The proof is an inductive one. If E is finite-dimensional, we can use induction on the dimension of F. If E is infinite-dimensional, we must appeal to the Axiom of Choice, using Zorn's lemma.

First we describe the inductive argument. Let $\mathcal{S}$ be the set of all pairs (G, g), where G is a linear subspace of E containing F, and g is a linear functional on G satisfying

$$g(x) = f(x) \ \text{for } x \in F \quad \text{and} \quad g(z) \leq p(z) \ \text{for } z \in G.$$

We give $\mathcal{S}$ a partial order by setting $(G_1, g_1) \leq (G_2, g_2)$ if $G_1 \subseteq G_2$ and $g_2(z) = g_1(z)$ for $z \in G_1$: that is, g_2 extends g_1. Every chain in $\mathcal{S}$ has an upper bound: the union of the linear subspaces occurring in the chain is a linear subspace K, say, and if $z \in K$ we define $k(z)$ to be the common value of the functionals in whose domain it lies. Then it is easy to check that (K, k) is an upper bound for the chain. Thus, by Zorn's lemma, there is a maximal element (G, g) of $\mathcal{S}$. In order to complete the proof, we must show that $G = E$.

Suppose not. Then there exists $y \in E \setminus G$. Let $G_1 = \text{span}(G, y)$. G_1 properly contains G, and we shall show that g can be extended to a linear functional g_1 on G_1 which satisfies the required inequality, giving the necessary contradiction.

Now any element $x \in G_1$ can be written uniquely as $x = z + \lambda y$, with $z \in G$, so that if g_1 is a linear functional that extends g then $g_1(x) = g(z) + \lambda g_1(y)$. Thus g_1 is determined by $g_1(y)$, and our task is to find a suitable value for $g_1(y)$. We need to consider the cases where λ is zero, positive or negative. There is no problem when $\lambda = 0$, for then $x \in G$, and $g_1(x) = g(x)$. Let us suppose then that $z + \alpha y$ and $w - \beta y$ are elements of G_1 with $\alpha > 0$ and $\beta > 0$. Then, using the sublinearity of p,

$$\alpha g(w) + \beta g(z) = g(\alpha w + \beta z) \leq p(\alpha w + \beta z)$$
$$\leq p(\alpha w - \alpha\beta y) + p(\beta z + \alpha\beta y)$$
$$= \alpha p(w - \beta y) + \beta p(z + \alpha y),$$

so that

$$\frac{g(w) - p(w - \beta y)}{\beta} \leq \frac{p(z + \alpha y) - g(z)}{\alpha}.$$

Thus if we set

$$\theta_0 = \sup\left\{ \frac{g(w) - p(w - \beta y)}{\beta} : w \in G, \beta > 0 \right\},$$
$$\theta_1 = \inf\left\{ \frac{p(z + \alpha y) - g(z)}{\alpha} : z \in G, \alpha > 0 \right\},$$

then $\theta_0 \leq \theta_1$. Let us choose $\theta_0 \leq \theta \leq \theta_1$, and let us set $g_1(y) = \theta$. Then

$$g_1(z + \alpha y) = g(z) + \alpha\theta \leq p(z + \alpha y),$$
$$g_1(w - \beta y) = g(w) - \beta\theta \leq p(w - \beta y)$$

for any $z, w \in G$ and any positive α, β, and so we have found a suitable extension. $\square$

Corollary 4.4.1 *Suppose that f is a convex function on a radially open convex subset C of a real vector space E and that $x \in C$. Then there exists an affine function a such that $a(x) = f(x)$ and $a(y) \leq f(y)$ for $y \in C$.*

Proof By the Hahn–Banach theorem there exists a linear functional g on E such that $g(z) \leq D_z(f)(x)$ for all $z \in E$ (take $F = \{0\}$ in the theorem). Let $a(z) = f(x) + g(z - x)$. This is affine, and if $y \in C$ then

$$a(y) = f(x) + g(y - x) \leq f(x) + D_{y-x}(f)(x) \leq f(y),$$

by Proposition 4.3.1. $\square$

We can also express the Hahn–Banach theorem as a separation theorem. We do this in three steps.

Theorem 4.4.2 (The separation theorem: I) *Suppose that U is a non-empty radially open convex subset of a real vector space E.*

(i) If $0 \notin U$ there exists a linear functional ϕ on E for which $\phi(x) > 0$ for $x \in U$.

(ii) If V is a non-empty convex subset of E disjoint from U there exists a linear functional ϕ on E and a real number λ for which $\phi(x) > \lambda$ for $x \in U$ and $\phi(y) \leq \lambda$ for $y \in V$.

(iii) If F is a linear subspace of E disjoint from U there exists a linear functional ϕ on E for which $\phi(x) > 0$ for $x \in U$ and $\phi(y) = 0$ for $y \in F$.

Proof (i) Choose x_0 in U and let $W = U - x_0$. W is radially open and $0 \in W$. Let p_W be the gauge of W. Then $-x_0 \notin W$, and so $p_W(-x_0) \geq 1$. Let $y_0 = -x_0/p_W(-x_0)$, so that $p_W(y_0) = 1$. If $\alpha y_0 \in$ span (y_0), let $f(\alpha y_0) = \alpha$. Then f is a linear functional on span (y_0) and $f(-x_0) = p_W(-x_0) \geq 1$. If $\alpha \geq 0$, then $f(\alpha y_0) = p_W(\alpha y_0)$ and if $\alpha < 0$ then $f(\alpha y_0) = -p_W(-\alpha y_0) \leq p_W(\alpha y_0)$, since $p_W(-\alpha y_0) + p_W(\alpha y_0) \geq p_W(0) = 0$. By the Hahn–Banach Theorem, f can be extended to a linear functional h on E for which $h(x) \leq p_W(x)$ for all $x \in E$. If $x \in U$ then, since $h(-x_0) = p_W(-x_0) \geq 1$ and $p_W(x - x_0) < 1$,

$$h(x) = h(x - x_0) - h(-x_0) \leq p_W(x - x_0) - p_W(-x_0) < 0;$$

now take $\phi = -h$.

(ii) Let $W = U - V$. Then W is radially open, and $0 \notin W$. By (i), there exists a linear functional ϕ on E such that $\phi(x) > 0$ for $x \in W$: that is, $\phi(x) > \phi(y)$ for $x \in U$, $y \in V$. Thus ϕ is bounded above on V: let $\lambda = \sup\{\phi(y) : y \in V\}$. The linear functional ϕ is non-zero: let z be a vector for which $\phi(z) = 1$. If $x \in U$ then, since U is radially open, there exists $\alpha > 0$ such that $x - \alpha z \in U$. Then $\phi(x) = \phi(x - \alpha z) + \phi(\alpha z) \geq \lambda + \alpha > \lambda$.

(iii) Take ϕ as in (ii) (with F replacing V). Since F is a linear subspace, $\phi(F) = \{0\}$ or $\mathbf{R}$. The latter is not possible, since $\phi(F)$ is bounded above. Thus $\phi(F) = \{0\}$, and we can take $\lambda = 0$. $\qquad\square$

4.5 Normed spaces, Banach spaces and Hilbert space

Theorem 4.4.1 is essentially a real theorem. There is however an important version which applies in both the real and the complex case. A real-valued function p on a real or complex vector space is a *semi-norm* if it is sub-additive and if $p(\alpha x) = |\alpha| p(x)$ for every scalar α and vector x. A semi-norm is necessarily non-negative, since $0 = p(0) \leq p(x) + p(-x) = 2p(x)$. A semi-norm p is a *norm* if in addition $p(x) \neq 0$ for $x \neq 0$.

A norm is often denoted by a symbol such as $\|x\|$. $(E, \|.\|)$ is then a *normed* space. The function $d(x, y) = \|x - y\|$ is a metric on E; if E is complete under this metric, then $(E, \|.\|)$ is called a *Banach* space.

Many of the inequalities that we shall establish involve normed spaces and Banach spaces, which are the building blocks of functional analysis. Let us give some important fundamental examples. We shall meet many more.

Let $B(S)$ denote the space of bounded functions on a set S. $B(S)$ is a Banach space under the *supremum norm* $\|f\|_\infty = \sup_{s \in S} |f(s)|$. It is not separable if S is infinite. We write l_∞ for $B(\mathbf{N})$. The space

$$c_0 = \{x \in l_\infty : x_n \to 0 \text{ as } n \to \infty\}$$

is a separable closed linear subspace of l_∞, and is therefore also a Banach space under the norm $\|.\|_\infty$. If (X, τ) is a topological space then the space $C_b(X)$ of bounded continuous functions on X is a closed linear subspace of $B(X)$ and is therefore also a Banach space under the norm $\|.\|_\infty$.

Suppose that $(E, \|.\|_E)$ and $(F, \|.\|_F)$ are normed spaces. It is a standard result of linear analysis that a linear mapping T from E to F is continuous if and only if

$$\|T\| = \sup_{\|x\|_E \leq 1} \|T(x)\|_F < \infty,$$

that $L(E, F)$, the set of all continuous linear mappings from E to F, is a vector space under the usual operations, and that $\|T\|$ is a norm on $L(E, F)$. Further, $L(E, F)$ is a Banach space if and only if F is. In particular E^*, the *dual* of E, the space of all continuous linear functionals on E (continuous linear mappings from E into the underlying field), is a Banach space under the norm $\|\phi\|^* = \sup\{|\phi(x)| : \|x\|_E \leq 1\}$.

Standard results about normed spaces and Banach spaces are derived in Exercises 4.9–4.13.

Suppose that $f, g \in L^1(\Omega, \Sigma, \mu)$. Integrating the inequality $|f(x) + g(x)| \leq |f(x)| + |g(x)|$ and the equation $|\alpha f(x)| = |\alpha| \cdot |f(x)|$, we see that $L^1(\Omega, \Sigma, \mu)$ is a vector space, and that the function $\|f\|_1 = \int |f| \, d\mu$ is a seminorm on it. But $\int |f| \, d\mu = 0$ only if $f = 0$ almost everywhere, and so $\|.\|_1$ is in fact a norm. We shall see later (Theorem 5.1.1) that L^1 is a Banach space under this norm.

If V is an inner-product space, then, as we have seen in Chapter 2, $\|x\| = \langle x, x \rangle^{1/2}$ is a norm on V. If V is complete under this norm, V is called a *Hilbert space*. Again, we shall see later (Theorem 5.1.1) that $L^2 = L^2(\Omega, \Sigma, \mu)$ is a Hilbert space. A large amount of analysis, including the mathematical theory of quantum mechanics, takes place on a Hilbert space. Let us establish two fundamental results.

Proposition 4.5.1 *Suppose that V is an inner-product space. If $x, y \in V$, let $l_y(x) = \langle x, y \rangle$. Then l_y is a continuous linear functional on V, and*

$$\|l_y\|^* = \sup\{|l_y(x)|: \|x\| \leq 1\} = \|y\|.$$

The mapping $l : y \to l_y$ is an antilinear isometry of V into the dual space V^*: that is $\|l_y\|^* = \|y\|$ for each $y \in V$.

Proof Since the inner product is sesquilinear, l_y is a linear functional on V. By the Cauchy–Schwarz inequality, $|l_y(x)| \leq \|x\| \cdot \|y\|$, so that l_y is continuous, and $\|l_y\|^* \leq \|y\|$. On the other hand, $l_0 = 0$, and if $y \neq 0$ and $z = y/\|y\|$ then $\|z\| = 1$ and $l_y(z) = \|y\|$, so that $\|l_y\|^* = \|y\|$. Finally, l is antilinear, since the inner product is sesquilinear. □

When V is complete, we can say more.

Theorem 4.5.1 (The Fréchet–Riesz representation theorem) *Suppose that ϕ is a continuous linear functional on a Hilbert space H. Then there is a unique element $y \in H$ such that $\phi(x) = \langle x, y \rangle$.*

Proof The theorem asserts that the antilinear map l of the previous proposition maps H *onto* its dual H^*. If $\phi = 0$, we can take $y = 0$. Otherwise, by scaling (considering $\phi/\|\phi\|^*$), we can suppose that $\|\phi\|^* = 1$. Then for each n there exists y_n with $\|y_n\| \leq 1$ such that $\phi(y_n)$ is real and $\phi(y_n) \geq 1 - 1/n$. Since $\phi(y_n + y_m) \geq 2 - 1/n - 1/m$, $\|y_n + y_m\| \geq 2 - 1/n - 1/m$. We now apply the parallelogram law:

$$\|y_n - y_m\|^2 = 2\|y_n\|^2 + 2\|y_m\|^2 - \|y_n + y_m\|^2$$
$$\leq 4 - (2 - 1/n - 1/m)^2 < 4(1/n + 1/m).$$

Thus (y_n) is a Cauchy sequence: since H is complete, y_n converges to some y. Then $\|y\| = \lim_{n \to \infty} \|y_n\| \leq 1$ and $\phi(y) = \lim_{n \to \infty} \phi(y_n) = 1$, so that $\|y\| = 1$. We claim that $\phi(x) = \langle x, y \rangle$, for all $x \in H$.

First, consider $z \neq 0$ for which $\langle z, y \rangle = 0$. Now $\|y + \alpha z\|^2 = 1 + |\alpha|^2 \|z\|^2$ and $\phi(y + \alpha z) = 1 + \alpha \phi(z)$, so that $|1 + \alpha \phi(z)|^2 \leq 1 + |\alpha|^2 \|z\|^2$ for all scalars α. Setting $\alpha = \overline{\phi(z)} / \|z\|^2$, we see that

$$\left(1 + \frac{|\phi(z)|^2}{\|z\|^2}\right)^2 \leq 1 + \frac{|\phi(z)|^2}{\|z\|^2},$$

so that $\phi(z) = 0$. Suppose that $x \in H$. Let $z = x - \langle x, y \rangle y$, so that $\langle z, y \rangle = 0$. Then $\phi(x) = \langle x, y \rangle \phi(y) + \phi(z) = \langle x, y \rangle$. Thus y has the required property. This shows that the mapping l of the previous proposition is surjective. Since l is an isometry, it is one-one, and so y is unique. □

We shall not develop the rich geometric theory of Hilbert spaces (see [DuS 88] or [Bol 90]), but Exercises 4.5–4.8 establish results that we shall use.

4.6 The Hahn–Banach theorem for normed spaces

Theorem 4.6.1 *Suppose that p is a semi-norm on a real or complex vector space E, that F is a linear subspace of E and that f is a linear functional on F satisfying $|f(x)| \leq p(x)$ for all $x \in F$. Then there is a linear functional h on E such that*

$$h(x) = f(x) \ \text{for } x \in F \quad \text{and} \quad |h(y)| \leq p(y) \ \text{for } y \in E.$$

Proof In the real case, p is a sublinear functional on E which satisfies $p(x) = p(-x)$. By Theorem 4.4.1, there is a linear functional h on E which satisfies $h(x) \leq p(x)$. Then

$$|h(x)| = \max(h(x), h(-x)) \leq \max(p(x), p(-x)) = p(x).$$

We use Theorem 4.4.1 to deal with the complex case, too. Let $f_{\mathbf{R}}(x)$ be the real part of $f(x)$. Then $f_{\mathbf{R}}$ is a real linear functional on E, when E is considered as a real space, and $|f_{\mathbf{R}}(x)| \leq p(x)$ for all $x \in F$, and so there exists a real linear functional k on E extending $f_{\mathbf{R}}$ and satisfying $k(x) \leq p(x)$ for all x. Set $h(x) = k(x) - ik(ix)$. We show that h has the required properties. First, h is a complex linear functional on E: $h(x+y) = h(x) + h(y)$, $h(\alpha x) = \alpha h(x)$ when α is real, and

$$h(ix) = k(ix) - ik(-x) = k(ix) + ik(x) = ih(x).$$

Next, if $y \in F$ and $f(y) = re^{i\theta}$, then $f(e^{-i\theta}y) = r = k(e^{-i\theta}y)$ and $f(ie^{-i\theta}y) = ir$ so that $k(ie^{-i\theta}y) = 0$; thus $h(e^{-i\theta}y) = r = f(e^{-i\theta}y)$, and so $h(y) = f(y)$: thus h extends f. Finally, if $h(x) = re^{i\theta}$ then

$$|h(x)| = r = h(e^{-i\theta}x) = k(e^{-i\theta}x) \le p(e^{-i\theta}x) = p(x).$$

$\square$

This theorem is the key to the duality theory of normed spaces (and indeed of locally convex spaces, though we won't discuss these).

Corollary 4.6.1 *Suppose that x is a non-zero vector in a normed space $(E, \|.\|)$. Then there exists a linear functional ϕ on E such that*

$$\phi(x) = \|x\|, \quad \|\phi\|^* = \sup_{\|y\| \le 1} |\phi(y)| = 1.$$

Proof Take $F = \text{span}(x)$, and set $f(\alpha x) = \alpha \|x\|$. Then f is a linear functional on F, and $|f(\alpha x)| = |\alpha| \|x\| = \|\alpha x\|$. Thus f can be extended to a linear functional ϕ on E satisfying $|\phi(y)| \le \|y\|$, for $y \in E$. Thus $\|\phi\|^* \le 1$. As $\phi(x/\|x\|) = 1$, $\|\phi\|^* = 1$. $\square$

The dual E^{**} of E^* is called the *bidual* of E. The next corollary is an immediate consequence of the preceding one, once the linearity properties have been checked.

Corollary 4.6.2 *Suppose that $(E, \|.\|)$ is a normed space. If $x \in E$ and $\phi \in E^*$, let $E_x(\phi) = \phi(x)$. Then $E_x \in E^{**}$ and the mapping $x \to E_x$ is a linear isometry of E into E^{**}.*

We now have a version of the separation theorem for normed spaces.

Theorem 4.6.2 (The separation theorem: II) *Suppose that U is a non-empty open convex subset of a real normed space $(E, \|.\|_E)$.*

(i) If $0 \notin U$ there exists a continuous linear functional ϕ on E for which $\phi(x) > 0$ for $x \in U$.

(ii) If V is a non-empty convex subset of E disjoint from U there exists a continuous linear functional ϕ on E and a real number λ for which $\phi(x) > \lambda$ for $x \in U$ and $\phi(y) \le \lambda$ for $y \in V$.

(iii) If F is a linear subspace of E disjoint from U there exists a continuous linear functional ϕ on E for which $\phi(x) > 0$ for $x \in U$ and $\phi(y) = 0$ for $y \in F$.

Proof U is radially open, and so by Theorem 4.4.2 there exists a linear functional ϕ on E for which $\phi(x) > 0$ for $x \in U$. We show that ϕ is continuous: inspection of the proof of Theorem 4.4.2 then shows that (ii) and (iii) are also satisfied.

Let $x_0 \in U$. Since U is open, there exists $r > 0$ such that if $\|x - x_0\|_E \leq r$ then $x \in U$. We show that if $\|x\|_E \leq 1$ then $|\phi(x)| < \phi(x_0)/r$. Suppose not, so that there exists x_1 with $\|x_1\|_E \leq 1$ and $|\phi(x_1)| \geq \phi(x_0)/r$. Let $y = x_0 - r(\phi(x_1)/|\phi(x_1)|)x_1$. Then $y \in U$ and $\phi(y) = \phi(x_0) - r|\phi(x_1)| \leq 0$, giving the required contradiction. $\qquad\square$

We also have the following metric result.

Theorem 4.6.3 (The separation theorem: III) *Suppose that A is a non-empty closed convex subset of a real normed space $(E, \|.\|_E)$, and that x_0 is a point of E not in A. Let $d = d(x_0, A) = \inf\{\|x_0 - a\| : a \in A\}$. Then there exists $\psi \in E^*$ with $\|\psi\|^* = 1$ such that $\psi(x_0) \geq \psi(a) + d$ for all $a \in A$.*

Proof We apply Theorem 4.6.2 (ii) to the disjoint convex sets $x_0 + dU$ and A, where $U = \{x \in E : \|x\| < 1\}$. There exists a continuous linear functional ϕ on E and a real number λ such that $\phi(a) \leq \lambda$ for $a \in A$ and $\phi(x_0 + x) > \lambda$ for $\|x\|_E < d$. Let $\psi = \phi/\|\phi\|^*$, so that $\|\psi\|^* = 1$. Suppose that $a \in A$ and that $0 < \theta < 1$. There exists $y \in E$ with $\|y\| < 1$ such that $\psi(y) > \theta$. Then $\psi(x_0) - d\theta > \psi(x_0 - dy) > \psi(a)$. Since this holds for all $0 < \theta < 1$, $\psi(x_0) \geq \psi(a) + d$. $\qquad\square$

We also have the following normed-space version of Corollary 4.4.1.

Corollary 4.6.3 *Suppose that f is a continuous convex function on an open convex subset C of a real normed space $(E, \|.\|)$ and that $x \in C$. Then there exists a continuous affine function a such that $a(x) = f(x)$ and $a(y) \leq f(y)$ for $y \in C$.*

Proof By Corollary 4.4.1, there exists an affine function a such that $a(x) = f(x)$ and $a(y) \leq f(y)$ for $y \in C$. We need to show that a is continuous. We can write $a(z) = f(x) + \phi(z - x)$, where ϕ is a linear functional on E. Given $\epsilon > 0$, there exists $\delta > 0$ such that if $\|z\| < \delta$ then $x + z \in C$ and $|f(x + z) - f(x)| < \epsilon$. Then if $\|z\| < \delta$,

$$f(x) + \phi(z) = a(x + z) \leq f(x + z) < f(x) + \epsilon,$$

so that $\phi(z) < \epsilon$. But also $\|-z\| < \delta$, so that $-\phi(z) = \phi(-z) < \epsilon$, and $|\phi(z)| < \epsilon$. Thus ϕ is continuous at 0, and is therefore continuous (Exercise 4.9); so therefore is a. $\qquad\square$

4.7 Barycentres and weak integrals

We now return to Jensen's inequality, and consider what happens on Banach spaces. Once again, we must first consider barycentres. Suppose that μ is a probability measure defined on the Borel sets of a real Banach space $(E, \|.\|)$. If $\phi \in E^*$ then ϕ is Borel measurable. Suppose that each $\phi \in E^*$ is in $L^1(\mu)$. Let $I_\mu(\phi) = \int_E \phi(x)\, d\mu(x)$. Then I_μ is a linear functional on E^*. If there exists $\bar{\mu}$ in E such that $I_\mu(\phi) = \phi(\bar{\mu})$ for all $\phi \in E^*$, then $\bar{\mu}$ is called the *barycentre* of μ.

A barycentre need not exist: but in fact if μ is a probability measure defined on the Borel sets of a real Banach space $(E, \|.\|)$, and μ is supported on a bounded closed set B (that is, $\mu(E \setminus B) = 0$), then μ has a barycentre in E.

Here is another version of Jensen's inequality.

Theorem 4.7.1 (Jensen's inequality: III) *Suppose that μ is a probability measure on the Borel sets of a separable real normed space E, and that μ has a barycentre $\bar{\mu}$. If f is a continuous convex function on E with $\int_E f^-\, d\mu < \infty$ then $f(\bar{\mu}) \leq \int_E f\, d\mu$. If f is strictly convex then equality holds if and only if $\mu = \delta_{\bar{\mu}}$.*

Proof The proof is exactly the same as Theorem 4.2.1. Proposition 4.6.3 ensures that the affine function that we obtain is continuous. $\qquad\square$

Besides considering measures defined on a Banach space, we shall also consider functions taking values in a Banach space. Let us describe here what we need to know.

Theorem 4.7.2 (Pettis' theorem) *Suppose that (Ω, Σ, μ) is a measure space, and that $g : \Omega \to (E, \|.\|)$ is a mapping of Ω into a Banach space $(E, \|.\|)$. The following are equivalent:*

(i) $g^{-1}(B) \in \Sigma$, for each Borel set B in E, and there exists a sequence g_n of simple E-valued measurable functions which converges pointwise almost everywhere to g.

(ii) g is weakly measurable – that is, $\phi \circ g$ is measurable for each ϕ in E^* – and there exists a closed separable subspace E_0 of E such that $g(\omega) \in E_0$ for almost all ω.

If these equivalent conditions hold, we say that g is *strongly measurable*.

Now suppose that g is strongly measurable and that $I \in E$. We say that g is *weakly integrable*, with weak integral I, if $\phi(g) \in L^1(\mu)$, and $\int_\Omega \phi(g) \, d\mu = \phi(I)$, for each $\phi \in E^*$. Note that when μ is a probability measure this simply states that I is the barycentre of the image measure $g(\mu)$, which is the Borel measure on E defined by $g(\mu)(B) = \mu(g^{-1}(B))$ for each Borel set B in E.

By contrast, we say that a measurable function g is *Bochner integrable* if there exists a sequence (g_n) of simple functions such that $\int_\Omega \|g - g_n\| \, d\mu \to 0$ as $n \to \infty$. Then $\int g_n \, d\mu$ (defined in the obvious way) converges in E, and we define the Bochner integral $\int g \, d\mu$ as the limit. A measurable function g is Bochner integrable if and only if $\int \|g\| \, d\mu < \infty$. A Bochner integrable function is weakly integrable, and the Bochner integral is then the same as the weak integral.

We conclude this chapter with the following useful mean-value inequality.

Proposition 4.7.1 (The mean-value inequality) *Suppose that* $g : (\Omega, \Sigma, \mu) \to (E, \|.\|)$ *is weakly integrable, with weak integral* I. *Then*

$$\|I\| \leq \int_\Omega \|g\| \, d\mu.$$

Proof There exists an element $\phi \in E^*$ with $\|\phi\|^* = 1$ such that

$$\|I\| = \phi(I) = \int \phi(g) \, d\mu.$$

Then since $|\phi(g)| \leq \|g\|$,

$$\|I\| \leq \int |\phi(g)| \, d\mu \leq \int \|g\| \, d\mu.$$

$\square$

4.8 Notes and remarks

Jensen proved versions of his inequality in [Jen 06], a landmark in convex analysis. He wrote: *It seems to me that the notion of 'convex function' is almost as fundamental as these: 'positive function', 'increasing function'. If I am not mistaken in this then the notion should take its place in elementary accounts of real functions.*

The Hahn–Banach theorem for real vector spaces, was proved independently by Hahn [Hah 27] and Banach [Ban 29]. The complex version was proved several years later, by Bohnenblust and Sobczyk [BoS 38].

Details of the results described in Section 4.7 are given in [DiU 77].

Exercises

4.1 (i) Use Jensen's inequality to show that if $x > 0$ then

$$\frac{2x}{2+x} < \log(1+x) < \frac{2x+x^2}{2+2x}.$$

Let $d_n = (n+1/2)\log(1+1/n) - 1$. Show that

$$0 < d_n < 1/4n(n+1).$$

Let $r_n = n!e^n/n^{n+1/2}$. Calculate $\log(r_{n+1}/r_n)$, and show that r_n decreases to a finite limit C. Show that $r_n \le e^{1/4n}C$.

(ii) Let $I_n + \int_0^{\pi/2} \sin^n \theta \, d\theta$. Show that I_n is a decreasing sequence of positive numbers, and show, by integration by parts, that $nI_n = (n-1)I_{n-2}$ for $n \ge 2$. Show that

$$\frac{I_{2n+1}}{I_{2n}} = \frac{2^{4n+1}(n!)^4}{\pi(2n)!(2n+1)!} \to 1$$

as $n \to \infty$, and deduce that $C = \sqrt{2\pi}$. Thus $n! \sim \sqrt{2\pi}n^{n+1/2}/e^n$.

This is *Stirling's formula*. Another derivation of the value of C will be given in Theorem 13.6.1.

4.2 Suppose that f is a convex function defined on an open interval I of the real line. Show that D^+f and D^-f are increasing functions, that D^+f is right-continuous and D^-f left-continuous, and that $D^-f(x) \ge D^+f(y)$ if $x > y$. Show that $D^+f(x)$ and $D^-f(x)$ are equal and continuous, except at a countable set of points where

$$\lim_{h\searrow 0} D^+f(x-h) = D^-f(x) < D^+f(x) = \lim_{h\searrow 0} D^-f(x+h).$$

Show that f is differentiable, except at this countable set of points.

4.3 Suppose that f is a real-valued function defined on an open interval I of the real line. Show that f is convex if and only if there exists an increasing function g on I such that

$$f(x) = \int_{x_0}^x g(t) \, dt + c,$$

where x_0 is a point of I and c is a constant.

4.4 Suppose that $(\Omega, \Sigma, \mathbf{P})$ is a probability space, and that f is a non-negative measurable function on Ω for which

$$\mathbf{E}(\log^+ f) = \int_\Omega \log^+ f \, d\mathbf{P} = \int_{(f>1)} \log f \, d\mathbf{P} < \infty,$$

so that $-\infty \leq \mathbf{E}(\log f) < \infty$. Let $G(f) = \exp(\mathbf{E}(\log f))$, so that $0 \leq G(f) < \infty$. $G(f)$ is the *geometric mean* of f. Explain this terminology. Show that $G(f) \leq \mathbf{E}(f)$.

4.5   This question, and the three following ones, establish results about Hilbert spaces that we shall use later. Suppose that A is a non-empty subset of a Hilbert space H. Show that $A^\perp = \{y \colon \langle a, y \rangle = 0 \text{ for } a \in A\}$ is a closed linear subspace of H.

4.6   Suppose that C is a non-empty closed convex subset of a Hilbert space H and that $x \in H$. Use an argument similar to that of Theorem 4.5.1 to show that there is a unique point $c \in C$ with $\|x - c\| = \inf\{\|x - y\| \colon y \in C\}$.

4.7   Suppose that F is a closed linear subspace of a Hilbert space H and that $x \in H$.

(i) Let $P(x)$ be the unique nearest point to x in F. Show that $x - P(x) \in F^\perp$, and that if $y \in F$ and $x - y \in F^\perp$ then $y = P(x)$.

(ii) Show that $P : H \to H$ is linear and that if $F \neq \{0\}$ then $\|P\| = 1$. P is the *orthogonal projection* of H onto F.

(iii) Show that $H = F \oplus F^\perp$, and that if P is the orthogonal projection of H onto F then $I - P$ is the orthogonal projection of H onto $F^\perp$.

4.8   Suppose that (x_n) is a linearly independent sequence of elements of a Hilbert space x.

(i) Let $P_0 = 0$, let P_n be the orthogonal projection of H onto span $(x_1, \ldots, x_n)$, and let $Q_n = I - P_n$. Let $y_n = Q_{n-1}(x_n)/\|Q_{n-1}(x_n)\|$. Show that (y_n) is an orthonormal sequence in H: $\|y_n)\| = 1$ for each n, and $\langle y_m, y_n \rangle = 0$ for $m \neq n$. Show that span $(y_1, \ldots, y_n) =$ span $(x_1, \ldots, x_n)$, for each n.

(ii) [Gram–Schmidt orthonormalization] Show that the sequence (y_n) can be defined recursively by setting

$$y_1 = x_1/\|x_1\|, \quad z_n = x_n - \sum_{i=1}^{n-1} \langle x_i, y_i \rangle \, y_i \text{ and } y_n = z_n/\|z_n\|.$$

4.9   This question, and the four following ones, establish fundamental properties about normed spaces. Suppose that $(E, \|.\|_E)$ and $(F, \|.\|_F)$ are normed spaces. Suppose that T is a linear mapping from E to F. Show that the following are equivalent:

(i) T is continuous at 0;

(ii) T is continuous at each point of E;

(iii) T is uniformly continuous;

(iv) T is Lipschitz continuous at 0;

(v) T is a Lipschitz function;

(vi) $\|T\| = \sup\{\|T(x)\|_F : \|x\|_E \le 1\} < \infty$.

4.10 Show that the set $L(E, F)$ of continuous linear mappings from E to F is a vector space under the usual operations. Show that $\|T\| = \sup\{\|T(x)\|_F : \|x\|_E \le 1\}$ is a norm (the *operator norm*) on $L(E, F)$. Show that if $(F, \|.\|_F)$ is complete then $L(E, F)$ is complete under the operator norm.

4.11 Suppose that $T \in L(E, F)$. If $\phi \in F^*$ and $x \in E$, let $T^*(\phi)(x) = \phi(T(x))$. Show that $T^*(\phi) \in E^*$ and that $\|T^*(\phi)\|_{E^*} \le \|T\| \cdot \|\phi\|_{F^*}$. Show that $T^* \in L(F^*, E^*)$ and that $\|T^*\| \le \|T\|$. Use Corollary 4.6.1 to show that $\|T^*\| = \|T\|$. T^* is the *transpose* or *conjugate* of T.

4.12 Suppose that T is a linear functional on a normed space $(E, \|.\|_E)$. Show that ϕ is continuous if and only if its null-space $\phi^{-1}(\{0\})$ is closed.

4.13 Suppose that F is a closed linear subspace of a normed space $(E, \|.\|_E)$, and that $q : E \to E/F$ is the quotient mapping. If $x \in E$, let $d(x, F) = \inf\{\|x - y\|_E : y \in F\}$. Show that if $q(x_1) = q(x_2)$ then $d(x_1, F) = d(x_2, F)$. If $z = q(x)$, let $\|z\|_{E/F} = d(x, F)$. Show that $\|.\|_{E/F}$ is a norm on E/F (the *quotient norm*). Show that if E is complete then $(E/F, \|.\|_{E/F})$ is.

4.14 Show that the vector space $B(S)$ of all bounded (real- or complex-valued) functions on a set S is complete under the norm $\|f\|_\infty = \sup\{|f(s)|: s \in S\}$, and that if (X, τ) is a topological space then the space $C_b(X)$ of bounded continuous functions on X is a closed linear subspace of $B(X)$ and is therefore also a Banach space under the norm $\|.\|_\infty$.

4.15 Suppose that f is a bounded convex function defined on an open convex subset of a normed space E. Show that f is Lipschitz continuous. Give an example of a convex function defined on an open convex subset of a normed space E which is not continuous.

4.16 Show that a sublinear functional is convex, and that a convex positive homogeneous function is sublinear.

4.17 Show that the closure and the interior of a convex subset of a normed space are convex.

4.18 Here is a version of the separation theorem for complex normed spaces. A convex subset A of a real or complex vector space is *absolutely convex* if whenever $x \in A$ then $\lambda x \in A$ for all λ with $|\lambda| \le 1$. Show that if A is a closed absolutely convex subset of a complex normed space

$(E, \|.\|_E)$ and $x_0 \notin A$ then there exists a continuous linear functional ψ on E with $\|\psi\|^* = 1$, $\psi(x_0)$ real and

$$\psi(x_0) \geq \sup_{a \in A} |\psi(a)| + d(x_0, A).$$

4.19 Let ϕ be the vector space of all infinite sequences with only finitely many non-zero terms, with the supremum norm. Let μ be defined by

$$\mu(A) = \sum \{2^{-n} : e_n \in A\},$$

where e_n is the sequence with 1 in the nth place, and zeros elsewhere. Show that μ is a probability measure on the Borel sets of ϕ which is supported on the unit ball of ϕ, and show that μ does not have a barycentre.

4.20 Let μ be the Borel probability measure on c_0 defined by

$$\mu(A) = \sum \{2^{-n} : 2^n e_n \in A\},$$

where e_n is the sequence with 1 in the nth place, and zeros elsewhere. Show that μ does not have a barycentre.

5

The L^p spaces

5.1 L^p spaces, and Minkowski's inequality

Our study of convexity led us to consider normed spaces. We are interested in inequalities between sequences and between functions, and this suggests that we should consider normed spaces whose elements are sequences, or (equivalence classes of) functions. We begin with the L^p spaces.

Suppose that (Ω, Σ, μ) is a measure space, and that $0 < p < \infty$. We define $\mathcal{L}^p(\Omega, \Sigma, \mu)$ to be the collection of those (real- or complex-valued) measurable functions for which

$$\int_\Omega |f|^p \, d\mu < \infty.$$

If $f = g$ almost everywhere, then $\int_\Omega |f - g|^p \, d\mu = 0$ and $\int_\Omega |f|^p \, d\mu = \int_\Omega |g|^p \, d\mu$. We therefore identify functions which are equal almost everywhere, and denote the resulting space by $L^p = L^p(\Omega, \Sigma, \mu)$.

If $f \in L^p$ and α is a scalar, then $\alpha f \in L^p$. Since $|a+b|^p \le 2^p \max(|a|^p, |b|^p) \le 2^p(|a|^p + |b|^p)$, $f + g \in L^p$ if $f, g \in L^p$. Thus f is a vector space.

Theorem 5.1.1 *(i) If $1 \le p < \infty$ then $\|f\|_p = (\int |f|^p \, d\mu)^{1/p}$ is a norm on L^p.*

(ii) If $0 < p < 1$ then $d_p(f, g) = \int |f - g|^p \, d\mu$ is a metric on L^p.

(iii) $(L^p, \|.\|_p)$ is a Banach space for $1 \le p < \infty$ and (L^p, d_p) is a complete metric space for $0 < p < 1$.

Proof The proof depends on the facts that the function t^p is convex on $[0, \infty)$ for $1 \le p < \infty$ and is concave for $0 < p < 1$.

(i) Clearly $\|\alpha f\|_p = |\alpha| \|f\|_p$. If f or g is zero then trivially $\|f + g\|_p \le \|f\|_p + \|g\|_p$. Otherwise, let $F = f/ \|f\|_p$, $G = g/ \|g\|_p$, so that $\|F\|_p = $

$\|G\|_p = 1$. Let $\lambda = \|g\|_p / (\|f\|_p + \|g\|_p)$, so that $0 < \lambda < 1$. Now

$$
\begin{aligned}
|f + g|^p &= (\|f\|_p + \|g\|_p)^p \, |(1 - \lambda)F + \lambda G|^p \\
&\leq (\|f\|_p + \|g\|_p)^p \, ((1 - \lambda)|F| + \lambda|G|)^p \\
&\leq (\|f\|_p + \|g\|_p)^p \, ((1 - \lambda)|F|^p + \lambda|G|^p) \, ,
\end{aligned}
$$

since t^p is convex, for $1 \leq p < \infty$. Integrating,

$$
\begin{aligned}
\int |f + g|^p \, d\mu &\leq (\|f\|_p + \|g\|_p)^p \left((1 - \lambda) \int |F|^p \, d\mu + \lambda \int |G|^p \, d\mu \right) \\
&= (\|f\|_p + \|g\|_p)^p.
\end{aligned}
$$

Thus we have established *Minkowski's inequality*

$$
\left(\int |f + g|^p \, d\mu \right)^{1/p} \leq \left(\int |f|^p \, d\mu \right)^{1/p} + \left(\int |g|^p \, d\mu \right)^{1/p},
$$

and shown that $\|.\|_p$ is a norm.

(ii) If $0 < p < 1$, the function t^{p-1} is decreasing on $(0, \infty)$, so that if a and b are non-negative, and not both 0, then

$$
(a + b)^p = a(a + b)^{p-1} + b(a + b)^{p-1} \leq a^p + b^p.
$$

Integrating,

$$
\int |f + g|^p \, d\mu \leq \int (|f| + |g|)^p \, d\mu \leq \int |f|^p \, d\mu + \int |g|^p \, d\mu;
$$

this is enough to show that d_p is a metric.

(iii) For this, we need *Markov's inequality*: if $f \in L^p$ and $\alpha > 0$ then $\alpha^p I_{(|f| > \alpha)} \leq |f|^p$; integrating, $\alpha^p \mu(|f| > \alpha) \leq \int |f|^p \, d\mu$. Suppose that (f_n) is a Cauchy sequence. Then it follows from Markov's inequality that (f_n) is locally Cauchy in measure, and so it converges locally in measure to a function f. By Proposition 1.2.2, there is a subsequence (f_{n_k}) which converges almost everywhere to f. Now, given $\epsilon > 0$ there exists K such that $\int |f_{n_k} - f_{n_l}|^p \, d\mu < \epsilon$ for $k, l \geq K$. Then, by Fatou's lemma, $\int |f_{n_k} - f|^p \, d\mu \leq \epsilon$ for $k \geq K$. This shows first that $f_{n_k} - f \in L^p$, for $k \geq K$, so that $f \in L^p$, and secondly that $f_{n_k} \to f$ in norm as $k \to \infty$. Since (f_n) is a Cauchy sequence, it follows that $f_n \to f$ in norm, as $n \to \infty$, so that L^p is complete. $\qquad\square$

In a similar way if E is a Banach space, and $0 < p < \infty$, then we denote by $L^p(\Omega; E) = L^p(E)$ the collection of (equivalence classes of) measurable E-valued functions for which $\int \|f\|^p \, d\mu < \infty$. The results of Theorem 5.1.1

carry over to these spaces, with obvious changes to the proof (replacing absolute values by norms).

Let us also introduce the space $L^\infty = L^\infty(\Omega, \Sigma, \mu)$. A measurable function f is *essentially bounded* if there exists a set B of measure 0 such that f is bounded on $\Omega \backslash B$. If f is essentially bounded, we define its essential supremum to be

$$\text{ess sup } f = \inf\{t \colon \lambda_{|f|}(t) = \mu(|f| > t) = 0\}.$$

If f is essentially bounded and $g = f$ almost everywhere then g is also essentially bounded, and ess sup f = ess sup g. We identify essentially bounded functions which are equal almost everywhere; the resulting space is L^∞. L^∞ is a vector space, $\|f\|_\infty$ = ess sup $|f|$ is a norm and straightforward arguments show that $(L^\infty, \|.\|_\infty)$ is a Banach space.

5.2 The Lebesgue decomposition theorem

As an important special case, L^2 is a Hilbert space. We now use the Fréchet–Riesz representation theorem to prove a fundamental theorem of measure theory.

Theorem 5.2.1 (The Lebesgue decomposition theorem) *Suppose that (Ω, Σ, μ) is a measure space, and that ν is a measure on Σ with $\nu(\Omega) < \infty$. Then there exists a non-negative $f \in L^1(\mu)$ and a set $B \in \Sigma$ with $\mu(B) = 0$ such that $\nu(A) = \int_A f \, d\mu + \nu(A \cap B)$ for each $A \in \Sigma$.*

If we define $\nu_B(A) = \nu(A \cap B)$ for $A \in \Sigma$, then ν_B is a measure. The measures μ and ν_B are mutually singular; we decompose Ω as $B \cup (\Omega \setminus B)$, where $\mu(B) = 0$ and $\nu_B(\Omega \setminus B) = 0$; μ and ν_B live on disjoint sets.

Proof Let $\pi(A) = \mu(A) + \nu(A)$; π is a measure on Σ. Suppose that $g \in L_{\mathbf{R}}^2(\pi)$. Let $L(g) = \int g \, d\nu$. Then, by the Cauchy–Schwarz inequality,

$$|L(g)| \leq (\nu(\Omega))^{1/2} \left(\int |g|^2 \, d\nu\right)^{1/2} \leq (\nu(\Omega))^{1/2} \|g\|_{L^2(\pi)},$$

so that L is a continuous linear functional on $L_{\mathbf{R}}^2(\pi)$. By the Fréchet–Riesz theorem, there exists an element $h \in L_{\mathbf{R}}^2(\pi)$ such that $L(g) = \langle g, h \rangle$, for each $g \in L^2(\pi)$; that is, $\int_\Omega g \, d\nu = \int_\Omega gh \, d\mu + \int_\Omega gh \, d\nu$, so that

$$\int_\Omega g(1 - h) \, d\nu = \int_\Omega gh \, d\mu. \quad (*)$$

Taking g as an indicator function I_A, we see that

$$\nu(A) = L(I_A) = \int_A h \, d\pi = \int_A h \, d\mu + \int_A h \, d\nu$$

for each $A \in \Sigma$.

Now let $N = (h < 0)$, $G_n = (0 \leq h \leq 1 - 1/n)$, $G = (0 \leq h < 1)$ and $B = (h \geq 1)$. Then

$$\nu(N) = \int_N h \, d\mu + \int_N h \, d\nu \leq 0, \text{ so that } \mu(N) = \nu(N) = 0,$$

and

$$\nu(B) = \int_B h \, d\mu + \int_B h \, d\nu \geq \nu(B) + \mu(B), \text{ so that } \mu(B) = 0.$$

Let $f(x) = h(x)/(1 - h(x))$ for $x \in G$, and let $h(x) = 0$ otherwise. Note that if $x \in G_n$ then $0 \leq f(x) \leq 1/(1 - h(x)) \leq n$. If $A \in \Sigma$, then, using $(*)$,

$$\nu(A \cap G_n) = \int_\Omega \frac{1 - h}{1 - h} I_{A \cap G_n} \, d\nu = \int_\Omega f I_{A \cap G_n} \, d\mu = \int_{A \cap G_n} f \, d\mu.$$

Applying the monotone convergence theorem, we see that $\nu(A \cap G) = \int_{A \cap G} f \, d\mu = \int_A f \, d\mu$. Thus

$$\nu(A) = \nu(A \cap G) + \nu(A \cap B) + \nu(A \cap N) = \int_A f \, d\mu + \nu(A \cap B).$$

Taking $A = \Omega$, we see that $\int_\Omega f \, d\mu < \infty$, so that $f \in L^1(\mu)$. □

This beautiful proof is due to von Neumann.

Suppose that (Ω, Σ, μ) is a measure space, and that ψ is a real-valued function on Σ. We say that ψ is *absolutely continuous* with respect to μ if, given $\epsilon > 0$, there exists $\delta > 0$ such that if $\mu(A) < \delta$ then $|\psi(A)| < \epsilon$.

Corollary 5.2.1 (The Radon–Nykodým theorem) *Suppose that (Ω, Σ, μ) is a measure space, and that ν is a measure on Σ with $\nu(\Omega) < \infty$. Then ν is absolutely continuous with respect to μ if and only if there exists a non-negative $f \in L^1(\mu)$ such that $\nu(A) = \int_A f \, d\mu$ for each $A \in \Sigma$.*

Proof Suppose first that ν is absolutely continuous with respect to μ. If $\mu(B) = 0$ then $\nu(B) = 0$, and so the measure ν_B of the theorem is zero. Conversely, suppose that the condition is satisfied. Let $B_n = (f > n)$. Then by the dominated convergence theorem, $\nu(B_n) = \int_{B_n} f \, d\mu \to 0$. Suppose that $\epsilon > 0$. Then there exists n such that $\nu(B_n) < \epsilon/2$. Let $\delta = \epsilon/2n$. Then if $\mu(A) < \delta$,

$$\nu(A) = \nu(A \cap B_n) + \int_{A \cap (0 \leq f \leq n)} f \, d\mu < \epsilon/2 + n\delta = \epsilon.$$

$\square$

We also need a 'signed' version of this corollary.

Theorem 5.2.2 *Suppose that* (Ω, Σ, μ) *is a measure space, with* $\mu(\Omega) < \infty$, *and that* ψ *is a bounded absolutely continuous real-valued function on* Σ *which is additive: if* A, B *are disjoint sets in* Σ *then* $\psi(A \cup B) = \psi(A) + \psi(B)$. *Then there exists* $f \in L^1$ *such that* $\psi(A) = \int_A f \, d\mu$, *for each* $A \in \Sigma$.

Proof If $A \in \Sigma$, let $\psi^+(A) = \sup\{\psi(B): B \subseteq A\}$. ψ^+ is a bounded additive non-negative function on Σ. We shall show that ψ^+ is countably additive. Suppose that A is the disjoint union of (A_i). Let $R_j = \cup_{i>j} A_i$. Then $R_j \searrow \emptyset$, and so $\mu(R_j) \to 0$ as $j \to \infty$. By absolute continuity, $\sup\{|\psi(B)|: B \subseteq R_j\} \to 0$ as $j \to \infty$, and so $\psi^+(R_j) \to 0$ as $j \to \infty$. This implies that ψ^+ is countably additive. Thus ψ^+ is a measure on Σ, which is absolutely continuous with respect to μ, and so it is represented by some $f^+ \in L^1(\mu)$. But now $\psi^+ - \psi$ is additive, non-negative and absolutely continuous with respect to μ, and so is represented by a function f^-. Let $f = f^+ - f^-$. Then $f \in L^1(\mu)$ and

$$\psi(A) = \psi^+(A) - (\psi^+(A) - \psi(A)) = \int_A f^+ \, d\mu - \int_A f^- \, d\mu = \int_A f \, d\mu.$$

$\square$

5.3 The reverse Minkowski inequality

When $0 < p < 1$ and L^p is infinite-dimensional then there is no norm on L^p which defines the topology on L^p. Indeed if (Ω, Σ, μ) is atom-free there are no non-trivial convex open sets, and so no non-zero continuous linear functionals (see Exercise 5.4). In this case, the inequality in Minkowski's inequality is reversed.

Proposition 5.3.1 (The reverse Minkowski inequality) *Suppose that* $0 < p < 1$ *and that* f *and* g *are non-negative functions in* L^p. *Then*

$$\left(\int f^p \, d\mu \right)^{1/p} + \left(\int g^p \, d\mu \right)^{1/p} \leq \left(\int (f+g)^p \, d\mu \right)^{1/p}.$$

Proof Let $q = 1/p$ and let $w = (u, v) = (f^q, g^q)$. Thus w takes values in $\mathbf{R}^2$, which we equip with the norm $\|(x, y)\|_q = (|x|^q + |y|^q)^{1/q}$. Let

$$I(w) = \int w \, d\mu = \left(\int u \, d\mu, \int v \, d\mu \right).$$

Then

$$\|I(w)\|_q^q = \left(\int u \, d\mu \right)^q + \left(\int v \, d\mu \right)^q = \left(\int f^p \, d\mu \right)^{1/p} + \left(\int g^p \, d\mu \right)^{1/p},$$

while

$$\left(\int \|w\|_q \, d\mu \right)^q = \left(\int (u^q + v^q)^{1/q} \, d\mu \right)^q = \left(\int (f + g)^p \, d\mu \right)^{1/p},$$

so that the result follows from the mean-value inequality (Proposition 4.7.1).   □

In the same way, the inequality in Proposition 4.7.1 is reversed.

Proposition 5.3.2 *Suppose that $0 < p < 1$ and that f and g are non-negative functions in L^1. Then*

$$\int (f^p + g^p)^{1/p} \, d\mu \leq \left(\left(\int f \, d\mu \right)^p + \left(\int g \, d\mu \right)^p \right)^{1/p}.$$

Proof As before, let $q = 1/p$ and let $u = f^p$, $v = g^p$. Then $u, v \in L^q$ and, using Minkowski's inequality,

$$\int (f^p + g^p)^{1/p} \, d\mu = \int (u + v)^q \, d\mu = \|u + v\|_q^q$$

$$\leq (\|u\|_q + \|v\|_q)^q = \left(\left(\int f \, d\mu \right)^p + \left(\int g \, d\mu \right)^p \right)^{1/p}.$$

    □

5.4 Hölder's inequality

If $1 < p < \infty$, we define the *conjugate index* p' to be $p' = p/(p - 1)$. Then $1/p + 1/p' = 1$, so that p is the conjugate index of p'. We also define ∞ to be the conjugate index of 1, and 1 to be the conjugate index of ∞.

Note that, by Proposition 4.1.3, if p and p' are conjugate indices, and t and u are non-negative, then

$$tu \leq \frac{t^p}{p} + \frac{u^{p'}}{p'},$$

with equality if and only if $t^p = u^{p'}$. We use this to prove *Hölder's inequality*. This inequality provides a natural and powerful generalization of the Cauchy–Schwarz inequality.

We define the *signum* $\operatorname{sgn}(z)$ of a complex number z as $z/|z|$ if $z \neq 0$, and 0 if $z = 0$.

Theorem 5.4.1 (Hölder's inequality) *Suppose that $1 < p < \infty$, that $f \in L^p$ and $g \in L^{p'}$. Then $fg \in L^1$, and*

$$\left| \int fg\,d\mu \right| \leq \int |fg|\,d\mu \leq \|f\|_p \|g\|_{p'}.$$

Equality holds throughout if and only if either $\|f\|_p \|g\|_{p'} = 0$, or $g = \lambda \overline{\operatorname{sgn}(f)}|f|^{p-1}$ almost everywhere, where $\lambda \neq 0$.

Proof The result is trivial if either f or g is zero. Otherwise, by scaling, it is enough to consider the case where $\|f\|_p = \|g\|_{p'} = 1$. Then by the inequality above $|fg| \leq |f|^p/p + |g|^{p'}/p'$; integrating,

$$\int |fg|\,d\mu \leq \int |f|^p/p\,d\mu + \int |g|^{p'}/p'\,d\mu = 1/p + 1/p' = 1.$$

Thus $fg \in L^1(\mu)$ and $|\int fg\,d\mu| \leq \int |fg|\,d\mu$.

If $g = \lambda \overline{\operatorname{sgn}(f)}|f|^{p-1}$ almost everywhere, then $fg = \lambda|fg| = \lambda|f|^p = \lambda|g|^{p'}$ almost everywhere, so that equality holds.

Conversely, suppose that

$$\left| \int fg\,d\mu \right| = \int |fg|\,d\mu = \|f\|_p \|g\|_{p'}.$$

Then, again by scaling, we need only consider the case where $\|f\|_p = \|g\|_{p'} = 1$. Since $|\int fg\,d\mu| = \int |fg|\,d\mu$, there exists θ such that $e^{i\theta}fg = |fg|$ almost everywhere. Since

$$\int |fg|\,d\mu = 1 = \int |f|^p/p\,d\mu + \int |g|^{p'}/p'\,d\mu \quad \text{and} \quad |f|^p/p + |g|^{p'}/p' \geq |fg|,$$

$|fg| = |f|^p/p + |g|^{p'}/p'$ almost everywhere, and so $|f|^p = |g|^{p'}$ almost everywhere. Thus $|g| = |f|^{p/p'} = |f|^{p-1}$ almost everywhere, and $g = e^{-i\theta}\overline{\operatorname{sgn}(f)}|f|^{p-1}$ almost everywhere. $\square$

Corollary 5.4.1 *if $f \in L^p$ then*

$$\|f\|_p = \sup\left\{\int |fg|\,d\mu\colon \|g\|_{p'} \le 1\right\} = \sup\left\{\left|\int fg\,d\mu\right|\colon \|g\|_{p'} \le 1\right\},$$

and the supremum is attained.

Proof The result is trivially true if $f = 0$; let us suppose that $f \ne 0$. Certainly

$$\|f\|_p \ge \sup\left\{\int |fg|\,d\mu\colon \|g\|_{p'} \le 1\right\} \ge \sup\left\{\left|\int fg\,d\mu\right|\colon \|g\|_{p'} \le 1\right\},$$

by Hölder's inequality. Let $h = |f|^{p-1}\overline{\operatorname{sgn} f}$. Then

$$fh = |fh| = |f|^p = |h|^{p'},$$

so that $h \in L^{p'}$ and $\|h\|_{p'} = \|f\|_p^{p/p'}$. Let $g = h/\|h\|_{p'}$, so that $\|g\|_{p'} = 1$. Then

$$\int fg\,d\mu = \int |fg|\,d\mu = \int \frac{|f|^p}{\|f\|_p^{p/p'}}\,d\mu = \|f\|_p^p / \|f\|_p^{p/p'} = \|f\|_p\,.$$

Thus

$$\|f\|_p = \sup\left\{\int |fg|\,d\mu\colon \|g\|_{p'} \le 1\right\} = \sup\left\{\left|\int fg\,d\mu\right|\colon \|g\|_{p'} \le 1\right\},$$

and the supremum is attained. $\qquad\qquad\square$

As an application of this result, we have the following important corollary.

Corollary 5.4.2 *Suppose that f is a non-negative measurable function on $(\Omega_1, \Sigma_1, \mu_1) \times (\Omega_2, \Sigma_2, \mu_2)$ and that $0 < p \le q < \infty$. Then*

$$\left(\int_{X_1}\left(\int_{X_2} f(x,y)^p\,d\mu_2(y)\right)^{q/p} d\mu_1(x)\right)^{1/q}$$

$$\le \left(\int_{X_2}\left(\int_{X_1} f(x,y)^q\,d\mu_1(x)\right)^{p/q} d\mu_2(y)\right)^{1/p}.$$

Proof Let $r = q/p$. Then

$$\left(\int_{X_1} \left(\int_{X_2} f(x,y)^p \, d\mu_2(y) \right)^{q/p} d\mu_1(x) \right)^{1/q}$$

$$= \left(\int_{X_1} \left(\int_{X_2} f(x,y)^p \, d\mu_2(y) \right)^r d\mu_1(x) \right)^{1/rp}$$

$$= \left(\int_{X_1} \left(\int_{X_2} f(x,y)^p \, d\mu_2(y) \right) g(x) \, d\mu_1(x) \right)^{1/p} \quad \text{for some } g \text{ with } \|g\|_{r'} = 1$$

$$= \left(\int_{X_2} \left(\int_{X_1} f(x,y)^p g(x) \, d\mu_1(x) \right) d\mu_2(y) \right)^{1/p} \quad \text{(by Fubini's theorem)}$$

$$\leq \left(\int_{X_2} \left(\int_{X_1} f(x,y)^{pr} \, d\mu_1(x) \right)^{1/r} d\mu_2(y) \right)^{1/p} \quad \text{(by Corollary 5.4.1)}$$

$$= \left(\int_{X_2} \left(\int_{X_1} f(x,y)^q \, d\mu_1(x) \right)^{p/q} d\mu_2(y) \right)^{1/p}.$$

$\square$

We can consider f as a vector-valued function $f(y)$ on Ω_2, taking values in $L^q(\Omega_1)$, and with $\int_{\Omega_2} \|f(y)\|_q^p \, d\mu_2 < \infty$: thus $f \in L^p_{\Omega_2}(L^q_{\Omega_1})$. The corollary then says that $f \in L^q_{\Omega_1}(L^p_{\Omega_2})$ and $\|f\|_{L^q_{\Omega_1}(L^p_{\Omega_2})} \leq \|f\|_{L^p_{\Omega_2}(L^q_{\Omega_1})}$.

Here is a generalization of Hölder's inequality.

Proposition 5.4.1 *Suppose that* $1/p_1 + \cdots + 1/p_n = 1$ *and that* $f_i \in L_{p_i}$ *for* $1 \leq i \leq n$. *Then* $f_1 \cdots f_n \in L_1$ *and*

$$\int |f_1 \cdots f_n| \, d\mu \leq \|f_1\|_{p_1} \cdots \|f_n\|_{p_n}.$$

Equality holds if and only if either the right-hand side is zero, or there exist $\lambda_{ij} > 0$ *such that* $|f_i|^{p_i} = \lambda_{ij} |f_j|^{p_j}$ *for* $1 \leq i, j \leq n$.

Proof By Proposition 4.1.3,

$$|f_1 \cdots f_n| \leq |f_1|^{p_1}/p_1 + \cdots + |f_n|^{p_n}/p_n.$$

We now proceed exactly as in Theorem 5.4.1. $\square$

It is also easy to prove this by induction on n, using Hölder's inequality.

5.5 The inequalities of Liapounov and Littlewood

Hölder's inequality shows that there is a natural scale of inclusions for the L^p spaces, when the underlying space has finite measure.

Proposition 5.5.1 *Suppose that (Ω, Σ, μ) is a measure space and that $\mu(\Omega) < \infty$. Suppose that $0 < p < q < \infty$. If $f \in L^q$ then $f \in L^p$ and $\|f\|_p \leq \mu(\Omega)^{1/p - 1/q} \|f\|_q$. If $f \in L^\infty$ then $f \in L^p$ and $\|f\|_p \leq \mu(\Omega)^{1/p} \|f\|_\infty$.*

Proof Let $r = q/(q - p)$, so that $p/q + 1/r = 1$ and $1/rp = 1/p - 1/q$. We apply Hölder's inequality to the functions 1 and $|f|^p$, using exponents r and q/p:

$$\int |f|^p \, d\mu \leq (\mu(\Omega))^{1/r} \left(\int |f|^q \, d\mu \right)^{p/q},$$

so that

$$\|f\|_p \leq (\mu(\Omega))^{1/rp} \left(\int |f|^q \, d\mu \right)^{1/q} = \mu(\Omega)^{1/p - 1/q} \|f\|_q.$$

When $f \in L^\infty$, $\int |f|^p \, d\mu \leq \|f\|_\infty^p \, \mu(\Omega)$, so that $\|f\|_p \leq \mu(\Omega)^{1/p} \|f\|_\infty$. $\square$

When the underlying space has counting measure, we denote the space $L^p(\Omega)$ by $l_p(\Omega)$ or l_p; when $\Omega = \{1, \ldots, n\}$ we write l_p^n. With counting measure, the inclusions go the other way.

Proposition 5.5.2 *Suppose that $0 < p < q \leq \infty$. If $f \in l_p$ then $f \in l_q$ and $\|f\|_q \leq \|f\|_p$.*

Proof The result is certainly true when $q = \infty$, and when $f = 0$. Otherwise, let $F = f/\|f\|_p$, so that $\|F\|_p = 1$. Thus if $i \in \Omega$ then $|F_i| \leq 1$ and so $|F_i|^q \leq |F_i|^p$. Thus $\sum_i |F_i|^q \leq \sum_i |F_i|^p = 1$, so that $\|F\|_q \leq 1$ and $\|f\|_q \leq \|f\|_p$. $\square$

For general measure spaces, if $p \neq q$ then L^p neither includes nor is included in L^q. On the other hand if $0 < p_0 < p < p_1 \leq \infty$ then

$$L^{p_0} \cap L^{p_1} \subseteq L^p \subseteq L^{p_0} + L^{p_1}.$$

More precisely, we have the following.

Theorem 5.5.1 *(i) (Liapounov's inequality) Suppose that $0 < p_0 < p_1 < \infty$ and that $0 < \theta < 1$. Let $p = (1 - \theta)p_0 + \theta p_1$. If $f \in L^{p_0} \cap L^{p_1}$ then $f \in L^p$ and $\|f\|_p^p \leq \|f\|_{p_0}^{(1-\theta)p_0} \|f\|_{p_1}^{\theta p_1}$.*

(ii) (Littlewood's inequality) Suppose that $0 < p_0 < p_1 < \infty$ and that $0 < \theta < 1$. Define p by $1/p = (1 - \theta)/p_0 + \theta/p_1$. If $f \in L^{p_0} \cap L^{p_1}$ then $f \in L^p$ and $\|f\|_p \le \|f\|_{p_0}^{1-\theta} \|f\|_{p_1}^{\theta}$.

(iii) Suppose that $0 < p_0 < p_1 \le \infty$ and that $0 < \theta < 1$. Define p by $1/p = (1 - \theta)/p_0 + \theta/p_1$. Then if $f \in L^p$ there exist functions $g \in L^{p_0}$ and $h \in L^{p_1}$ such that $f = g + h$ and $\|g\|_{p_0}^{1-\theta} \|h\|_{p_1}^{\theta} \le \|f\|_p$.

Proof (i) We use Hölder's inequality with exponents $1/(1 - \theta)$ and $1/\theta$:

$$\|f\|_p^p = \int |f|^p \, d\mu = \int |f|^{(1-\theta)p_0} |f|^{\theta p_1} \, d\mu$$

$$\le \left(\int |f|^{p_0} \, d\mu \right)^{1-\theta} \left(\int |f|^{p_1} \, d\mu \right)^{\theta} = \|f\|_{p_0}^{(1-\theta)p_0} \|f\|_{p_1}^{\theta p_1}.$$

(ii) Let $1 - \gamma = (1 - \theta)p/p_0$, so that $\gamma = \theta p/p_1$. We apply Hölder's inequality with exponents $1/(1 - \gamma)$ and $1/\gamma$:

$$\|f\|_p = \left(\int |f|^p \, d\mu \right)^{1/p} = \left(\int |f|^{(1-\theta)p} |f|^{\theta p} \, d\mu \right)^{1/p}$$

$$\le \left(\int |f|^{(1-\theta)p/(1-\gamma)} \, d\mu \right)^{(1-\gamma)/p} \left(\int |f|^{\theta p/\gamma} \, d\mu \right)^{\gamma/p}$$

$$= \left(\int |f|^{p_0} \, d\mu \right)^{(1-\theta)/p_0} \left(\int |f|^{p_1} \, d\mu \right)^{\theta/p_1} = \|f\|_{p_0}^{1-\theta} \|f\|_{p_1}^{\theta}.$$

(iii) Let $g = f I_{(|f|>1)}$ and let $h = f - g$. Then $|g|^{p_0} \le |f|^p$, and so $\|g\|_{p_0} \le \|f\|_p^{p/p_0}$. On the other hand, $|h| \le 1$, so that $|h|^{p_1} \le |h|^p \le |f|^p$, and $\|h\|_{p_1} \le \|f\|_p^{p/p_1}$. Thus

$$\|g\|_{p_0}^{1-\theta} \|h\|_{p_1}^{\theta} \le \|f\|_p^{p((1-\theta)/p_0 + \theta/p_1)} = \|f\|_p.$$

$\square$

Liapounov's inequality says that $\log \|f\|_p$ is a convex function of p, and Littlewood's inequality says that $\log \|f\|_{1/t}$ is a convex function of t.

5.6 Duality

We now consider the structure of the L^p spaces, and their duality properties.

Proposition 5.6.1 *The simple functions are dense in L_p, for $1 \le p < \infty$.*

Proof Suppose that $f \in L_p$. Then there exists a sequence (f_n) of simple functions with $|f_n| \le |f|$ which converges pointwise to f. Then $|f - f_n|^p \le |f|^p$, and $|f - f_n|^p \to 0$ pointwise, and so by the theorem of dominated convergence, $\|f - f_n\|_p^p = \int |f - f_n|^p \, d\mu \to 0$. $\qquad\square$

This result holds for L^∞ if and only if $\mu(\Omega) < \infty$.

Proposition 5.6.2 *Suppose that $1 \le p < \infty$. A measurable function f is in L^p if and only if $fg \in L^1$ for all $g \in L^{p'}$.*

Proof The condition is certainly necessary, by Hölder's inequality. It is trivially sufficient when $p = 1$ (take $g = 1$). Suppose that $1 < p < \infty$ and that $f \notin L^p$. There exists an increasing sequence (k_n) of non-negative simple functions which increases pointwise to $|f|$. By the monotone convergence theorem, $\|k_n\|_p \to \infty$; extracting a subsequence if necessary, we can suppose that $\|k_n\|_p \ge 4^n$, for each n. Let $h_n = k_n^{p-1}$. Then as in Corollary 5.4.1, $\|h_n\|_{p'} = \|k_n\|_p^{p/p'}$; setting $g_n = h_n / \|h_n\|_{p'}$, $\|g_n\|_{p'} = 1$ and

$$\int |f| g_n \, d\mu \ge \int k_n g_n \, d\mu = \|k_n\|_p^{-p/p'} \int k_n^p \, d\mu = \|k_n\|_p \ge 4^n.$$

If we set $s = \sum_{n=1}^\infty g_n / 2^n$, then $\|s\|_{p'} \le 1$, so that $s \in L^{p'}$, while $\int |f| s \, d\mu = \infty$. $\qquad\square$

Suppose that $1 \le p < \infty$ and that $g \in L^{p'}$. If $f \in L^p$, let $l_g(f) = \int fg \, d\mu$. Then it follows from Hölder's inequality that the mapping $g \to l_g$ is a linear isometry of $L^{p'}$ into $(L^p)^*$. In fact, we can say more.

Theorem 5.6.1 *If $1 < p < \infty$, the mapping $g \to l_g$ is a linear isometric isomorphism of $L^{p'}$ onto $(L^p)^*$.*

Proof We shall prove this in the real case: the extension to the complex case is given in Exercise 5.11. We must show that the mapping is surjective. There are several proofs of this; the proof that we give here appeals to measure theory. First, suppose that $\mu(\Omega) < \infty$. Suppose that $\phi \in (L^p)^*$ and that $\phi \ne 0$. Let $\psi(E) = \phi(I_E)$, for $E \in \Sigma$. Then ψ is an additive function on Σ. Further, $|\psi(E)| \le \|\phi\|^* . (\mu(E))^{1/p}$, so that ψ is absolutely continuous with respect to μ. By Theorem 5.2.2 there exists $g \in L^1$ such that $\phi(I_E) = \nu(E) = \int_E g \, d\mu$ for all $E \in \Sigma$. Now let $\phi^+(f) = \phi(f.I_{g \ge 0})$ and $\phi^-(f) = \phi(f.I_{g < 0})$: ϕ^+ and ϕ^- are continuous linear functionals on L^p, and $\phi = \phi^+ - \phi^-$. If f is a simple function then $\phi^+(f) = \int fg^+ \, d\mu$. We now show that $g^+ \in L^{p'}$. There exists a sequence (g_n) of non-negative simple functions which increase pointwise to g^+. Let $f_n = g_n^{p'-1}$. Then

$$\int g_n^{p'} \, d\mu \leq \int g_n^{p'-1} g^+ \, d\mu = \phi^+(f_n) \leq \left\| \phi^+ \right\|^* \|f_n\|_p$$

$$= \left\| \phi^+ \right\|^* \left(\int g_n^{p(p'-1)} \, d\mu \right)^{1/p} = \left\| \phi^+ \right\|^* \left(\int g_n^{p'} \, d\mu \right)^{1/p},$$

so that $\int g_n^{p'} \, d\mu \leq (\| \phi^+ \|^*)^{p'}$. It now follows from the monotone convergence theorem that $\int (g^+)^{p'} \, d\mu \leq (\| \phi^+ \|^*)^{p'}$, and so $g^+ \in L^{p'}$. Similarly $g^- \in L^{p'}$, and so $g \in L^{p'}$. Now $\phi(f) = l_g(f)$ when f is a simple function, and the simple functions are dense in L_p, and so $\phi = l_g$.

In the general case, we can write $\Omega = \cup_n \Omega_n$, where the sets Ω_n are disjoint sets of finite measure. Let ϕ_n be the restriction of ϕ to $L^{p'}(\Omega_n)$. Then by the above result, for each n there exists $g_n \in L^{p'}(\Omega_n)$ such that $\phi_n = l_{g_n}$. Let g be the function on Ω whose restriction to Ω_n is g_n, for each n. Then straightforward arguments show that $g \in L^{p'}(\Omega)$ and that $\phi = l_g$. $\qquad \square$

The theorem is also true for $p = 1$ (see Exercise 5.8), but is not true for $p = \infty$, unless L^∞ is finite dimensional. This is the first indication of the fact that the L^p spaces, for $1 < p < \infty$, are more well-behaved than L^1 and L^∞.

A Banach space $(E, \|.\|)$ is *reflexive* if the natural isometry of E into E^{**} maps E *onto* E^{**}: thus we can identify the bidual of E with E.

Corollary 5.6.1 L^p *is reflexive, for* $1 < p < \infty$.

The proof of Theorem 5.6.1 appealed to measure theory. In Chapter 9 we shall establish some further inequalities, concerning the geometry of the unit ball of L^p, which lead to a very different proof.

5.7 The Loomis–Whitney inequality

The spaces L^1 and L^∞ are clearly important, and so is L^2, which provides an important example of a Hilbert space. But why should we be interested in L^p spaces for other values of p? The next few results begin to give an answer to this question.

First we need to describe the setting in which we work, and the notation which we use. This is unfortunately rather complicated. It is well worth writing out the proof for the case $d = 3$. Suppose that $(\Omega_1, \Sigma_1, \mu_1), \ldots,$ $(\Omega_d, \Sigma_d, \mu_d)$ are measure spaces; let (Ω, Σ, μ) be the product measure space

$\prod_{i=1}^{d}(\Omega_i, \Sigma_i, \mu_i)$. We want to consider products with one or two factors omitted. Let $(\Omega^j, \Sigma^j, \mu^j) = \prod_{i \neq j}(\Omega_i, \Sigma_i, \mu_i)$. Similarly, if j, k are distinct indices, let $(\Omega^{j,k}, \Sigma^{j,k}, \mu^{j,k}) = \prod_{i \neq j,k}(\Omega_i, \Sigma_i, \mu_i)$. If $\omega \in \Omega$, we write $\omega = (\omega_j, \omega^j)$, where $\omega_j \in \Omega_j$ and $\omega^j \in \Omega^j$, and if $\omega^j \in \Omega^j$, where $j \neq 1$ we write $\omega^j = (\omega_1, \omega^{1,j})$, where $\omega_1 \in \Omega_1$ and $\omega^{1,j} \in \Omega^{1,j}$.

Theorem 5.7.1 *Suppose that h_j is a non-negative function in L^{d-1} $(\Omega^j, \Sigma^j, \mu^j)$, for $1 \leq j \leq d$. Let $g_j(\omega_j, \omega^j) = h_j(\omega^j)$ and let $g = \prod_{j=1}^{d} g_j$. Then*

$$\int_{\Omega} g \, d\mu \leq \prod_{j=1}^{d} \|h_j\|_{d-1}.$$

Proof The proof is by induction on d. The result is true for $d = 2$, since we can write $g(\omega_1, \omega_2) = h_1(\omega_2)h_2(\omega_1)$, and then

$$\int_{\Omega} g \, d\mu = \left(\int_{\Omega^1} h_1 \, d\mu^1 \right) \left(\int_{\Omega^2} h_2 \, d\mu^2 \right).$$

Suppose that the result holds for $d - 1$. Suppose that $\omega_1 \in \Omega_1$. We define the function g_{ω_1} on Ω^1 by setting

$$g_{\omega_1}(\omega^1) = g(\omega_1, \omega^1);$$

similarly if $2 \leq j \leq d$ we define the function h_{j,ω_1} on $\Omega^{1,j}$ by setting

$$h_{j,\omega_1}(\omega^{1,j}) = h_j(\omega_1, \omega^{1,j})$$

and define the function g_{j,ω_1} on Ω^1 by setting

$$g_{j,\omega_1}(\omega^1) = g_j(\omega_1, \omega^1).$$

Then by Hölder's inequality, with indices $d - 1$ and $(d - 1)/(d - 2)$,

$$\int_{\Omega^1} g_{\omega_1} \, d\mu^1 = \int_{\Omega^1} h_1 \left(\prod_{j=2}^{d} g_{j,\omega_1} \right) d\mu^1$$

$$\leq \|h_1\|_{d-1} \left(\int_{\Omega^1} \left(\prod_{j=2}^{d} g_{j,\omega_1} \right)^{(d-1)/(d-2)} d\mu^1 \right)^{(d-2)/(d-1)}.$$

But now by the inductive hypothesis,

$$\left(\int_{\Omega^1}(\prod_{j=2}^d g_{j,\omega_1})^{(d-1)/(d-2)}\,d\mu^1\right)^{(d-2)/(d-1)}$$

$$=\left(\int_{\Omega^1}(\prod_{j=2}^d g_{j,\omega_1}^{(d-1)/(d-2)})\,d\mu^1\right)^{(d-2)/(d-1)}$$

$$\leq\left(\prod_{j=2}^d\left\|h_{j,\omega_1}^{(d-1)/(d-2)}\right\|_{d-2}\right)^{(d-2)/(d-1)}$$

$$=\left(\prod_{j=2}^d\int h_{j,\omega_1}^{d-1}\,d\mu^{1,j}\right)^{1/(d-1)}$$

$$=\prod_{j=2}^d\|h_{j,\omega_1}\|_{d-1}.$$

Consequently, integrating over Ω_1, and using the generalized Hölder inequality with indices $(d-1,\ldots,d-1)$,

$$\int_\Omega g\,d\mu\leq\|h_1\|_{d-1}\int_{\Omega_1}\left(\prod_{j=2}^d\|h_{j,\omega_1}\|_{d-1}\right)d\mu_1(\omega_1)$$

$$\leq\|h_1\|_{d-1}\prod_{j=2}^d\left(\int_{\Omega_1}\|h_{j,\omega_1}\|_{d-1}^{d-1}\,d\mu_1\right)^{1/(d-1)}$$

$$=\|h_1\|_{d-1}\prod_{j=2}^d\left(\int_{\Omega_1}\left(\int_{\Omega^{1,j}}h_{j,\omega_1}^{d-1}\,d\mu^{1,j}\right)d\mu_1\right)^{1/(d-1)}$$

$$=\prod_{j=1}^d\|h_j\|_{d-1}.$$

Corollary 5.7.1 *Suppose that* $h_j\in L^{\alpha_j}(\Omega^j,\Sigma^j,\mu^j)$ *for* $1\leq j\leq d$, *where* $\alpha_j\geq 1$. *If* f *is a measurable function on* Ω *satisfying* $|f(\omega_j,\omega^j)|\leq|h_j(\omega^j)|$ *for all* $\omega=(\omega_j,\omega^j)$, *for* $1\leq j\leq d$, *then*

$$\|f\|_{\alpha/(d-1)}\leq\left(\prod_{j=1}^d\|h_j\|_{\alpha_j}^{\alpha_j}\right)^{1/\alpha}\leq(1/\alpha)\sum_{j=1}^d\alpha_j\,\|h_j\|_{\alpha_j},$$

where $\alpha=\alpha_1+\cdots+\alpha_d$.

Proof For $|f(\omega)|^{\alpha/(d-1)} \le \prod_{j=1}^d |h_j(\omega^j)|^{\alpha_j/(d-1)}$. The second inequality follows from the generalized AM–GM inequality. □

Corollary 5.7.2 (The Loomis–Whitney inequality) *Suppose that K is a compact subset of $\mathbf{R}^d$. Let K_j be the image of K under the orthogonal projection onto the subspace orthogonal to the j-th axis. Then*

$$\lambda_d(K) \le \left(\prod_{j=1}^d \lambda_{d-1}(K_j) \right)^{1/(d-1)}.$$

[Here λ_d denotes d-dimensional Borel measure, and λ_{d-1} $(d-1)$-dimensional measure.]

Proof Apply the previous corollary to the characteristic functions of K and the K_j, taking $\alpha_j = 1$ for each j. □

5.8 A Sobolev inequality

In the theory of partial differential equations, it is useful to estimate the size of a function in terms of its partial derivatives. Such estimates are called *Sobolev inequalities*. We use Corollary 5.7.1 to prove the following fundamental Sobolev inequality.

Theorem 5.8.1 *Suppose that f is a continuously differentiable function of compact support on $\mathbf{R}^d$, where $d > 1$. If $1 \le p < d$ then*

$$\|f\|_{pd/(d-p)} \le \frac{p(d-1)}{2(d-p)} \left(\prod_{j=1}^d \left\| \frac{\partial f}{\partial x_j} \right\|_p \right)^{1/d} \le \frac{p(d-1)}{2d(d-p)} \left(\sum_{j=1}^d \left\| \frac{\partial f}{\partial x_j} \right\|_p^p \right)^{1/p}.$$

Proof We first consider the case when $p = 1$. Let us write $x = (x_j, x^j)$. Then

$$f(x) = \int_{-\infty}^{x_j} \frac{\partial f}{\partial x_j}(t, x^j)\, dt = \int_{x_j}^{\infty} \frac{\partial f}{\partial x_j}(t, x^j)\, dt,$$

so that

$$|f(x)| \le \tfrac{1}{2} \int_{-\infty}^{\infty} \left| \frac{\partial f}{\partial x_j}(t, x^j) \right| dt.$$

Then, applying Corollary 5.7.1 with $\alpha_j = 1$ for each j,

$$\|f\|_{d/(d-1)} \le \frac{1}{2} \left(\prod_{j=1}^{d} \left\| \frac{\partial f}{\partial x_j} \right\|_1 \right)^{1/d} \le \frac{1}{2d} \left(\sum_{j=1}^{d} \left\| \frac{\partial f}{\partial x_j} \right\|_1 \right).$$

Next suppose that $1 < p < d$. Let $s = p(d-1)/(d-p)$. Then $(s-1)p' = sd/(d-1) = pd/(d-p)$; we shall see why this is useful shortly. Now

$$|f(x)|^s = \left| \int_{-\infty}^{x_j} \frac{\partial}{\partial x_j} \left(|f(t, x^j)|^s \right) dt \right|$$

$$\le s \int_{-\infty}^{x_j} |f(t, x^j)|^{s-1} \left| \frac{\partial f}{\partial x_j}(t, x^j) \right| dt;$$

similarly

$$|f(x)|^s \le s \int_{x_j}^{\infty} |f(t, x^j)|^{s-1} \left| \frac{\partial f}{\partial x_j}(t, x^j) \right| dt,$$

so that

$$|f(x)| \le \left(\frac{s}{2} \int_{-\infty}^{\infty} |f(t, x^j)|^{s-1} \left| \frac{\partial f}{\partial x_j}(t, x^j) \right| dt \right)^{1/s}.$$

Now take $\alpha_j = s$ for each j: by Corollary 5.7.1,

$$\|f\|_{sd/(d-1)}^s \le \frac{s}{2} \left(\prod_{j=1}^{d} \left\| |f|^{s-1} \left| \frac{\partial f}{\partial x_j} \right| \right\|_1 \right)^{1/d}.$$

Now

$$\left\| |f|^{s-1} \left| \frac{\partial f}{\partial x_j} \right| \right\|_1 \le \left\| |f|^{s-1} \right\|_{p'} \left\| \frac{\partial f}{\partial x_j} \right\|_p = \|f\|_{(s-1)p'}^{s-1} \left\| \frac{\partial f}{\partial x_j} \right\|_p,$$

so that

$$\|f\|_{sd/(d-1)}^s \le \frac{s}{2} \|f\|_{(s-1)p'}^{s-1} \left(\prod_{j=1}^{d} \left\| \frac{\partial f}{\partial x_j} \right\|_p \right)^{1/d}.$$

Thus, bearing in mind that $(s - 1)p' = sd/(d - 1) = pd/(d - p)$,

$$\|f\|_{pd/(d-p)} \leq \frac{p(d - 1)}{2(d - p)} \left(\prod_{j=1}^{d} \left\| \frac{\partial f}{\partial x_j} \right\|_p \right)^{1/d} \leq \frac{p(d - 1)}{2d(d - p)} \left(\sum_{j=1}^{d} \left\| \frac{\partial f}{\partial x_j} \right\|_p^p \right)^{1/p}.$$

$\square$

This theorem illustrates strongly the way in which the indices and constants depend upon the dimension d. This causes problems if we wish to let d increase to infinity. We return to this point in Chapter 13.

5.9 Schur's theorem and Schur's test

We end this chapter with two results of Schur, which depend upon Hölder's inequality. The first of these is an interpolation theorem. Although the result is a remarkable one, it is a precursor of more powerful and more general results that we shall prove later. Suppose that (Ω, Σ, μ) and (Φ, T, ν) are σ-finite measure spaces, and that K is a measurable function on $\Omega \times \Phi$ for which there are constants M and N such that

$$\int \left(\text{ess} \sup_{y \in \Phi} |K(x, y)| \right) d\mu(x) \leq M,$$

and

$$\int |K(x, y)| \, d\nu(y) \leq N, \quad \text{for almost all } x \in \Omega.$$

If $f \in L^1(\nu)$, then

$$\left| \int K(x, y) f(y) \, d\nu(y) \right| \leq \left(\text{ess} \sup_{y \in \Phi} |K(x, y)| \right) \int |f(y)| \, d\nu(y),$$

so that, setting $T(f)(x) = \int K(x, y) f(y) \, d\nu(y)$,

$$\|T(f)\|_1 \leq \int \left(\text{ess} \sup_{y \in \Phi} |K(x, y)| \right) d\mu(x) \|f\|_1 \leq M \|f\|_1.$$

Thus $T \in L(L^1(\nu), L^1(\mu))$, and $\|T\| \leq M$.

On the other hand, if $f \in L^\infty(\nu)$, then

$$|T(f)(x)| \leq \int |K(x, y)| |f(y)| \, d\nu(y) \leq \|f\|_\infty \int |K(x, y)| \, d\nu(y) \leq N \|f\|_\infty,$$

so that $T \in L(L^\infty(\nu), L^\infty(\mu))$, and $\|T\| \leq N$.

Hölder's inequality enables us to interpolate these results. By Theorem 5.5.1, if $1 < p < \infty$ then $L^p \subseteq L^1 + L^\infty$, and so we can define $T(f)$ for $f \in L^p$.

Theorem 5.9.1 (Schur's theorem) *Suppose that* (Ω, Σ, μ) *and* (Φ, T, ν) *are* σ*-finite measure spaces, and that* K *is a measurable function on* $\Omega \times \Phi$ *for which there are constants* M *and* N *such that*

$$\int \left(\operatorname*{ess\,sup}_{y \in \Phi} |K(x,y)| \right) d\mu(x) \leq M,$$

and

$$\int |K(x,y)| \, d\nu(y) \leq N, \quad \text{for almost all } x \in \Omega.$$

Let $T(f) = \int K(x,y) f(y) \, d\nu(y)$. *If* $1 < p < \infty$ *and* $f \in L^p(\nu)$ *then* $T(f) \in L^p(\mu)$ *and* $\|T(f)\|_p \leq M^{1/p} N^{1/p'} \|f\|_p$.

Proof Applying Hölder's inequality,

$$|T(f)(x)| \leq \int |K(x,y)| |f(y)| \, d\nu(y)$$

$$= \int |K(x,y)|^{1/p} |f(y)| |K(x,y)|^{1/p'} \, d\nu(y)$$

$$\leq \left(\int |K(x,y)| |f(y)|^p \, d\nu(y) \right)^{1/p} \left(\int |K(x,y)| \, d\nu(y) \right)^{1/p'}$$

$$\leq N^{1/p'} \left(\int |K(x,y)| |f(y)|^p \, d\nu(y) \right)^{1/p} \qquad x\text{-almost everywhere.}$$

Thus

$$\int |T(f)(x)|^p \, d\mu(x) \leq N^{p/p'} \int \left(\int |K(x,y)| |f(y)|^p \, d\nu(y) \right) d\mu(x)$$

$$= N^{p/p'} \int \left(\int |K(x,y)| \, d\mu(x) \right) |f(y)|^p \, d\nu(y)$$

$$\leq N^{p/p'} M \|f\|_p^p.$$

$\square$

The next result remains a powerful tool.

Theorem 5.9.2 (Schur's test) *Suppose that* $k = k(x,y)$ *is a non-negative measurable function on a product space* $(X, \Sigma, \mu) \times (Y, T, \nu)$, *and that* $1 < p < \infty$. *Suppose also that there exist strictly positive measurable functions* s *on* (X, Σ, μ) *and* t *on* (Y, T, ν), *and constants* A *and* B *such that*

$$\int_Y k(x,y) (t(y))^{p'} \, d\nu(y) \leq (As(x))^{p'} \quad \text{for almost all } x,$$

and

$$\int_X (s(x))^p k(x,y)\, d\mu(x) \le (Bt(y))^p \quad \text{for almost all } y.$$

Then if $f \in L^p(Y)$, $T(f)(x) = \int_Y k(x,y)f(y)\, d\nu(y)$ exists for almost all x, $T(f) \in L^p(X)$ and $\|T(f)\|_p \le AB\,\|f\|_p$.

Proof Hölder's inequality shows that it is enough to prove that if h is a non-negative function in $L^{p'}(X)$ and g is a non-negative function in $L^p(Y)$ then

$$\int_X \int_Y h(x)k(x,y)g(y)\, d\nu(y)\, d\mu(x) \le AB\,\|h\|_{p'}\,\|g\|_p\,.$$

Now, using Hölder's inequality,

$$\int_Y k(x,y)g(y)\, d\nu(y)$$

$$= \int_Y (k(x,y))^{1/p'} t(y) \frac{(k(x,y))^{1/p} g(y)}{t(y)}\, d\nu(y)$$

$$\le \left(\int_Y k(x,y)(t(y))^{p'}\, d\nu(y) \right)^{1/p'} \left(\int_Y \frac{k(x,y)(g(y))^p}{(t(y))^p}\, d\nu(y) \right)^{1/p}$$

$$\le As(x) \left(\int_Y \frac{k(x,y)(g(y))^p}{(t(y))^p}\, d\nu(y) \right)^{1/p}.$$

Thus, using Hölder's inequality again,

$$\int_X \int_Y h(x)k(x,y)g(y)\, d\nu(y)\, d\mu(x)$$

$$\le A \int_X h(x)s(x) \left(\int_Y \frac{k(x,y)(g(y))^p}{(t(y))^p}\, d\nu(y) \right)^{1/p} d\mu(x)$$

$$\le A\,\|h\|_{p'} \left(\int_X (s(x))^p \left(\int_Y \frac{k(x,y)(g(y))^p}{(t(y))^p}\, d\nu(y) \right) d\mu(x) \right)^{1/p}$$

$$= A\,\|h\|_{p'} \left(\int_Y \left(\int_X (s(x))^p k(x,y)\, d\mu(x) \right) \frac{(g(y))^p}{(t(y))^p}\, d\nu(y) \right)^{1/p}$$

$$\le AB\,\|h\|_{p'} \left(\int_Y (g(y))^p\, d\nu(y) \right)^{1/p} = AB\,\|h\|_{p'}\,\|g\|_p\,.$$

$\square$

5.10 Hilbert's absolute inequality

Let us apply Schur's test to the kernel $k(x, y) = 1/(x+y)$ on $[0, \infty) \times [0, \infty)$. We take $s(x) = t(x) = 1/x^{pp'}$. Then

$$\int_0^\infty (s(x))^p k(x, y)\, dx = \int_0^\infty \frac{1}{(x+y)x^{1/p'}}\, dy = \frac{\pi}{\sin(\pi/p')} \frac{1}{y^{1/p'}}$$
$$= \frac{\pi}{\sin(\pi/p)}(t(x))^p,$$

and similarly

$$\int_0^\infty k(x, y)(t(y))^{p'}\, dy = \frac{\pi}{\sin(\pi/p)}(s(x))^{p'},$$

Here we use the formula

$$\int_0^\infty \frac{1}{(1+y)y^\alpha}\, dy = \frac{\pi}{\sin \alpha\pi} \quad \text{for } 0 < \alpha < 1,$$

which is a familiar exercise in the calculus of residues (Exercise 5.13).

Thus we have the following version of Hilbert's inequality for the kernel $k(x, y) = 1/(x+y)$. (There is another more important inequality, also known as Hilbert's inequality, for the kernel $k(x, y) = 1/(x-y)$: we consider this in Chapter 11. To distinguish the inequalities, we refer to the present inequality as *Hilbert's absolute inequality*.)

Theorem 5.10.1 (Hilbert's absolute inequality: the continuous case) *If $f \in L^p[0, \infty)$ and $g \in L^{p'}[0, \infty)$, where $1 < p < \infty$, then*

$$\int_0^\infty \int_0^\infty \frac{|f(x)g(y)|}{x+y}\, dx\, dy \le \frac{\pi}{\sin(\pi/p)} \|f\|_p \|g\|_{p'},$$

and the constant $\pi/\sin(\pi/p)$ is best possible.

Proof It remains to show that the constant $\pi/\sin(\pi/p)$ is the best possible. Suppose that $1 < \lambda < 1 + 1/2p'$. Let

$$f_\lambda(x) = (\lambda - 1)^{1/p} x^{-\lambda/p} I_{[1,\infty)}(x) \text{ and } g_\lambda(y) = (\lambda - 1)^{1/p'} y^{-\lambda/p'} I_{[1,\infty)}(y).$$

Then $\|f_\lambda\|_p = \|g_\lambda\|_{p'} = 1$. Also

$$\int_0^\infty \int_0^\infty \frac{f_\lambda(x)g_\lambda(y)}{x+y}\,dx\,dy = (\lambda - 1)\int_1^\infty \left(\int_1^\infty \frac{dx}{x^{\lambda/p}(x+y)}\right)\frac{dy}{y^{\lambda/p'}}$$

$$= (\lambda - 1)\int_1^\infty \left(\int_{1/y}^\infty \frac{du}{u^{\lambda/p}(1+u)}\right)\frac{dy}{y^\lambda}$$

$$= (\lambda - 1)\int_1^\infty \left(\int_0^\infty \frac{du}{u^{\lambda/p}(1+u)}\right)\frac{dy}{y^\lambda}$$

$$- (\lambda - 1)\int_1^\infty \left(\int_0^{1/y} \frac{du}{u^{\lambda/p}(1+u)}\right)\frac{dy}{y^\lambda}$$

$$\geq \frac{\pi}{\sin(\lambda\pi/p)} - (\lambda - 1)\int_1^\infty \left(\int_0^{1/y} \frac{du}{u^{\lambda/p}}\right)\frac{dy}{y^\lambda}.$$

Now $\int_0^{1/y} u^{-\lambda/p}\,du = 1/(\beta y^\beta)$, where $\beta = 1 - \lambda/p = 1/p' - (\lambda-1)/p \geq 1/2p'$, and so

$$\int_1^\infty \left(\int_0^{1/y} \frac{du}{u^{\lambda/p}}\right)\frac{dy}{y^\lambda} = \frac{1}{\beta}\int_1^\infty \frac{dy}{y^{\beta+\lambda}} = \frac{1}{\beta(\beta+\lambda-1)} \leq 4p'^2.$$

Thus

$$\int_0^\infty \int_0^\infty \frac{f_\lambda(x)g_\lambda(y)}{x+y}\,dx\,dy \geq \frac{\pi}{\sin(\lambda\pi/p)} - 4p'^2(\lambda - 1).$$

Letting $\lambda \to 1$, we obtain the result. $\qquad\square$

Similar arguments establish the following discrete result.

Theorem 5.10.2 (Hilbert's absolute inequality: the discrete case)
If $a \in l_p(\mathbf{Z}^+)$ and $b \in l_p(\mathbf{Z}^+)$, where $1 < p < \infty$, then

$$\sum_{m=0}^\infty \sum_{n=0}^\infty \frac{|a_m b_n|}{m+n+1} \leq \frac{\pi}{\sin(\pi/p)}\|a\|_p\|b\|_{p'},$$

and the constant $\pi/\sin(\pi/p)$ is best possible.

Let us give an application to the theory of analytic functions. The *Hardy space* $H^1(\mathbf{D})$ is the space of analytic functions f on the unit disc $\mathbf{D} = \{z: |z| < 1\}$ which satisfy

$$\|f\|_{H^1} = \sup_{0 < r < 1}\frac{1}{2\pi}\int_{-\pi}^\pi |f(re^{i\theta})|\,d\theta < \infty.$$

Theorem 5.10.3 (Hardy) *If* $f(z) = \sum_{n=0}^{\infty} a_n z^n \in H^1$ *then*

$$\sum_{n=0}^{\infty} \frac{|a_n|}{n+1} < \pi \, \|f\|_{H^1}.$$

We need the fact that we can write $f = bg$, where b is an analytic function on $\mathbf{D}$ for which $|b(re^{i\theta})| \to 1$ as $r \to 1$ for almost all θ, and g is a function in $H^1(\mathbf{D})$ with no zeros in $\mathbf{D}$. (See [Dur 70], Theorem 2.5.) Then $\|g\|_{H^1} = \|f\|_{H^1}$. Since $0 \notin g(\mathbf{D})$, there exists an analytic function h on $\mathbf{D}$ such that $h^2 = g$. Let

$$h(z) = \sum_{j=0}^{\infty} h_n z^n, \quad b(z)h(z) = \sum_{j=0}^{\infty} c_n z^n.$$

Then

$$\sum_{n=0}^{\infty} |h_n|^2 = \sup_{0<r<1} \frac{1}{2\pi} \int_{-\pi}^{\pi} |h(re^{i\theta})|^2 \, d\theta = \|f\|_{H^1},$$

$$\sum_{n=0}^{\infty} |c_n|^2 = \sup_{0<r<1} \frac{1}{2\pi} \int_{-\pi}^{\pi} |b(re^{i\theta})h(re^{i\theta})|^2 \, d\theta = \|f\|_{H^1},$$

and $a_n = \sum_{j=0}^{n} h_j c_{n-j}$. Thus, using Hilbert's inequality with $p = 2$,

$$\sum_{n=0}^{\infty} \frac{|a_n|}{n+1} \le \sum_{n=0}^{\infty} \sum_{j=0}^{n} \frac{|h_j c_{n-j}|}{n+1} = \sum_{n=0}^{\infty} \sum_{k=0}^{\infty} \frac{|h_n c_k|}{n+k+1} \le \pi \, \|f\|_{H^1}.$$

5.11 Notes and remarks

Hölder's inequality was proved in [Höl 89], and Minkowski's in [Min 96]. The systematic study of the L^p spaces was inaugurated by F. Riesz [Ri(F) 10], as part of his programme investigating integral equations.

Exercises

5.1   When does equality hold in Minkowski's inequality?

5.2   (Continuation of Exercise 4.4.)

   (i) Suppose that $(\Omega, \Sigma, \mathbf{P})$ is a probability space, and that f is a non-negative measurable function on Ω for which $\mathbf{E}(\log^+ f) < \infty$. Show that if $0 < r < \infty$ then $G(f) = \exp(\mathbf{E}(\log f)) \le \|f\|_r = (\mathbf{E}(f^r))^{1/r}$.

   (ii) Suppose that $t > 1$. Show that $(t^r - 1)/r$ is an increasing function of r on $(0, \infty)$, and that $(t^r - 1)/r \to \log t$ as $r \searrow 0$.

(iii) Suppose that $\|f\|_{r_0} < \infty$ for some $r_0 > 0$. Show that $\log(\|f\|_r) \le \mathbf{E}((|f|^r - 1)/r)$ for $0 < r \le r_0$. Use the theorem of dominated convergence to show that $\|f\|_r \searrow G(f)$ as $r \searrow 0$.

5.3   Let f^+ and f^- be the functions defined in Theorem 5.2.2. Show that $\mu((f^+ > 0) \cap (f^- > 0)) = 0$.

5.4   Suppose that $f \in L^p(0,1)$, where $0 < p < 1$. Choose $0 = t_0 < t_1 < \cdots t_n = 1$ so that $\int_{t_{j-1}}^{t_j} |f(x)|^p \, dx = (1/n) \int_0^1 |f(x)|^p \, dx$ for $1 \le j \le n$. Let $f_j = nf I_{(t_{j-1}, t_j]}$. Calculate $d_p(f_j, 0)$. Show that if U is a non-empty convex open subset of $L^p(0,1)$ then $U = L^p(0,1)$.

5.5   Show that $(L^\infty, \|.\|_\infty)$ is a Banach space.

5.6   Show that the simple functions are dense in $(L^\infty, \|.\|_\infty)$ if and only if $\mu(\Omega) < \infty$.

5.7   Give an inductive proof of Proposition 5.4.1.

5.8   Prove the following:

(i) If $f \in L^1$ and $g \in L^\infty$ then $fg \in L^1$ and $|l_g(f)| = |\int fg \, d\mu| \le \|f\|_1 \|g\|_\infty$.

(ii) l is a norm-decreasing linear mapping of L^∞ into $(L^1)^*$.

(iii) If g is a non-zero element of L^∞ and $0 < \epsilon < 1$ there exists a set A_ϵ of finite positive measure such that $|g(\omega)| > (1 - \epsilon)\|g\|_\infty$ for $\omega \in A_\epsilon$.

(iv) Show that $\|l_g\|_1^* = \|g\|_\infty$. (Consider $\overline{\operatorname{sgn} g} I_{A_\epsilon}$.)

(v) By following the proof of Theorem 5.6.1, show that l is an isometry of L^∞ onto $(L^1)^*$. (Find g, and show that $\mu(|g| > \|\phi\|^*) = 0$.)

5.9   Show that there is a natural isometry l of L^1 into $(L^\infty)^*$.

5.10   It is an important fact that the mapping l of the preceding question is not surjective when L^1 is infinite-dimensional: L^1 is not reflexive.

(i) Let $c = \{x = (x_n): x_n \to l \text{ for some } l, \text{ as } n \to \infty\}$. Show that c is a closed linear subspace of l_∞. If $x \in c$, let $\phi(x) = \lim_{n\to\infty} x_n$. Show that $\phi \in c^*$, and that $\|\phi\|^* = 1$. Use the Hahn–Banach theorem to extend ϕ to $\psi \in l_\infty^*$. Show that $\psi \notin l(l_1)$.

(ii) Use the Radon–Nykodým theorem, and the idea of the preceding example, to show that $l(L^1(0,1)) \ne (L^\infty(0,1))^*$.

5.11   Suppose that ϕ is a continuous linear functional on the complex Banach space $L^p_{\mathbf{C}}(\Omega, \Sigma, \mu)$, where $1 \le p < \infty$. If $f \in L^p_{\mathbf{R}}(\Omega, \Sigma, \mu)$, we can consider f as an element of $L^p_{\mathbf{C}}(\Omega, \Sigma, \mu)$. Let $\psi(f)$ be the real part of $\phi(f)$ and $\chi(f)$ the imaginary part. Show that ϕ and χ are continuous linear functionals on $L^p_{\mathbf{R}}(\Omega, \Sigma, \mu)$. Show that ϕ is represented by an element g of $L^{p'}_{\mathbf{C}}(\Omega, \Sigma, \mu)$. Show that $\|g\|_{p'} = \|\phi\|^*$.

5.12 Suppose that (Ω, Σ, μ) is a σ-finite measure space, that E is a Banach space and that $1 \le p < \infty$.

(i) If $\phi = \sum_{j=1}^{k} \phi_j I_{A_j}$ is a simple measurable E^*-valued function and $f \in L^p(E)$, let

$$j(\phi)(f) = \sum_{j=1}^{k} \int_{A_k} \phi_j(f) \, d\mu.$$

Show that $j(\phi) \in (L^p(E))^*$ and that $\|j(\phi)\|_{L^p(E)}^* = \|\phi\|_{L^{p'}(E^*)}$.

(ii) Show that j extends to an isometry of $L^{p'}(E^*)$ into $(L^p(E))^*$. [It is an important fact that j need not be surjective: this requires the so-called Radon–Nikodym property. See [DiU 77] for details; this is an invaluable source of information concerning vector-valued functions.]

(iii) Show that

$$p \, \|f\|_{L^p(E)} = \sup\{j(\phi)(f) : \phi \text{ simple}, \ \|\phi\|_{L^{p'}(E^*)} \le 1\}.$$

5.13 Prove that

$$\int_0^\infty \frac{1}{1+y} \frac{1}{y^\alpha} \, dy = \frac{\pi}{\sin \alpha \pi} \quad \text{for } 0 < \alpha < 1,$$

by contour integration, or otherwise.

5.14 Write out a proof of Theorem 5.10.2.

6

Banach function spaces

6.1 Banach function spaces

In this chapter, we introduce the idea of a Banach function space; this provides a general setting for most of the spaces of functions that we consider. As an example, we introduce the class of *Orlicz spaces*, which includes the L^p spaces for $1 < p < \infty$. As always, let (Ω, Σ, μ) be a σ-finite measure space, and let $M = M(\Omega, \Sigma, \mu)$ be the space of (equivalence classes of) measurable functions on Ω.

A *function norm* on M is a function $\rho : M \rightarrow [0, \infty]$ (note that ∞ is allowed) satisfying the following properties:

(i) $\rho(f) = 0$ if and only if $f = 0$; $\rho(\alpha f) = |\alpha|\rho(f)$ for $\alpha \neq 0$; $\rho(f + g) \leq \rho(f) + \rho(g)$.
(ii) If $|f| \leq |g|$ then $\rho(f) \leq \rho(g)$.
(iii) If $0 \leq f_n \nearrow f$ then $\rho(f) = \lim_{n\to\infty} \rho(f_n)$.
(iv) If $A \in \Sigma$ and $\mu(A) < \infty$ then $\rho(I_A) < \infty$.
(v) If $A \in \Sigma$ and $\mu(A) < \infty$ there exists C_A such that $\int_A |f|\, d\mu \leq C_A \rho(f)$ for any $f \in M$.

If ρ is a function norm, the space $E = \{f \in M : \rho(f) < \infty\}$ is called a *Banach function space*. If $f \in E$, we write $\|f\|_E$ for $\rho(f)$. Then condition (i) ensures that E is a vector space and that $\|.\|_E$ is a norm on it. We denote the closed unit ball $\{x : \rho(x) \leq 1\}$ of E by B_E. As an example, if $1 \leq p < \infty$, let $\rho_p(f) = (\int |f|^p\, d\mu)^{1/p}$. Then ρ_p is a Banach function norm, and the corresponding Banach function space is L^p. Similarly, L^∞ is a Banach function space.

Condition (ii) ensures that E is a lattice, and rather more: if $g \in E$ and $|f| \leq |g|$ then $f \in E$ and $\|f\|_E \leq \|g\|_E$. Condition (iv) ensures that the simple functions are in E, and condition (v) ensures that we can integrate functions in E over sets of finite measure. In particular, if $\mu(\Omega) < \infty$ then

$L^\infty \subseteq E \subseteq L^1$, and the inclusion mappings are continuous. Condition (iii) corresponds to the monotone convergence theorem for L^1, and has similar uses, as the next result shows.

Proposition 6.1.1 (Fatou's lemma) *Suppose that (f_n) is a sequence in a Banach function space $(E, \|.\|_E)$, that $f_n \to f$ almost everywhere and that $\liminf \|f_n\|_E < \infty$. Then $f \in E$ and $\|f\|_E \leq \liminf \|f_n\|_E$.*

Proof Let $h_n = \inf_{m \geq n} |f_m|$; note that $h_n \leq |f_n|$. Then $0 \leq h_n \nearrow |f|$, so that

$$\rho(f) = \rho(|f|) = \lim_{n \to \infty} \|h_n\|_E \leq \lim_{n \to \infty} \inf \|f_n\|_E.$$

$\square$

Suppose that $A \in \Sigma$. Then if E is a Banach function space, we set $E_A = \{f \in E : f = fI_A\}$. E_A is the linear subspace of E consisting of those functions which are zero outside A.

Proposition 6.1.2 *If E is a Banach function space and $\mu(A) < \infty$ then $\{f \in E_A : \|f\|_E \leq 1\}$ is closed in L_A^1.*

Proof Suppose that (f_n) is a sequence in $\{f \in E_A : \|f\|_E \leq 1\}$ which converges in L_A^1 norm to f_A, say. Then there is a subsequence (f_{n_k}) which converges almost everywhere to f_A. Then f_A is zero outside F, and it follows from Fatou's lemma that $\rho(f_A) \leq 1$. $\square$

Theorem 6.1.1 *If $(E, \|.\|_E)$ is a Banach function space, then it is norm complete.*

Proof Suppose that (f_n) is a Cauchy sequence. Then if $\mu(A) < \infty$, $(f_n I_A)$ is a Cauchy sequence in L_A^1, and so it converges in L_A^1 norm to f_A, say. Further, there is a subsequence $(f_{n_k} I_A)$ which converges almost everywhere to f_A. Since (Ω, Σ, μ) is σ-finite, we can use a diagonal argument to show that there exists a subsequence $(g_k) = (f_{d_k})$ which converges almost everywhere to a function f. It will be enough to show that $f \in E$ and that $\|f - g_k\|_E \to 0$.

First, $\rho(f) \leq \sup_k \|g_k\|_E < \infty$, by Fatou's lemma, so that $f \in E$. Second, given $\epsilon > 0$ there exists k_0 such that $\|g_l - g_k\|_E < \epsilon$ for $l > k \geq k_0$. Since $g_l - g_k \to f - g_k$ almost everywhere as $l \to \infty$, another application of Fatou's lemma shows that $\|f - g_k\|_E \leq \epsilon$ for $k \geq k_0$. $\square$

It is convenient to characterize function norms and Banach function spaces in terms of the unit ball.

Proposition 6.1.3 *Let B_E be the unit ball of a Banach function space. Then*

(i) B_E *is convex.*

(ii) *If $|f| \leq |g|$ and $g \in B_E$ then $f \in B_E$.*

(iii) *If $0 \leq f_n \nearrow f$ and $f_n \in B_E$ then $f \in B_E$.*

(iv) *If $A \in \Sigma$ and $\mu(A) < \infty$ then $I_A \in \lambda B_E$ for some $0 \leq \lambda < \infty$.*

(v) *If $A \in \Sigma$ and $\mu(A) < \infty$ then there exists $0 < C_A < \infty$ such that $\int_A |f| \, d\mu \leq C_A$ for any $f \in B_E$.*

Conversely, suppose that B satisfies these conditions. Let

$$\rho(f) = \inf\{\lambda > 0 \colon f \in \lambda B\}.$$

[The infimum of the empty set is ∞.]

Then ρ is a function norm, and $B = \{f \colon \rho(f) \leq 1\}$.

Proof This is a straightforward but worthwhile exercise. □

6.2 Function space duality

We now turn to function space duality.

Proposition 6.2.1 *Suppose that ρ is a function norm. If $f \in M$, let*

$$\rho'(f) = \sup \left\{ \int |fg| \, d\mu \colon g \in B_E \right\}.$$

Then ρ' is a function norm.

Proof This involves more straightforward checking. Let us just check two of the conditions. First, suppose that $\rho'(f) = 0$. Then $\rho'(|f|) = 0$, and by condition (iv), $\int_F |f| \, d\mu = 0$ whenever $\mu(F) < \infty$, and this ensures that $f = 0$.

Second, suppose that $0 \leq f_n \nearrow f$ and that $\sup \rho'(f_n) = \alpha < \infty$. If $\rho(g) \leq 1$ then $\int f_n |g| \, d\mu \leq \alpha$, and so $\int f|g| \, d\mu \leq \alpha$, by the monotone convergence theorem. Thus $\rho(f) \leq \alpha$. □

ρ' is the *associate function norm*, and the corresponding Banach function space $(E', \|.\|_{E'})$ is the *associate function space*. If $f \in E'$ then the mapping $g \to \int fg \, d\mu$ is an isometry of $(E' \, \|.\|_{E'})$ into the dual space E^* of all continuous linear functionals on $(E, \|.\|_E)$, and we frequently identify E' with a subspace of E^*.

Theorem 6.2.1 *If ρ is a function norm then $\rho'' = \rho$.*

Proof This uses the Hahn–Banach theorem, and also uses the fact that the dual of L^1 can be identified with L^∞ (Exercise 5.8). It follows from the definitions that $\rho'' \le \rho$, so that we must show $\rho'' \ge \rho$. For this it is enough to show that if $\rho(f) > 1$ then $\rho''(f) > 1$. There exist simple functions f_n such that $0 \le f_n \nearrow |f|$. Then $\rho(f_n) \to \rho(|f|) = \rho(f)$. Thus there exists a simple function g such that $0 \le g \le |f|$ and $\rho(g) > 1$.

Suppose that g is supported on A, where $\mu(A) < \infty$. Then g is disjoint from $\{hI_A : h \in B_E\}$, and this set is a closed convex subset of L^1_A. By the separation theorem (Theorem 4.6.3) there exists $k \in L^\infty_A$ such that

$$\int_A gk\,d\mu > 1 \ge \sup\left\{\left|\int_A hk\,d\mu\right| : hI_A \in B_E\right\} = \sup\left\{\left|\int hk\,d\mu\right| : h \in B_E\right\}.$$

This implies first that $\rho'(k) \le 1$ and second that $\rho''(g) > 1$. Thus $\rho''(f) \ge \rho''(g) > 1$. □

6.3 Orlicz spaces

Let us give an example of an important class of Banach function spaces, the *Orlicz spaces*. A *Young's function* Φ is a non-negative convex function on $[0, \infty)$, with $\Phi(0) = 0$, for which $\Phi(t)/t \to \infty$ as $t \to \infty$. Let us consider

$$B_\Phi = \left\{f \in M \colon \int \Phi(|f|)\,d\mu \le 1\right\}.$$

Then B_Φ satisfies the conditions of Proposition 6.1.3; the corresponding Banach function space L_Φ is called the *Orlicz space* defined by Φ. The norm

$$\|f\|_\Phi = \inf\left\{\lambda > 0 \colon \int \Phi(\lambda|f|)\,d\mu \le 1\right\}$$

is known as the *Luxemburg norm* on L_Φ.

The most important, and least typical, class of Orlicz spaces occurs when we take $\Phi(t) = t^p$, where $1 < p < \infty$; in this case we obtain L^p.

[The spaces L^1 and L^∞ are also Banach function spaces, although, according to our definition, they are not Orlicz spaces.]

Let us give some examples of Orlicz spaces.

- $\Phi(t) = e^t - 1$. We denote the corresponding Orlicz space by $(L_{\exp}, \|\cdot\|_{\exp})$. Note that if $\mu(\Omega) < \infty$ then $L_{\exp} \subseteq L^p$ for $1 \le p < \infty$, and $\|f\|_{\exp} \le 1$ if and only if $\int e^{|f|}\,d\mu \le 1 + \mu(\Omega)$.
- $\Phi(t) = e^{t^2} - 1$. We denote the corresponding Orlicz space by $(L_{\exp^2}, \|\cdot\|_{\exp^2})$. Note that $L_{\exp^2} \subseteq L_{\exp}$.

- $\Phi(t) = t \log^+ t$, where $\log^+ t = \max(\log t, 0)$. We denote the corresponding Orlicz space by $(L_{L \log L}, \|.\|_{L \log L})$.

We now turn to duality properties. First we consider Young's functions more carefully. As Φ is convex, it has a left-derivative $D^- \Phi$ and a right-derivative $D^+ \Phi$. We choose to work with the right-derivative, which we denote by ϕ, but either will do. ϕ is a non-negative increasing right-continuous function on $[0, \infty)$, and $\phi(t) \to \infty$ as $t \to \infty$, since $D^+ \Phi(t) \geq \Phi(t)/t$.

Proposition 6.3.1 *Suppose that Φ is a Young's function with right-derivative ϕ. Then $\Phi(t) = \int_0^t \phi(s) \, ds$.*

Proof Suppose that $\epsilon > 0$. There exists a partition $0 = t_0 < t_1 < \cdots < t_n = t$ such that

$$\sum_{i=1}^n \phi(t_i)(t_i - t_{i-1}) - \epsilon \leq \int_0^t \phi(s) \, ds \leq \sum_{i=1}^n \phi(t_{i-1})(t_i - t_{i-1}) + \epsilon.$$

But $\phi(t_{i-1})(t_i - t_{i-1}) \leq \Phi(t_i) - \Phi(t_{i-1})$ and

$$\phi(t_i)(t_i - t_{i-1}) \geq D^- f(t_i)(t_i - t_{i-1}) \geq \Phi(t_i) - \Phi(t_{i-1}),$$

so that

$$\Phi(t) - \epsilon \leq \int_0^t \phi(s) \, ds \leq \Phi(t) + \epsilon.$$

Since ϵ is arbitrary, the result follows. $\square$

The function ϕ is increasing and right-continuous, but it need not be strictly increasing, and it can have jump discontinuities. Nevertheless, we can define an appropriate inverse function: we set

$$\psi(u) = \sup\{t : \phi(t) \leq u\}.$$

Then ψ is increasing and right-continuous, and $\psi(u) \to \infty$ as $u \to \infty$. The functions ϕ and ψ have symmetric roles.

Proposition 6.3.2 $\phi(t) = \sup\{u : \psi(u) \leq t\}$.

Proof Let us set $\gamma(t) = \sup\{u : \psi(u) \leq t\}$. Suppose that $\psi(u) \leq t$. Then if $t' > t$, $\phi(t') > u$. Since ϕ is right-continuous, $\phi(t) \geq u$, and so $\gamma(t) \leq \phi(t)$. On the other hand, if $u < \phi(t)$, then $\psi(u) \leq t$, so that $\gamma(t) \geq u$. Thus $\gamma(t) \geq \phi(t)$. $\square$

We now set $\Psi(u) = \int_0^u \psi(v)\, dv$. Ψ is a Young's function, the Young's function *complementary to* Φ.

Theorem 6.3.1 (Young's inequality) *Suppose that* $\Phi(t) = \int_0^s \phi(s)\, ds$ *and* $\Psi(u) = \int_0^u \psi(v)\, dv$ *are complementary Young's functions. Then* $tu \leq \Phi(t) + \Psi(u)$, *with equality if and only if* $u = \phi(t)$ *or* $t = \psi(u)$.

Proof We consider the integrals as 'areas under the curve'. First suppose that $\phi(t) = u$. Then if $0 \leq s < t$ and $0 \leq v < u$, then either $v \leq \phi(s)$ or $s < \psi(v)$, but not both. Thus the rectangle $[0, t) \times [0, u)$ is divided into two disjoint sets with measures $\int_0^t \phi(s)\, ds$ and $\int_0^u \psi(v)\, dv$. [Draw a picture!]

Next suppose that $\phi(t) < u$. Then, since ϕ is right continuous, it follows from the definition of ψ that $\psi(v) > t$ for $\phi(t) < v \leq u$. Thus

$$tu = t\phi(t) + t(u - \phi(t))$$
$$< (\Phi(t) + \Psi(\phi(t))) + \int_{\phi(t)}^u \psi(v)\, dv \leq \Phi(t) + \Psi(u).$$

Finally, if $\phi(t) > u$ then $\psi(u) \leq t$, and we obtain the result by interchanging ϕ and ψ. $\qquad\square$

Corollary 6.3.1 *If* $f \in L_\Phi$ *and* $g \in L_\Psi$ *then* $fg \in L^1$ *and*

$$\int |fg|\, d\mu \leq 2\,\|f\|_\Phi \cdot \|g\|_\Psi\,.$$

Proof Suppose that $\alpha > \|f\|_\Phi$ and $\beta > \|g\|_\Psi$. Then

$$\frac{|fg|}{\alpha\beta} \leq \Phi\left(\frac{f}{\alpha}\right) + \Psi\left(\frac{g}{\beta}\right);$$

integrating, $\int |fg|\, d\mu \leq 2\alpha\beta$, which gives the result. $\qquad\square$

Thus $L_\Psi \subseteq (L_\Phi)'$, and $\|g\|_\Phi' \leq 2\,\|g\|_\Psi$ (where $\|.\|_\Phi'$ is the norm associate to $\|.\|_\Phi$). In fact, we can say more.

Theorem 6.3.2 $L_\Psi = (L_\Phi)'$ *and*

$$\|g\|_\Psi \leq \|g\|_\Phi' \leq 2\,\|g\|_\Psi\,.$$

Proof We have seen that $L_\Psi \subseteq (L_\Phi)'$ and that $\|g\|_\Phi' \leq 2\,\|g\|_\Psi$. Suppose that $g \in L_\Phi'$ and that $\|g\|_\Phi' \leq 1$. Then there exists a sequence (g_n) of simple functions such that $0 \leq g_n \nearrow |g|$. Since $\rho_\Psi(g) = \rho_\Psi(|g|) = \sup_n \|g_n\|_\Psi$, it is therefore enough to show that if g is a non-negative simple function with $\|g\|_\Psi = 1$ then $\|g\|'_\Phi \geq 1$.

Let $h = \psi(g)$. Then the conditions for equality in Young's inequality hold pointwise, and so $hg = \Phi(h) + \Psi(g)$. Thus

$$\int hg\,d\mu = \int \Phi(h)\,d\mu + \int \Psi(g)\,d\mu = \int \Phi(h)\,d\mu + 1.$$

If $\|h\|_\Phi \leq 1$, this implies that $\|g\|'_\Phi \geq 1$. On the other hand, if $\|h\|_\Phi = \lambda > 1$ then

$$\|h\|_\Phi = \lambda \int \Phi(h/\lambda)\,d\mu \leq \int \Phi(h)\,d\mu,$$

by the convexity of Φ. Thus $\int hg\,d\mu \geq \|h\|_\Phi$, and so $\|g\|'_\Phi \geq 1$. $\square$

We write $\|.\|_{(\Psi)}$ for the norm $\|.\|'_\Phi$ on L_Ψ: it is called the *Orlicz norm*. Theorem 6.3.2 then states that the Luxemburg norm and the Orlicz norm are equivalent.

Finally, let us observe that we can also consider vector-valued function spaces. If (X, ρ) is a Banach function space and $(E, \|.\|_E)$ is a Banach space, we set $X(E)$ to be the set of E-valued strongly measurable functions, for which $\rho(\|f\|_E) < \infty$. It is a straightforward matter to verify that $X(E)$ is a vector space, that $\|f\|_{X(E)} = \rho(\|f\|_E)$ is a norm on $X(E)$, and that under this norm $X(E)$ is a Banach space.

6.4 Notes and remarks

A systematic account of Banach function spaces was given by Luxemburg [Lux 55] in his PhD thesis, and developed in a series of papers with Zaanen [LuZ 63]. Orlicz spaces were introduced in [Orl 32]. The definition of these spaces can be varied (for example to include L^1 and L^∞): the simple definition that we have given is enough to include the important spaces $L_{\exp}$, $L_{\exp^2}$ and $L_{L\log L}$. A fuller account of Banach function spaces, and much else, is given in [BeS 88].

Exercises

6.1 Write out a proof of Proposition 6.1.3 and the rest of Proposition 6.2.1.

6.2 Suppose that the step functions are dense in the Banach function space E. Show that the associate space E' can be identified with the Banach space dual of E.

6.3 Suppose that E_1 and E_2 are Banach function spaces, and that $E_1 \subseteq E_2$. Use the closed graph theorem to show that the inclusion mapping is continuous. Give a proof which does not depend on the closed graph theorem. [The closed graph theorem is a fundamental theorem of functional analysis: if you are not familiar with it, consult [Bol 90] or [TaL 80].]

6.4 Suppose that E is a Banach function space and that $fg \in L^1$ for all $g \in E$. Show that $g \in E'$.

6.5 Suppose that E is a Banach function space. Show that the associate space E' can be identified with the dual E^* of E if and only if whenever (f_n) is an increasing sequence of non-negative functions in E which converges almost everywhere to $f \in E$ then $\|f - f_n\|_E \to 0$.

6.6 Calculate the functions complementary to $e^t - 1$, $e^{t^2} - 1$ and $t \log^+ t$.

6.7 Suppose that Φ is an Orlicz function with right derivative ϕ. Show that $\rho_\Phi(f) = \int_0^\infty \phi(u)\mu(|f| > u)\,du$.

6.8 Suppose that Φ is a Young's function. For $s \geq 0$ and $t \geq 0$ let $f_s(t) = st - \Phi(t)$. Show that $f_s(t) \to -\infty$ as $t \to \infty$. Let $\Psi(s) = \sup\{f_s(t) : t \geq 0\}$. Show that Ψ is the Young's function conjugate to Φ.

　　The formula $\Psi(s) = \sup\{st - \Phi(t) : t \geq 0\}$ expresses Ψ as the *Legendre–Fenchel transform* of Φ.

7

Rearrangements

7.1 Decreasing rearrangements

Suppose that $(E, \|.\|_E)$ is a Banach function space and that $f \in E$. Then $\|f\|_E = \|\,|f|\,\|_E$, so that the norm of f depends only on the absolute values of f. For many important function spaces we can say more. Suppose for example that $f \in L^p$, where $1 < p < \infty$. By Proposition 1.3.4, $\|f\|_p = (p \int t^{p-1} \mu(|f| > t)\, dt)^{1/p}$, and so $\|f\|_p$ depends only on the distribution of $|f|$. The same is true for functions in Orlicz spaces. In this chapter, we shall consider properties of functions and spaces of functions with this property.

In order to avoid some technical difficulties which have little real interest, we shall restrict our attention to two cases:

(i) (Ω, Σ, μ) is an atom-free measure space;

(ii) $\Omega = \mathbf{N}$ or $\{1, \ldots, n\}$, with counting measure.

In the second case, we are concerned with sequences, and the arguments are usually, but not always, easier. We shall begin by considering case (i) in detail, and shall then describe what happens in case (ii), giving details only when different arguments are needed.

Suppose that we are in the first case, so that (Ω, Σ, μ) is atom-free. We shall then make use of various properties of the measure space, which follow from the fact that if $A \in \Sigma$ and $0 < t < \mu(A)$ then there exists a subset B of A with $\mu(B) = t$ (Exercise 7.1). If $f \geq 0$ then the distribution function λ_f takes values in $[0, \infty]$. The fact that λ_f can take the value ∞ is a nuisance. For example, if $\Omega = \mathbf{R}$, with Lebesgue measure, and $f(x) = \tan^2 x$, then $\lambda_f(t) = \infty$ for all $t > 0$, which does not give us any useful information about f; similarly, if $f(x) = \sin^2 x$, then $\lambda_f(t) = \infty$ for $0 < t < 1$ and $\lambda_f(t) = 0$ for $t \geq 1$. We shall frequently restrict attention to functions in

$$M_1(\Omega, \Sigma, \mu) = \{f \in M(\Omega, \Sigma, \mu) \colon \lambda_{|f|}(u) < \infty, \text{ for some } u > 0\}.$$

Thus M_1 contains $\sin^2 x$, but does not contain $\tan^2 x$. If $f \in M_1$, let $C_f = \inf\{u\colon \lambda_{|f|}(u) < \infty\}$. Let us also set

$$M_0 = \{f \in M_1\colon C_f = 0\} = \{f \in M\colon \lambda_{|f|}(u) < \infty, \text{ for all } u > 0\},$$

and at the other extreme, let M_∞ denote the space of (equivalence classes) of measurable functions, taking values in $(-\infty, \infty]$. Thus $M_0 \subseteq M_1 \subseteq M \subseteq M_\infty$. Note that $L^p \subseteq M_0$ for $0 < p < \infty$ and that $L^\infty \subseteq M_1$.

Suppose that $f \in M_1$. Then the distribution function $\lambda_{|f|}$ is a decreasing right-continuous function on $[0, \infty)$, taking values in $[0, \infty]$ (Proposition 1.3.3). We now consider the distribution function f^* of $\lambda_{|f|}$.

Proposition 7.1.1 *If $f \in M_1$, f^* is a decreasing right-continuous function on $[0, \infty)$, taking values in $[0, \infty]$, and $f^*(t) = 0$ if $t > \mu(\Omega)$. If $\mu(\Omega) = \infty$ then $f^*(t) \to C_f$ as $t \to \infty$.*

The functions $|f|$ and f^* are equidistributed: $\mu(|f| > u) = \lambda(f^* > u)$ for $0 \le u < \infty$.

Proof The statements in the first paragraph follow from the definitions, and Proposition 1.3.3.

If $\mu(|f| > u) = \infty$, then certainly $\mu(|f| > u) \ge \lambda(f^* > u)$. If $\lambda_{|f|}(u) = \mu(|f| > u) = t < \infty$, then $f^*(t) \le u$, so that $\lambda(f^* > u) \le t = \mu(|f| > u)$.

If $\lambda(f^* > u) = \infty$, then certainly $\mu(|f| > u) \le \lambda(f^* > u)$. If $\lambda(f^* > u) = t < \infty$, then $f^*(t) \le u$: that is, $\lambda(\lambda_{|f|} > t) \le u$. Thus if $v > u$, $\lambda_{|f|}(v) \le t$. But $\lambda_{|f|}$ is right-continuous, and so $\mu(|f| > u) = \lambda_{|f|}(u) \le t = \lambda(f^* > u)$. $\qquad\square$

The function f^* is called the *decreasing rearrangement* of f: it is a right-continuous decreasing function on $[0, \infty)$ with the same distribution as $|f|$.

Two applications of Proposition 1.3.3 also give us the following result.

Proposition 7.1.2 *If $0 \le f_n \nearrow f$ and $f \in M_1$ then $0 \le f_n^* \nearrow f^*$.*

This proposition is very useful, since it allows us to work with simple functions.

Proposition 7.1.3 *If $f \in M_1$ and E is a measurable set, then $\int_E |f|\,d\mu \le \int_0^{\mu(E)} f^*\,d\mu$.*

Proof Let $h = |f|I_E$. Since $0 \leq h \leq |f|$, $h^* \leq f^*$, and $h^*(t) = 0$ for $t > \mu(E)$. Since h and h^* are equidistributed,

$$\int_E |f| \, d\mu = \int h \, d\mu = \int_0^{\mu(E)} h^* \, d\mu \leq \int_0^{\mu(E)} f^* \, d\mu.$$

$\square$

Proposition 7.1.4 *If* $f, g \in M_1$ *then* $\int |fg| \, d\mu \leq \int_0^\infty f^* g^* \, dt$.

Proof We can suppose that $f, g \geq 0$. Let (f_n) be an increasing sequence of non-negative simple functions, increasing to f. Then $f_n^* g^* \nearrow f^* g^*$, by Proposition 7.1.2. By the monotone convergence theorem, $\int fg \, d\mu = \lim_{n \to \infty} \int f_n g \, d\mu$ and $\int f^* g^* \, d\mu = \lim_{n \to \infty} \int f_n^* g^* \, dt$. It is therefore suffi- cient to prove the result for simple f. We can write $f = \sum_{i=1}^n a_i I_{F_i}$, where $a_i \geq 0$ and $F_1 \subseteq F_2 \subseteq \cdots \subseteq F_n$. (Note that we have an increasing sequence of sets here, rather than a disjoint sequence, so that $f^* = \sum_{i=1}^n a_i I_{[0,\mu(F_i))}$.) Then, using Proposition 7.1.3,

$$\int fg \, d\mu = \sum_{i=1}^n a_i \left(\int_{F_i} g \, d\mu \right) \leq \sum_{i=1}^n a_i \left(\int_0^{\mu(F_i)} g^* \, dt \right)$$

$$= \int_0^\infty \left(\sum_{i=1}^n a_i I_{[0,\mu(F_i))} \right) g^* \, dt = \int_0^\infty f^* g^* \, dt.$$

$\square$

7.2 Rearrangement-invariant Banach function spaces

We say that a Banach function space $(X, \|.\|_X)$ is *rearrangement-invariant* if whenever $f \in X$ and $|f|$ and $|g|$ are equidistributed then $g \in X$ and $\|f\|_X = \|g\|_X$. Suppose that $(X, \|.\|_X)$ is rearrangement-invariant and ϕ is a measure-preserving map of (Ω, Σ, μ) onto itself (that is, $\mu(\phi^{-1}(A)) = \mu(A)$ for each $A \in \Sigma$). If $f \in X$ then f and $f \circ \phi$ have the same distribution, and so $f \circ \phi \in X$ and $\|f \circ \phi\|_X = \|f\|_X$; this explains the terminology.

Theorem 7.2.1 *Suppose that* $(X, \|.\|_X)$ *is a rearrangement-invariant func- tion space. Then* $(X' \|.\|_{X'})$ *is also a rearrangement-invariant function space, and*

$$\|f\|_X = \sup \left\{ \int f^* g^* \, dt \colon \|g\|_{X'} \leq 1 \right\}$$

$$= \sup \left\{ \int f^* g^* \, dt \colon g \text{ simple}, \ \|g\|_{X'} \leq 1 \right\}.$$

Proof By Proposition 7.1.4

$$\|f\|_X = \sup\left\{\int |fg|\,d\mu \colon \|g\|_{X'} \le 1\right\} \le \sup\left\{\int f^* g^*\,dt \colon \|g\|_{X'} \le 1\right\}.$$

On the other hand, if $f \in X$ and $g \in X'$ with $\|g\|_{X'} \le 1$, there exist increasing sequences (f_n) and (g_n) of simple functions which converge to $|f|$ and $|g|$ respectively. Further, for each n, we can take f_n and g_n of the form

$$f_n = \sum_{j=1}^{k} a_j \chi_{E_j}, \quad g_n = \sum_{j=1}^{k} b_j \chi_{E_j},$$

where $E_1, \ldots, E_k$ are disjoint sets of equal measure (here we use the special properties of (Ω, Σ, μ); see Exercise 7.7) and where $b_1 \ge \cdots \ge b_k$. Now there exists a permutation σ of $(1, \ldots, n)$ such that $a_{\sigma(1)} \ge \cdots \ge a_{\sigma(k)}$. Let $f_n^\sigma = \sum_{j=1}^{k} a_\sigma(j) \chi_{E_j}$. Then f_n and f_n^σ are equidistributed, so that

$$\|f\|_X \ge \|f_n\|_X = \|f_n^\sigma\|_X \ge \int f_n^\sigma g_n\,d\mu = \int f_n^* g_n^*\,dt.$$

Letting $n \to \infty$, we see that $\|f\|_X \ge \int f^* g^*\,dt$.

Finally, suppose that $g \in X'$ and that $|g|$ and $|h|$ are equidistributed. Then if $f \in X$ and $\|f\|_X \le 1$,

$$\int |fh|\,d\mu \le \int f^* h^*\,dt = \int f^* g^*\,dt \le \|g\|_{X'}.$$

This implies that $h \in X'$ and that $\|h\|_{X'} \le \|g\|_{X'}$; similarly $\|g\|_{X'} \le \|h\|_{X'}$. $\quad\square$

7.3 Muirhead's maximal function

In Section 4.3 we introduced the notion of a sublinear functional; these functionals play an essential role in the Hahn–Banach theorem. We now extend this notion to more general mappings.

A mapping T from a vector space E into a space $M_\infty(\Omega, \Sigma, \mu)$ is *subadditive* if $T(f + g) \le T(f) + T(g)$ for $f, g \in E$, is *positive homogeneous* if $T(\lambda f) = \lambda T(f)$ for $f \in E$ and λ real and positive, and is *sublinear* if it is both subadditive and positive homogeneous. The mapping $f \to f^*$ gives good information about f, but it is not subadditive: if A and B are disjoint sets of positive measure t, then $I_A^* + I_B^* = 2I_{[0,t)}$, while $(I_A + I_B)^* = I_{[0,2t)}$. We now introduce a closely related mapping, of great importance, which is sublinear. Suppose that $f \in M_1$ and that $t > 0$. We define *Muirhead's maximal function* as

$$f^\dagger(t) = \sup\left\{\frac{1}{t}\int_E |f|\, d\mu\colon \mu(E) \le t\right\},$$

for $0 < t < \mu(\Omega)$.

Theorem 7.3.1 *The mapping $f \to f^\dagger$ is sublinear, and if $|f| \le |g|$ then $f^\dagger \le g^\dagger$. If $|f_n| \nearrow |f|$ then $f_n^\dagger \nearrow f^\dagger$. Further,*

$$f^\dagger(t) = \frac{1}{t}\int_0^t f^*(s)\, ds.$$

Proof It follows from the definition that the mapping $f \to f^\dagger$ is sublinear, and that if $|f| \le |g|$ then $f^\dagger \le g^\dagger$. Thus if $|f_n| \nearrow |f|$ then $\lim_{n\to\infty} f_n^\dagger \le f^\dagger$. On the other hand, if $\mu(B) \le t$ then, by the monotone convergence theorem, $\int_B |f_n|\, d\mu \to \int_B |f|\, d\mu$. Thus $\lim_{n\to\infty} f_n^\dagger(t) \ge (1/t)\int_B |f|\, d\mu$. Taking the supremum over B, it follows that $\lim_{n\to\infty} f_n^\dagger(t) \ge f^\dagger(t)$.

If $f \in M_1$, then $f^\dagger(t) \le (1/t)\int_0^t f^*(s)\, ds$, by Proposition 7.1.3. It follows from Proposition 7.1.2 and the monotone convergence theorem that if $|f_n| \nearrow |f|$ then $(1/t)\int_0^t f_n^*(s)\, ds \nearrow (1/t)\int_0^t f^*(s)\, ds$. It is therefore sufficient to prove the converse inequality for non-negative simple functions.

Suppose then that $f = \sum_{i=1}^n \alpha_i I_{F_i}$ is a simple function, with $\alpha_i > 0$ for $1 < i < n$ and $F_1 \subseteq F_2 \subseteq \cdots \subseteq F_n$. If $\mu(F_n) \le t$, choose $G \supset F_n$ with $\mu(G) = t$. If $t < \mu(F_n)$ there exists j such that $\mu(F_{j-1}) \le t < \mu(F_j)$. Choose G with $F_{j-1} \subseteq G \subseteq F_j$ and $\mu(G) = t$. Then $(1/t)\int_0^t f^*(s)\, ds = (1/t)\int_G f\, d\mu$, and so $(1/t)\int_0^t f^*(s)\, ds \le f^\dagger(t)$. $\square$

Corollary 7.3.1 *If $f \in M_1$ then either $f^\dagger(t) = \infty$ for all $0 < t < \mu(\Omega)$ or $0 \le f^*(t) \le f^\dagger(t) < \infty$ for all $0 < t < \mu(\Omega)$. In the latter case, $f^\dagger$ is a continuous decreasing function on $(0, \mu(\Omega))$, and $t f^\dagger(t)$ is a continuous increasing function on $(0, \mu(\Omega))$.*

Proof If $\int_0^t f^*(s)\, ds = \infty$ for all $0 < t < \mu(\Omega)$, then $f^\dagger(t) = \infty$ for all $0 < t < \mu(\Omega)$. If there exists $0 < t < \mu(\Omega)$ for which $\int_0^t f^*(s)\, ds < \infty$, then $\int_0^t f^*(s)\, ds < \infty$ for all $0 < t < \mu(\Omega)$, and so $0 \le f^*(t) \le f^\dagger(t) < \infty$ for all $0 < t < \mu(\Omega)$. The function $t f^\dagger(t) = \int_0^t f^*(s)\, ds$ is then continuous and increasing. Thus $f^\dagger$ is continuous. Finally, if $0 < t < u < \mu(\Omega)$ then, setting $\lambda = (u-t)/u$,

$$f^\dagger(u) = (1-\lambda)f^\dagger(t) + \frac{\lambda}{u-t}\int_t^u f^*(s)\,ds \le (1-\lambda)f^\dagger(t) + \lambda f^*(t) \le f^\dagger(t).$$

$\square$

Here is another characterization of Muirhead's maximal function.

Theorem 7.3.2 *Suppose that* $0 < t < \mu(\Omega)$. *The map* $f \to f^\dagger(t)$ *is a function norm, and the corresponding Banach function space is* $L^1 + L^\infty$. *If* $f \in L^1 + L^\infty$ *then*

$$f^\dagger(t) = \inf\{\|h\|_1/t + \|k\|_\infty : f = h + k\}.$$

Further the infimum is attained: if $f \in L^1 + L^\infty$ *there exist* $h \in L^1$ *and* $k \in L^\infty$ *with* $\|h\|_1/t + \|k\|_\infty = f^\dagger(t)$.

Proof We need to check the conditions of Section 6.1. Conditions (i) and (ii) are satisfied, and (iii) follows from Theorem 7.3.1. If A is measurable, then $I_A^\dagger(t) \le 1$, so that condition (iv) is satisfied. If $\mu(A) < \infty$ there exist measurable sets $A_1, \ldots, A_k$, with $\mu(A_i) = t$ for $1 \le i \le k$, whose union contains A. Then if $f \in M$,

$$\int_A |f|\,d\mu \le \sum_{i=1}^k \int_{A_i} |f|\,d\mu \le kt f^\dagger(t),$$

and so condition (v) is satisfied. Thus $f^\dagger$ is a function norm.

First, suppose that $f = h + k$, with $h \in L^1$ and $k \in L^\infty$. If $\mu(A) \le t$ then $\int_A |h|\,d\mu \le \|h\|_1$, and so $h^\dagger(t) \le \|h\|_1/t$. Similarly, $\int_A |k|\,d\mu \le t\|k\|_\infty$, and so $k^\dagger(t) \le \|k\|_\infty$. Thus f is in the corresponding Banach function space, and

$$f^\dagger(t) \le h^\dagger(t) + k^\dagger(t) \le \|h\|_1/t + \|k\|_\infty.$$

Conversely suppose that $f^\dagger(t) < \infty$. First we observe that $f \in M_1$. For if not, then for each $u > 0$ there exists a set of measure t on which $|f| > u$, and so $f^\dagger(t) > u/t$, for all $u > 0$, giving a contradiction. Let $B = (|f| > f^*(t))$. Thus $f^*(s) > f^*(t)$ for $0 < s < \mu(B)$, and $f^*(s) \le f^*(t)$ for $\mu(B) \le s < \mu(\Omega)$. Since $|f|$ and f^* are equidistributed, $\mu(B) = \lambda(f^* > f^*(t)) \le t$. Now let $h = \operatorname{sgn} f(|f| - f^*(t))I_B$, and let $k = f - h$. Then $h^*(s) = f^*(s) - f^*(t)$ for $0 < s < \mu(B)$, and $h^*(s) = 0$ for $\mu(B) \le s < \mu(\Omega)$, so that

$$\frac{1}{t}\int |h|\, d\mu = \frac{1}{t}\int_0^{\mu(B)} f^*(s) - f^*(t)\, ds$$

$$= \frac{1}{t}\int_0^t f^*(s) - f^*(t)\, ds = f^\dagger(t) - f^*(t).$$

On the other hand, $|k(\omega)| = f^*(t)$ for $\omega \in B$, and $|k(\omega)| = |f(\omega)| \le f^*(t)$ for $\omega \notin B$, so that $\|k\|_\infty \le f^*(t)$. Thus $\|h\|_1 / t + \|k\|_\infty \le f^\dagger(t)$. $\qquad\square$

Theorem 7.3.3 *Suppose that $t > 0$. Then $L^1 \cap L^\infty$ is the associate space to $L^1 + L^\infty$ and the function norm*

$$\rho_{\{t\}}(g) = \max(\|g\|_1, t\|g\|_\infty)$$

is the associate norm to $f^\dagger(t)$.

Proof It is easy to see that $L^1 \cap L^\infty$ is the associate space to $L^1 + L^\infty$. Let $\|.\|'$ denote the associate norm. Suppose that $g \in L^1 \cap L^\infty$.

If $\|f\|_1 \le 1$ then $f^\dagger(t) \le 1/t$, and so $|\int fg\, d\mu| \le \|g\|' / t$. Thus

$$\|g\|_\infty = \sup\left\{ \left|\int fg\, d\mu\right| : \|f\|_1 \le 1\right\} \le \|g\|' / t.$$

Similarly, if $\|f\|_\infty \le 1$ then $f^\dagger(t) \le 1$, and so $|\int fg\, d\mu| \le \|g\|'$. Thus

$$\|g\|_1 = \sup\left\{ \left|\int fg\, d\mu\right| : \|f\|_\infty \le 1\right\} \le \|g\|'.$$

Consequently, $\rho_{\{t\}}(g) \le \|g\|'$.

Conversely, if $f^\dagger(t) \le 1$ we can write $f = h + k$ with $\|h\|_1 / t + \|k\|_\infty \le 1$. Then

$$\left|\int fg\, d\mu\right| \le \int |hg|\, d\mu| + \int |kg|\, d\mu$$

$$\le (\|h\|_1 / t) \cdot (t\|g\|_\infty) + \|k\|_\infty \cdot \|g\|_1$$

$$\le \rho_{\{t\}}(g).$$

Thus $\|g\|' \le \rho_{\{t\}}(g)$. $\qquad\square$

7.4 Majorization

We use Muirhead's maximal function to define an order relation on $L^1 + L^\infty$: we say that g *weakly majorizes* f, and write $f \prec_w g$, if $f^\dagger(t) \le g^\dagger(t)$ for all $t > 0$. If in addition f and g are non-negative functions in L^1 and $\int_\Omega f \, d\mu = \int_\Omega g \, d\mu$, we say that g *majorizes* f and write $f \prec g$. We shall however principally be concerned with weak majorization.

The following theorem begins to indicate the significance of this ordering. For $c \geq 0$, let us define the *angle function* a_c by $a_c(t) = (t - c)^+$.

Theorem 7.4.1 *Suppose that f and g are non-negative functions in $L^1 + L^\infty$. The following are equivalent:*

(i) $f \prec_w g$;

(ii) $\int_0^\infty f^*(t)h(t) \, dt \leq \int_0^\infty g^*(t)h(t) \, dt$ for every decreasing non-negative function h on $[0, \infty)$;

(iii) $\int a_c(f) \, d\mu \leq \int a_c(g) \, d\mu$ for each $c \geq 0$;

(iv) $\int \Phi(f) \, d\mu \leq \int \Phi(g) \, d\mu$ for every convex increasing function Φ on $[0, \infty)$ with $\Phi(0) = 0$.

Proof We first show that (i) and (ii) are equivalent. Since $tf^\dagger(t) = \int_0^\infty f^*(s)I_{[0,t)} \, ds$, (ii) implies (i). For the converse, if h is a decreasing non-negative step function on $[0, \infty)$, we can write $h = \sum_{i=1}^j \alpha_i I_{[0,t_i)}$, with $\alpha_i > 0$ and $0 < t_1 < \cdots < t_j$, so that if $f \prec_w g$ then

$$\int f^*(t)h(t) \, dt = \sum_{i=1}^j \alpha_i t_i f^\dagger(t_i)$$

$$\leq \sum_{i=1}^j \alpha_i t_i g^\dagger(t_i) = \int g^*(t)h(t) \, dt.$$

For general decreasing non-negative h, let (h_n) be an increasing sequence of decreasing non-negative step functions which converges pointwise to h. Then, by the monotone convergence theorem,

$$\int f^*(t)h(t) \, dt = \lim_{n \to \infty} \int f^*(t)h_n(t) \, dt$$

$$\leq \lim_{n \to \infty} \int g^*(t)h_n(t) \, dt = \int g^*(t)h(t) \, dt.$$

Thus (i) and (ii) are equivalent.

Next we show that (i) and (iii) are equivalent. Suppose that $f \prec_w g$ and that $c > 0$. Let

$$t_f = \inf\{s : f^*(s) \leq c\} \quad \text{and} \quad t_g = \inf\{s : g^*(s) \leq c\}.$$

If $t_f \leq t_g$, then

$$\int a_c(f)\, d\mu = \int_{(f>c)} (f-c)\, d\mu = \int_0^{t_f} f^*(s)\, ds - ct_f$$

$$\leq \int_0^{t_f} g^*(s)\, ds - ct_f + \left(\int_{t_f}^{t_g} g^*(s)\, ds - c(t_g - t_f) \right)$$

$$= \int_0^{t_g} g^*(s)\, ds - ct_g = \int_{(g>c)} (g-c)\, d\mu,$$

since $g^*(s) > c$ on $[t_f, t_g)$.

On the other hand, if $t_f > t_g$, then

$$\int a_c(f)\, d\mu = \int_{(f>c)} (f-c)\, d\mu = \int_0^{t_f} f^*(s)\, ds - ct_f$$

$$\leq \int_0^{t_f} g^*(s)\, ds - ct_f$$

$$= \int_0^{t_g} g^*(s)\, ds + \int_{t_g}^{t_f} g^*(s)\, ds - ct_f$$

$$\leq \int_0^{t_g} g^*(s)\, ds + c(t_f - t_g) - ct_f$$

$$= \int_{(g>c)} (g-c)\, d\mu,$$

since $g^*(s) \leq c$ on $[t_g, t_f)$. Thus (i) implies (iii).

Conversely, suppose that (iii) holds. By monotone convergence, the inequality also holds when $c = 0$. Suppose that $t > 0$, and let $c = g^*(t)$. Let t_f and t_g be defined as above. Note that $t_g \leq t$.

If $t_f \leq t$, then

$$\int_0^t f^*(s)\, ds \leq \int_0^{t_f} f^*(s)\, ds + (t - t_f)c$$

$$= \int_{(f>c)} (f-c)\, d\mu + tc$$

$$\leq \int_{(g>c)} (g-c)\, d\mu + tc$$

$$= \int_0^{t_g} g^*(s)\, ds + (t - t_g)c$$

$$= \int_0^t g^*(s)\, ds,$$

since $g^*(s) = c$ on $[t_g, t)$.

On the other hand, if $t_f > t$, then

$$
\begin{aligned}
\int_0^t f^*(s)\, ds &= \int_0^t (f^*(s) - c)\, ds + ct \\
&\leq \int_0^{t_f} (f^*(s) - c)\, ds + ct \\
&= \int_{(f>c)} (f - c)\, d\mu + ct \\
&\leq \int_{(g>c)} (g - c)\, d\mu + ct \\
&= \int_0^t (g^*(s) - c)\, ds + ct \\
&= \int_0^t g^*(s)\, ds.
\end{aligned}
$$

Thus $f \prec_w g$, and (iii) implies (ii).

We finally show that (iii) and (iv) are equivalent. Since a_c is a non-negative increasing convex function on $[0, \infty)$, (iv) implies (iii). Suppose that (iii) holds. Then $\int \Phi(f)\, d\mu \leq \int \Phi(g)\, d\mu$ when $\Phi = \sum_{i=1}^j \alpha_i a_{c_i}$, where $\alpha_i > 0$ and a_{c_i} is an angle function for $1 \leq i \leq j$. As any convex increasing non-negative function Φ with $\Phi(0) = 0$ can be approximated by an increasing sequence of such functions (Exercise 7.8), the result follows from the monotone convergence theorem. $\qquad\square$

Corollary 7.4.1 *Suppose that $(X, \|.\|_X)$ is a rearrangement-invariant Banach function space. If $f \in X$ and $h \prec_w f$ then $h \in X$ and $\|h\|_X \leq \|f\|_X$.*

Proof By Theorem 7.2.1, and (ii),

$$
\begin{aligned}
\|h\|_X &= \sup \left\{ \int h^* g^*\, dt \colon \|g^*\|_{X'} \leq 1 \right\} \\
&\leq \sup \left\{ \int f^* g^*\, dt \colon \|g^*\|_{X'} \leq 1 \right\} = \|f\|_X.
\end{aligned}
$$

$\qquad\square$

Theorem 7.4.2 *Suppose that $(X, \|.\|_X)$ is a rearrangement-invariant function space. Then $L^1 \cap L^\infty \subseteq X \subseteq L^1 + L^\infty$, and the inclusions are continuous.*

Proof Let $0 < t < \mu(\Omega)$, and let E be a set of measure t. Set $C_t = \|I_E\|'_X / t$. Since X' is rearrangement-invariant, C_t does not depend on the choice of E. Suppose that $f \in X$ and that $\mu(F) \le t$. Then

$$\frac{1}{t} \int_F |f| \, d\mu \le \|f\|_X \|I_F\|'_X / t \le C_t \|f\|_X < \infty,$$

so that $f^\dagger(t) \le C_t \|f\|_X$. Thus $f \in L^1 + L^\infty$, and the inclusion: $X \to L^1 + L^\infty$ is continuous. Similarly $X' \subseteq L^1 + L^\infty$, with continuous inclusion; considering associates, we see that $L^1 \cap L^\infty \subseteq X$, with continuous inclusion.

$\square$

7.5 Calderón's interpolation theorem and its converse

We now come to the first of several interpolation theorems that we shall prove.

Theorem 7.5.1 (Calderón's interpolation theorem) *Suppose that T is a sublinear mapping from $L^1 + L^\infty$ to itself which is norm-decreasing on L^1 and norm-decreasing on L^∞. If $f \in L^1 + L^\infty$ then $T(f) \prec_w f$.*

If $(X, \|.\|_X)$ is a rearrangement-invariant function space, then $T(X) \subseteq X$ and $\|T(f)\|_X \le \|f\|_X$ for $f \in X$.

Proof Suppose that $f \in L^1 + L^\infty$ and that $0 < t < \mu(\Omega)$. By Theorem 7.3.2,

$$T(f)^\dagger(t) \le \inf\{\|T(h)\|_1 / t + \|T(k)\|_\infty : f = h + k\}$$
$$\le \inf\{\|h\|_1 / t + \|k\|_\infty : f = h + k\} = f^\dagger(t),$$

and so $T(f) \prec_w f$. The second statement now follows from Corollary 7.4.1.

$\square$

Here is an application of Calderón's interpolation theorem. We shall state it for $\mathbf{R}^d$, but it holds more generally for a locally compact group with Haar measure (see Section 9.5).

Proposition 7.5.1 *Suppose that ν is a probability measure on $\mathbf{R}^d$ and that $(X, \|.\|_X)$ is a rearrangement-invariant function space on $\mathbf{R}^d$. If $f \in X$, then the convolution product $f \star \nu$, defined by*

$$(f \star \nu)(x) = \int f(x - y) \, d\nu(y),$$

is in X, and $\|f \star \nu\|_X \le \|f\|_X$.

Proof If $f \in L^1$ then

$$\int |f \star \nu| \, d\lambda \leq \int \left(\int |f(x - y)| \, d\lambda(x) \right) d\nu(y) = \int \|f\|_1 \, d\nu = \|f\|_1,$$

while if $g \in L^\infty$ then

$$|(g \star \nu)(x)| \leq \int |g| \, d\nu \leq \|g\|_\infty.$$

Thus we can apply Calderón's interpolation theorem.   □

As a consequence, if $h \in L^1(\mathbf{R}^d)$ then, since $|f \star h| \leq |f| \star |h|$, $f \star h \in X$ and

$$\|f \star h\|_X \leq \||f| \star |h|\|_X \leq \|f\|_X \|h\|_1.$$

The first statement of Calderón's interpolation theorem has an interesting converse. We shall prove this in the case where Ω has finite measure (in which case we may as well suppose that $\mu(\Omega) = 1$), and μ is *homogeneous*: that is, if we have two partitions $\Omega = A_1 \cup \cdots \cup A_n = B_1 \cup \cdots \cup B_n$ into sets of equal measure then there is a measure-preserving transformation R of Ω such that $R(A_i) = B_i$ for $1 \leq i \leq n$. Neither of these requirements is in fact necessary.

Theorem 7.5.2 *Suppose that $\mu(\Omega) = 1$ and μ is homogeneous. If $f, g \in L^1$ and $f \prec_w g$ then there exists a linear mapping T from L^1 to itself which is norm-decreasing on L^1 and norm-decreasing on L^∞ and for which $T(g) = f$. If g and f are non-negative, we can also suppose that T is a positive operator (that is, $T(h) \geq 0$ if $h \geq 0$).*

Proof The proof that we shall give is based on that given by Ryff [Ryf 65]. It is a convexity proof, using the separation theorem.

First we show that it is sufficient to prove the result when f and g are both non-negative. If $f \prec_w g$ then $|f| \prec_w |g|$. We can write $f = \theta|f|$, with $|\theta(\omega)| = 1$ for all ω, and $g = \phi|f|$, with $|\phi(\omega)| = 1$ for all ω. If there exists a suitable S with $S(|g|) = |f|$, let $T(k) = \theta.S(k/\phi)$. Then $T(g) = f$, and T is norm-decreasing on L^1 and on L^∞. We can therefore suppose that f and g are both non-negative, and restrict attention to real-valued functions.

We begin by considering the set

$$\Delta = \{T : T \in L(L^1),\ T \geq 0,\ \|T(f)\|_1 \leq \|f\|_1,\ \|T(f)\|_\infty \leq \|f\|_\infty \text{ for } f \in L^\infty\}.$$

If $T \in \Delta$, the transposed mapping T^* is norm-decreasing on L^∞. Also, T^* extends by continuity to a norm-decreasing linear map on L^1. Thus the extension of T^* to L^1, which we again denote by T^*, is in Δ.

Δ is a semi-group, and is a convex subset of

$$B^+ = \{T \in L(L^\infty): T \geq 0, \|T\| \leq 1\}.$$

Now B^+ is compact under the weak operator topology defined by the semi-norms $p_{h,k}(T) = \int (T(h)k \, d\mu$, where $h \in L^\infty, k \in L^1$. [This is a consequence of the fact that if E and F are Banach spaces then $L(E, F^*)$ can be identified with the dual of the tensor product $E \hat{\otimes} F$ with the projective norm, and of the Banach–Alaoglu theorem [DiJT 95, p. 120]. We shall show that Δ is closed in B^+ in this topology, so that Δ is also compact in the weak operator topology.

Suppose that $h, k \in L^\infty$ and that $\|h\|_1 \leq 1, \|k\|_\infty \leq 1$. Then if $T \in \Delta$, $|\int T(h)k \, d\mu| \leq 1$. Thus if $S \in \bar{\Delta}$, $|\int S(h)k \, d\mu| \leq 1$. Since this holds for all $k \in L^\infty$ with $\|k\|_\infty \leq 1$, $\|S(h)\|_1 \leq 1$. Thus $S \in \Delta$.

As we have observed, we can consider elements of Δ as norm-decreasing operators on L^1. We now consider the orbit

$$O(g) = \{T(g): T \in \Delta\} \subseteq L^1.$$

The theorem will be proved if we can show that $O(g) \supseteq \{f: f \geq 0, f \prec_w g\}$. $O(g)$ is convex. We claim that $O(g)$ is also closed in L^1. Suppose that $k \in \overline{O(g)}$. There exists a sequence (T_n) in Δ such that $T_n(g) \to k$ in L^1 norm. Let S be a limit point, in the weak operator topology, of the sequence (T_n^*). Then S and S^* are in Δ. If $h \in L^\infty$, then

$$\int kh \, d\mu = \lim_{n \to \infty} \int T_n(g)h \, d\mu = \lim_{n \to \infty} \int gT_n^*(h) \, d\mu$$
$$= \int gS(h) \, d\mu = \int S^*(g)h \, d\mu.$$

Since this holds for all $h \in L^\infty$, $k = S^*(g) \in O(g)$. Thus $O(g)$ is closed.

Now suppose that $f \prec_w g$, but that $f \notin O(g)$. Then by the separation theorem (Theorem 4.6.3) there exists $h \in L^\infty$ such that

$$\int fh \, d\mu > \sup\left\{\int kh \, d\mu: k \in O(g)\right\}.$$

Let $A = (h > 0)$, so that $h^+ = hI_A$. Then if $k \in O(g)$, $I_A k \in O(g)$, since multiplication by I_A is in Δ, and Δ is a semigroup. Thus

$$\int fh^+ \, d\mu \geq \int fh \, d\mu > \sup \left\{ \int I_A k h \, d\mu \colon k \in O(g) \right\}$$

$$= \sup \left\{ \int kh^+ \, d\mu \colon k \in O(g) \right\}.$$

In other words, we can suppose that $h \geq 0$. Now $\int fh \, d\mu \leq \int_0^1 f^* h^* \, ds$, and so we shall obtain the required contradiction if we show that

$$\sup \left\{ \int kh \, d\mu \colon k \in O(g) \right\} \geq \int_0^1 g^* h^* \, ds.$$

We can find increasing sequences (g_n), (h_n) of simple non-negative functions converging to g and h respectively, of the form

$$g_n = \sum_{j=1}^{J_n} a_j \chi_{A_j}, \quad h_n = \sum_{j=1}^{J_n} b_j \chi_{B_j},$$

with $\mu(A_j) = \mu(B_j) = 1/J_n$ for each j. There exists a permutation σ_n of $\{1, \ldots, J_n\}$ such that

$$\frac{1}{J_n} \sum_{j=1}^{J_n} a_{\sigma(j)} b_j = \frac{1}{J_n} \sum_{j=1}^{J_n} a_j^* b_j^* = \int_0^1 g_n^* h_n^* \, ds.$$

By homogeneity, there exists a measure-preserving transformation R_n of Ω such that $R_n(B_{\sigma(j)}) = A_j$ for each j. If $l \in L^\infty$, let $T_n(l)(\omega) = l(R_n(\omega))$; then $T_n \in \Delta$. Then

$$\int T_n(g) h \, d\mu \geq \int T_n(g_n) h_n \, d\mu = \int g_n^* h_n^* \, ds.$$

Since $\int_0^1 g^* h^* \, ds = \sup \int_0^1 g_n^* h_n^* \, ds$, this finishes the proof. $\qquad\square$

7.6 Symmetric Banach sequence spaces

We now turn to the case where $\Omega = \mathbf{N}$, with counting measure. Here we are considering sequences, and spaces of sequences. The arguments are often technically easier, but they are no less important. Note that

$$L^1 = l_1 = \{x = (x_i): \|x\|_1 = \sum_{i=0}^{\infty} |x_i| < \infty\},$$

$M_0 = c_0 = \{x = (x_i): x_i \to 0\}$ with $\|x\|_{c_0} = \|x\|_\infty = \max |x_i|$, and
$M_1 = l_\infty$.

It is easy to verify that a Banach sequence space $(X, \|.\|_X)$ is rearrangement invariant if and only whenever $x \in X$ and σ is a permutation of $\mathbf{N}$ then $x_\sigma \in X$ and $\|x\|_X = \|x_\sigma\|_X$ (where x_σ is the sequence defined by $(x_\sigma)_i = x_{\sigma(i)}$). Let e_i denote the sequence with 1 in the i-th place, and zeros elsewhere. If $(X, \|.\|_X)$ is a rearrangement-invariant Banach sequence space then $\|e_i\|_X = \|e_j\|_X$: we scale the norm so that $\|e_i\|_X = 1$: the resulting space is called a *symmetric Banach sequence space*. If $(X, \|.\|_X)$ is a symmetric Banach sequence space, then $l_1 \subseteq X$, and the inclusion is norm-decreasing. By considering associate spaces, it follows that $X \subseteq l_\infty$, and the inclusion is norm-decreasing.

Proposition 7.6.1 *If $(X, \|.\|_X)$ is a symmetric Banach sequence space then either $l_1 \subseteq X \subseteq c_0$ or $X = l_\infty$.*

Proof Certainly $l_1 \subseteq X \subseteq l_\infty$. If $x \in X \setminus c_0$, then there exists a permutation σ and $\epsilon > 0$ such that $|x_{\sigma(2n)}| \geq \epsilon$ for all n; it follows from the lattice property and scaling that the sequence $(0, 1, 0, 1, 0, \ldots) \in X$. Similarly, the sequence $(1, 0, 1, 0, 1, \ldots) \in X$, and so $(1, 1, 1, 1, \ldots) \in X$; it follows again from the lattice property and scaling that $X \supseteq l_\infty$. $\square$

If $x \in c_0$, the decreasing rearrangement x^* is a sequence, which can be defined recursively by taking x_1^* as the absolute value of the largest term, x_2^* as the absolute value of the next largest, and so on. Thus there exists a one-one mapping $\tau : \mathbf{N} \to \mathbf{N}$ such that $x_n^* = |x_{\tau(n)}|$. x_n^* can also be described by a minimax principle:

$$x_n^* = \min\{\max\{|x_j|: j \notin E\}: |E| < n\}.$$

We then have the following results, whose proofs are the same as before, or easier.

Proposition 7.6.2 *(i) $|x|$ and x^* are equidistributed.*
(ii) If $0 \leq x^{(n)} \nearrow x$ then $0 \leq x^{(n)*} \nearrow x^*$.
(iii) If $x \geq 0$ and $A \subset \mathbf{N}$ then $\sum_{i \in A} x_i \leq \sum_{i=1}^{|A|} x_i^*$.
(iv) If $x, y \in c_0$ then $\sum_{i=1}^{\infty} |x_i y_i| \leq \sum_{i=1}^{\infty} x_i^* y_i^*$.

We define *Muirhead's maximal sequence* as

$$x_i^\dagger = \frac{1}{i} \sup \left\{ \sum_{j \in A} |x_j| : |A| = i \right\}.$$

Then $x_i^\dagger$ is a norm on c_0 equivalent to $\|x\|_\infty = \max_n |x_n|$, and $x_i^\dagger = (\sum_{j=1}^i x_j^*)/i$, so that $x^\dagger = (x^*)^\dagger \geq x^*$.

Again, we define $x \prec_w y$ if $x^\dagger \leq y^\dagger$. The results corresponding to those of Theorems 7.4.1, 7.2.1 and 7.5.1 all hold, with obvious modifications.

Let us also note the following multiplicative result, which we shall need when we consider linear operators.

Proposition 7.6.3 *Suppose that (x_n) and (y_n) are decreasing sequences of positive numbers, and that $\prod_{n=1}^N x_n \leq \prod_{n=1}^N y_n$, for each N. If ϕ is an increasing function on $[0, \infty)$ for which $\phi(e^t)$ is a convex function of t then $\sum_{n=1}^N \phi(x_n) \leq \sum_{n=1}^N \phi(y_n)$ for each N. In particular, $\sum_{n=1}^N x_n^p \leq \sum_{n=1}^N y_n^p$ for each N, for $0 < p < \infty$.*

If $(X, \|.\|_X)$ is a symmetric Banach sequence space, and $(y_n) \in X$, then $(x_n) \in X$ and $\|(x_n)\|_X \leq \|(y_n)\|_X$.

Proof Let $a_n = \log x_n - \log x_N$ and $b_n = \log y_n - \log x_N$ for $1 \leq n \leq N$. Then $(a_n) \prec_w (b_n)$. Let $\psi(t) = \phi(x_N e^t) - \phi(x_N)$. Then ψ is a convex increasing function on $[0, \infty)$ with $\psi(0) = 0$, and so by Theorem 7.4.1

$$\sum_{n=1}^N \phi(x_n) = \sum_{n=1}^N \psi(a_n) + N\phi(x_N)$$

$$\leq \sum_{n=1}^N \psi(b_n) + N\phi(x_N) = \sum_{n=1}^N \phi(y_n).$$

The second statement is just a special case, since e^{tp} is a convex function of t. In particular, $x_n^\dagger \leq y_n^\dagger$, and so the last statement follows from Corollary 7.4.1. $\square$

7.7 The method of transference

What about the converse of Calderón's interpolation theorem? Although it is a reasonably straightforward matter to give a functional analytic proof of the corresponding theorem along the lines of Theorem 7.5.2, we give a more direct proof, since this proof introduces important ideas, with useful applications. Before we do so, let us consider how linear operators are represented by infinite matrices.

Suppose that $T \in L(c_0)$ and that $T(x) = y$. Then $y_i = \sum_{j=1}^{\infty} t_{ij} x_j$, where $t_{ij} = (T(e_j))_i$, so that

$$t_{ij} \to 0 \text{ as } i \to \infty \text{ for each } j, \quad \text{and} \quad \|T\| = \sup_i \left(\sum_{j=1}^{\infty} |t_{ij}| \right) < \infty.$$

Conversely if (t_{ij}) is a matrix which satisfies these conditions then, setting $T(x)_i = \sum_{j=1}^{\infty} t_{ij} x_j$, $T \in L(c_0)$ and $\|T\| = \sup_i (\sum_{j=1}^{\infty} |t_{ij}|)$.

Similarly if $S \in L(l_1)$, then S is represented by a matrix (s_{ij}) which satisfies

$$\|S\| = \sup_j \left(\sum_{i=1}^{\infty} |s_{ij}| \right) < \infty,$$

and any such matrix defines an element of $L(l_1)$.

If $T \in L(c_0)$ or $T \in L(l_1)$ then T is positive if and only if $t_{ij} \geq 0$ for each i and j. A matrix is *doubly stochastic* if its terms are all non-negative and

$$\sum_{i=1}^{\infty} t_{ij} = 1 \text{ for each } j \quad \text{and} \quad \sum_{j=1}^{\infty} t_{ij} = 1 \text{ for each } i.$$

A doubly stochastic matrix defines an operator which is norm-decreasing on c_0 and norm-decreasing on l_1, and so, by Calderón's interpolation theorem, it defines an operator which is norm-decreasing on each symmetric sequence space. Examples of doubly stochastic matrices are provided by *permutation matrices*; $T = (t_{ij})$ is a permutation matrix if there exists a permutation σ of $\mathbf{N}$ for which $t_{\sigma(j)j} = 1$ for each j and $t_{\sigma(i)j} = 0$ for $i \neq j$. In other words, each row and each column of T contains exactly one 1, and all the other entries are 0. If T is a permutation matrix then $(T(x))_i = x_{\sigma(i)}$, so that T permutes the coordinates of a vector. More particularly, a *transposition matrix* is a permutation matrix that is defined by a transposition – a permutation that exchanges two elements, and leaves the others fixed.

Theorem 7.7.1 *Suppose that x and y are non-negative decreasing sequences in c_0 with $x \prec_w y$. There exists a doubly stochastic matrix $P = (p_{ij})$ such that $x_i \leq \sum_{j=1}^{\infty} p_{ij} y_j$ for $1 \leq i < \infty$.*

Proof We introduce the idea of a *transfer matrix*. Suppose that $\tau = \tau_{ij}$ is the transposition of $\mathbf{N}$ which exchanges i and j and leaves the other integers fixed, and let π_τ be the corresponding transposition matrix. Then if $0 < \lambda \leq 1$ the *transfer matrix* $T = T_{\tau,\lambda}$ is defined as

$$T = T_{\tau,\lambda} = (1 - \lambda)I + \lambda \pi_\tau.$$

Thus

$$T_{ii} = T_{jj} = 1 - \lambda,$$
$$T_{kk} = 1 \quad \text{for } k \neq i, j,$$
$$T_{ij} = T_{ji} = \lambda$$
$$T_{kl} = 0 \quad \text{otherwise.}$$

If $T(z) = z'$, then $z_k = z'_k$ for $k \neq i, j$, and

$$z'_i + z'_j = ((1 - \lambda)z_i + \lambda z_j) + (\lambda z_i + (1 - \lambda)z_j) = z_i + z_j,$$

so that some of z_i is transferred to z'_j (or conversely). Note also that T is an averaging procedure; if we write $z_i = m + d$, $z_j = m - d$, then $z'_i = m + \mu d$, $z'_j = m - \mu d$, where $-1 \leq \mu = 1 - 2\lambda \leq 1$. Since T is a convex combination of I and π_τ, T is doubly stochastic, and so it is norm-decreasing on c_0 and on l_1. Note that transposition matrices are special cases of transfer matrices (with $\lambda = 1$).

We shall build P up as an infinite product of transfer matrices. We use the fact that if $k < l$ and $y_k > x_k$, $y_l < x_l$ and $y_j = x_j$ for $k < j < l$, and if we transfer an amount $\min(y_k - x_k, x_l - y_l)$ from y_k to y_l then the resulting sequence z is still decreasing, and $x \prec_w z$. We also use the fact that if $x_l > y_l$ then there exists $k < l$ such that $y_k > x_k$.

It may happen that $y_i \geq x_i$ for all i, in which case we take P to be the identity matrix. Otherwise, there is a least l such that $y_l < x_l$. Then there exists a greatest $k < l$ such that $y_k > x_k$. We transfer the amount $\min(y_k - x_k, x_l - y_l)$ from y_k to y_l, and iterate this procedure until we obtain a sequence $y^{(1)}$ with $y_l^{(1)} = x_l$. Composing the transfer matrices that we have used, we obtain a doubly stochastic matrix $P^{(1)}$ for which $P^{(1)}(y) = y^{(1)}$.

We now iterate this procedure. If it finishes after a finite number of steps, we are finished. If it continues indefinitely, there are two possibilities. First, for each k for which $y_k > x_k$, only finitely many transfers are made from y_k. In this case, if $P^{(n)}$ is the matrix obtained by composing the transfers used in the first n steps, then as n increases, each row and each column of $P^{(n)}$ is eventually constant, and we can take P as the term-by-term limit of $P^{(n)}$.

The other possibility is that infinitely many transfers are made from y_k, for some k. There is then only one k for which this happens. In this case, we start again. First, we follow the procedure described above, omitting the transfers from y_k, whenever they should occur. As a result, we obtain a doubly stochastic matrix P such that if $z = P(y)$ then $z_i \geq x_i$ for $1 \leq i < k$, $z_k = y_k > x_k$, there exists an infinite sequence $k < l_1 < l_2 < \cdots$ such that $x_{l_j} > z_{l_j}$ for each j, and $z_i = x_i$ for all other i. Let $\delta = x_{l_1} - z_{l_1}$.

Note that $\sum_{j=1}^{\infty}(x_{l_j} - z_{l_j}) \leq z_k - x_k$. We now show that there is a doubly stochastic matrix Q such that $Q(z) \geq x$. Then $QP(y) \geq x$, and QP is doubly stochastic. To obtain Q, we transfer an amount $x_{l_1} - z_{l_1}$ from z_k to z_{l_1}, then transfer an amount $x_{l_2} - z_{l_2}$ from z_k to z_{l_2}, and so on. Let $Q^{(n)}$ be the matrix obtained after n steps, and let $w^{(n)} = Q^{(n)}(z)$. It is easy to see that every row of $Q^{(n)}$, except for the k-th, is eventually constant. Let λ_n be the parameter for the nth transfer, and let $p_n = \prod_{i=1}^{n}(1 - \lambda_i)$. Then easy calculations show that

$$Q_{kk}^{(n)} = p_n, \text{ and } Q_{kl_i}^{(n)} = (\lambda_i/p_i)p_n.$$

Then

$$w_k^{(n+1)} = (1 - \lambda_{n+1})w_k^{(n)} + \lambda_{n+1}z_{l_{n+1}} = w_k^{(n)} - (x_{l_{n+1}} - z_{l_{n+1}}),$$

so that $\lambda_{n+1}(w_k^{(n)} - z_{l_{n+1}}) = x_{l_{n+1}} - z_{l_{n+1}}$. But

$$w_k^{(n)} - z_{l_{n+1}} \geq x_k - z_{l_1} \geq x_{l_1} - z_{l_1} = \delta,$$

so that $\sum_{n=1}^{\infty} \lambda_n < \infty$. Thus p_n converges to a positive limit p. From this it follows easily that if Q is the term-by-term limit of $Q^{(n)}$ then Q is doubly stochastic, and $Q(z) \geq x$. □

Corollary 7.7.1 *If $x, y \in c_0$ and $x \prec_w y$ then there is a matrix Q which defines norm-decreasing linear mappings on l_1 and c_0 and for which $Q(y) = x$.*

Proof Compose P with suitable permutation and multiplication operators. □

Corollary 7.7.2 *If x and y are non-negative elements of l_1 and $x \prec y$ then there exists a doubly stochastic matrix P such that $P(y) = x$.*

Proof By composing with suitable permutation operators, it is sufficient to consider the case where x and y are decreasing sequences. If P satisfies the conclusions of Theorem 7.7.1 then

$$\sum_{j=1}^{\infty} y_j = \sum_{i=1}^{\infty} x_i \leq \sum_{i=1}^{\infty}\left(\sum_{j=1}^{\infty} p_{ij}y_j\right) = \sum_{j=1}^{\infty}\left(\sum_{i=1}^{\infty} p_{ij}\right)y_j = \sum_{j=1}^{\infty} y_j.$$

Thus we must have equality throughout, and so $x_i = \sum_{j=1}^{\infty} p_{ij}y_j$ for each j. □

7.8 Finite doubly stochastic matrices

We can deduce corresponding results for the case when $\Omega = \{1, \ldots, n\}$. In particular, we have the following.

Theorem 7.8.1 *Suppose that $x, y \in \mathbf{R}^n$ and that $x \prec_w y$. Then there exists a matrix $T = (t_{ij})$ with*

$$\sum_{j=1}^n |t_{ij}| \leq 1 \text{ for } 1 \leq i \leq n \quad \text{and} \quad \sum_{i=1}^n |t_{ij}| \leq 1 \text{ for } 1 \leq j \leq n$$

such that $x_i = \sum_{j=1}^n t_{ij} y_j$.

Theorem 7.8.2 *Suppose that $x, y \in \mathbf{R}^n$ and that $x \geq 0$ and $y \geq 0$. The following are equivalent:*

(i) $x \prec y$.

(ii) There exists a doubly stochastic matrix P such that $P(y) = x$.

(iii) There exists a finite sequence $(T^{(1)}, \ldots, T^{(n)})$ of transfer matrices such that $x = T^{(n)} T^{(n-1)} \cdots T^{(1)} y$.

(iv) x is a convex combination of $\{y_\sigma \colon \sigma \in \Sigma_n\}$.

Proof The equivalence of the first three statements follows as in the infinite-dimensional case. That (iii) implies (iv) follows by writing each $T^{(j)}$ as $(1 - \lambda_j)I + \lambda_j \tau^{(j)}$, where $\tau^{(j)}$ is a transposition matrix, and expanding. Finally, the fact that (iv) implies (i) follows immediately from the sublinearity of the mapping $x \to x^\dagger$. $\square$

The set $\{x \colon x \prec y\}$ is a bounded closed convex subset of $\mathbf{R}^n$. A point c of a convex set C is an *extreme point* of C if it cannot be written as a convex combination of two other points of C: if $c = (1 - \lambda)c_0 + \lambda c_1$, with $0 < \lambda < 1$ then $c = c_0 = c_1$.

Corollary 7.8.1 *The vectors $\{y_\sigma \colon \sigma \in \Sigma_n\}$ are the extreme points of $\{x \colon x \prec y\}$.*

Proof It is easy to see that each y_σ is an extreme point, and the theorem ensures that there are no other extreme points. $\square$

Theorem 7.8.2 and its corollary suggests the following theorem. It does however require a rather different proof.

Theorem 7.8.3 *The set $\mathbf{P}$ of doubly stochastic $n \times n$ matrices is a bounded closed convex subset of $\mathbf{R}^{n \times n}$. A doubly stochastic matrix is an extreme* *point of* **P** *if and only if it is a permutation matrix. Every doubly stochastic matrix can be written as a convex combination of permutation matrices.*

Proof It is clear that **P** is a bounded closed convex subset of $\mathbf{R}^{n \times n}$, and that the permutation matrices are extreme points of **P**. Suppose that $P = (p_{ij})$ is a doubly stochastic matrix which is not a permutation matrix. Then there is an entry p_{ij} with $0 < p_{ij} < 1$. Then the i-th row must have another entry strictly between 0 and 1, and so must the j-th column. Using this fact repeatedly, we find a circuit of entries with this property: there exist distinct indices $i_1, \ldots, i_r$ and distinct indices $j_1, \ldots, j_r$ such that, setting $j_{r+1} = j_1$,

$$0 < p_{i_s j_s} < 1 \text{ and } 0 < p_{i_s j_{s+1}} < 1 \quad \text{for } 1 \le s \le r.$$

We use this to define a matrix $D = (d_{ij})$, by setting

$$d_{i_s j_s} = 1 \text{ and } d_{i_s j_{s+1}} = -1 \quad \text{for } 1 \le s \le r.$$

Let

$$a = \inf_{1 \le s \le r} p_{i_s j_s}, \qquad b = \inf_{1 \le s \le r} p_{i_s j_{s+1}}.$$

Then $P + \lambda D \in \mathbf{P}$ for $-a \le \lambda \le b$, and so P is not an extreme point of **P**.

We prove the final statement of the theorem by induction on the number of non-zero entries, using this construction. The result is certainly true when this number is n, for then P is a permutation matrix. Suppose that it is true for doubly stochastic matrices with less than k non-zero entries, and that P has k non-zero entries. Then, with the construction above, $P - aD$ and $P + bD$ have fewer than k non-zero entries, and so are convex combinations of permutation matrices. Since P is a convex combination of $P - aD$ and $P + bD$, P has the same property. $\qquad \square$

7.9 Schur convexity

Schur [Sch 23] investigated majorization, and raised the following problem: for what functions on $(\mathbf{R}^n)^+$ is it true that if $x \ge 0$, $y \ge 0$ and $x \prec y$ then $\phi(x) \le \phi(y)$? Such functions are now called *Schur convex*. [If $\phi(x) \ge \phi(y)$, ϕ is *Schur concave*.] Since $x_\sigma \prec x \prec x_\sigma$ for any permutation σ, a Schur convex function must be symmetric: $\phi(x_\sigma) = \phi(x)$. We have seen in Theorem 7.4.1 that if Φ is a convex increasing non-negative function on $[0, \infty)$ then the function $x \to \sum_{i=1}^n \Phi(x_i)$ is Schur convex. Theorem 7.8.2 has the following immediate consequence.

Theorem 7.9.1 *A function ϕ on $(\mathbf{R}^n)^+$ is Schur convex if and only if $\phi(T(x)) \leq \phi(x)$ for each $x \in (\mathbf{R}^n)^+$ and each transfer matrix T.*

Let us give one example. This is the original example of Muirhead [Mui 03], where the method of transfer was introduced.

Theorem 7.9.2 (Muirhead's theorem) *Suppose that $t_1, \ldots, t_n$ are positive. If $x \in (\mathbf{R}^n)^+$, let*

$$\phi(x) = \frac{1}{n!} \sum_{\sigma \in \Sigma_n} t_{\sigma(1)}^{x_1} \cdots t_{\sigma(n)}^{x_n}.$$

Then ϕ is Schur convex.

Proof Suppose that $T = T_{\tau, \lambda}$, where $\tau = \tau_{ij}$ and $0 \leq \lambda \leq 1$. Let us write

$$x_i = m + d, \quad x_j = m - d, \quad T(x)_i = m + \mu d, \quad T(x)_j = m - \mu d,$$

where $-1 \leq \mu = 1 - 2\lambda \leq 1$. Then

$$\phi(x) = \frac{1}{2(n!)} \left[\sum_{\sigma \in \Sigma_n} t_{\sigma(1)}^{x_1} \cdots t_{\sigma(n)}^{x_n} + \sum_{\sigma \in \Sigma_n} t_{\sigma(\tau(1))}^{x_1} \cdots t_{\sigma(\tau(n))}^{x_n} \right]$$

$$= \frac{1}{2(n!)} \sum_{\sigma \in \Sigma_n} \left(\prod_{k \neq i,j} t_{\sigma(k)}^{x_k} \right) \left(t_{\sigma(i)}^{x_i} t_{\sigma(j)}^{x_j} + t_{\sigma(i)}^{x_j} t_{\sigma(j)}^{x_i} \right)$$

$$= \frac{1}{2(n!)} \sum_{\sigma \in \Sigma_n} \left(\prod_{k \neq i,j} t_{\sigma(k)}^{x_k} \right) \left(t_{\sigma(i)}^{m+d} t_{\sigma(j)}^{m-d} + t_{\sigma(i)}^{m-d} t_{\sigma(j)}^{m+d} \right),$$

and similarly

$$\phi(T(x)) = \frac{1}{2(n!)} \sum_{\sigma \in \Sigma_n} \left(\prod_{k \neq i,j} t_{\sigma(k)}^{x_k} \right) \left(t_{\sigma(i)}^{m+\mu d} t_{\sigma(j)}^{m-\mu d} + t_{\sigma(i)}^{m-\mu d} t_{\sigma(j)}^{m+\mu d} \right).$$

Consequently

$$\phi(x) - \phi(T(x)) = \frac{1}{2(n!)} \sum_{\sigma \in \Sigma_n} \left(\prod_{k \neq i,j} t_{\sigma(k)}^{x_k} \right) \theta(\sigma),$$

where

$$\theta(\sigma) = t_{\sigma(i)}^m t_{\sigma(j)}^m \left(t_{\sigma(i)}^d t_{\sigma(j)}^{-d} + t_{\sigma(i)}^{-d} t_{\sigma(j)}^d - t_{\sigma(i)}^{\mu d} t_{\sigma(j)}^{-\mu d} - t_{\sigma(i)}^{-\mu d} t_{\sigma(j)}^{\mu d} \right)$$
$$= t_{\sigma(i)}^m t_{\sigma(j)}^m \left((a_\sigma^d + a_\sigma^{-d}) - (a_\sigma^{\mu d} + a_\sigma^{-\mu d}) \right),$$

and $a_\sigma = t_{\sigma(i)}/t_{\sigma(j)}$. Now if $a > 0$ the function $f(s) = a^s + a^{-s}$ is even, and increasing on $[0, \infty)$, so that $\theta(\sigma) \geq 0$, and $\phi(x) \geq \phi(T(x))$. □

Note that this theorem provides an interesting generalization of the arithmetic-mean geometric mean inequality: if $x \in (\mathbf{R}^n)^+$ and $\sum_{i=1}^n x_i = 1$, then

$$\left(\prod_{i=1}^n x_i \right)^{1/n} \leq \phi(x) \leq \frac{1}{n} \sum_{i=1}^n x_i,$$

since $(1/n, \ldots, 1/n) \prec x \prec (1, 0, \ldots, 0)$.

7.10 Notes and remarks

Given a finite set of numbers (the populations of cities or countries, the scores a cricketer makes in a season), it is natural to arrange them in decreasing order. It was Muirhead [Mui 03] who showed that more useful information could be obtained by considering the running averages of the numbers, and it is for this reason that the term 'Muirhead function' has been used for $f^\dagger$ (which is denoted by other authors as f^{**}). It was also Muirhead who showed how effective the method of transference could be.

Doubly stochastic matrices occur naturally in the theory of stationary Markov processes. A square matrix $P = (p_{ij})$ is *stochastic* if all of its terms are non-negative, and $\sum_j p_{ij} = 1$, for each i: p_{ij} is the probability of transitioning from state i to state j at any stage of the Markov process. The matrix is doubly stochastic if and only if the probability distribution where all states are equally probable is an invariant distribution for the Markov process.

Minkowski showed that every point of a compact convex subset of $\mathbf{R}^n$ can be expressed as a convex combination of the set's extreme points, and Carathéodory showed that it can be expressed as a convex combination of at most $n + 1$ extreme points. The extension of these ideas to the infinite-dimensional case is called *Choquet theory*: excellent accounts have been given by Phelps [Phe 66] and Alfsen [Alf 71].

Exercises

7.1   Suppose that (Ω, Σ, μ) is an atom-free measure space, that $A \in \Sigma$ and that $0 < t < \mu(A) < \infty$. Let $l = \sup\{\mu(B): B \subseteq A, \mu(B) \leq t\}$ and $u = \inf\{\mu(B): B \subseteq A, \mu(B) \geq t\}$. Show that there exist measurable subsets L and U of A with $\mu(L) = l$, $\mu(U) = u$. Deduce that $l = u$, and that there exists a measurable subset B of A with $\mu(B) = t$.

7.2   Suppose that $f \in M_1(\Omega, \Sigma, \mu)$, that $0 < q < \infty$ and that $C \geq 0$. Show that the following are equivalent:

(i) $\lambda_{|f|}(u) = \mu(|f| > u) \leq C^q/u^q$ for all $u > 0$;

(ii) $f^*(t) \leq C/t^{1/q}$ for $0 < t < \mu(\Omega)$.

7.3   Suppose that $f \in M_1$. What conditions are necessary and sufficient for $\lambda_{|f|}$ to be (a) continuous, and (b) strictly decreasing? If these conditions are satisfied, what is the relation between $\lambda_{|f|}$ and f^*?

7.4   Show that a rearrangement-invariant function space is either equal to $L^1 + L^\infty$ or is contained in M_0.

7.5   Suppose that $1 < p < \infty$. Show that

$$L^p + L^\infty = \left\{ f \in M: \int_0^t (f^*(s))^p \, ds < \infty \text{ for all } t > 0 \right\}.$$

7.6   Suppose that f and g are non-negative functions on (Ω, Σ, μ) for which $\int \log^+ f \, d\mu < \infty$ and $\int \log^+ g \, d\mu < \infty$. Let

$$G_t(f) = \exp\left(\frac{1}{t} \int_0^t \log f^*(s) \, ds\right),$$

and let $G_t(g)$ be defined similarly. Suppose that $G_t(f) \leq G_t(g)$ for all $0 < t < \mu(\Omega)$. Show that $\int_\Omega \Phi(f) \, d\mu \leq \int_\Omega \Phi(g) \, d\mu$ for every increasing function Φ on $[0, \infty)$ with $\Phi(e^t)$ a convex function of t: in particular, $\int f^r \, d\mu \leq \int g^r \, d\mu$ for each $0 < r < \infty$. What about $r = \infty$?

Formulate and prove a corresponding result for sequences. (In this case, the results are used to prove Weyl's inequality (Corollary 15.8.1).)

7.7   Suppose that f is a non-negative measurable function on an atom-free measure space (Ω, Σ, μ). Show that there exists an increasing sequence (f_n) of non-negative simple functions, where each f_n is of the form $f_n = \sum_{j=1}^{k_n} a_{jn} I_{E_{jn}}$, where, for each n, the sets E_{jn} are disjoint, and have equal measure, such that $f_n \nearrow f$.

7.8 Suppose that Φ is a convex increasing non-negative function on $[0, \infty)$ with $\Phi(0) = 0$. Let

$$\Phi_n(x) = D^+ f(0)x + \sum_{j=1}^{4^n}(D^+ f(\frac{j}{2^n}) - D^+ f(\frac{j-1}{2^n}))(x - \frac{j}{2^n})^+.$$

Show that Φ_n increases pointwise to Φ.

7.9 Show that the representation of a doubly stochastic $n \times n$ matrix as a convex combination of permutation matrices need not be unique, for $n \geq 3$.

7.10 Let $\Delta_d = \{x \in \mathbf{R}^d \colon x = x^*\}$. Let $s(x) = (\sum_{j=1}^{i} x_i)_{i=1}^{d}$, and let $\delta = s^{-1} : s(\Delta_d) \to \Delta_d$. Suppose that ϕ is a symmetric function on $(\mathbf{R}^d)^+$. Find a condition on $\phi \circ \delta$ for ϕ to be Schur convex. Suppose that ϕ is differentiable, and that

$$0 \leq \partial\phi/\partial x_d \leq \partial\phi/\partial x_{d-1} \leq \cdots \leq \partial\phi/\partial x_1$$

on Δ_d. Show that ϕ is Schur convex.

7.11 Suppose that $1 \leq k \leq d$. Let

$$e_k(x) = \sum\{x_{i_1}x_{i_2}\ldots x_{i_k} \colon i_1 < i_2 < \cdots < i_k\}$$

be the k-th elementary symmetric polynomial. Show that e_k is Schur concave.

7.12 Let $X_1, \ldots, X_k$ be independent identically distributed random variables taking values $v_1, \ldots, v_d$ with probabilities $p_1, \ldots, p_d$. What is the probability π that $X_1, \ldots, X_k$ take distinct values? Show that π is a Schur concave function of $p = (p_1, \ldots, p_d)$. What does this tell you about the 'matching birthday' story?

7.13 Suppose that X is a discrete random variable taking values $v_1, \ldots, v_d$ with probabilities $p_1, \ldots, p_d$. The *entropy* h of the distribution is $\sum_{\{j : p_j \neq 0\}} p_j \log_2(1/p_j)$. Show that h is a Schur concave function of $p = (p_1, \ldots, p_d)$. Show that $h \leq \log_2 d$.

7.14 Let

$$s(x) = \frac{1}{d-1}\sum_{i=1}^{d}(x_i - \bar{x})^2$$

be the *sample variance* of $x \in \mathbf{R}^d$, where $\bar{x} = (x_1 + \cdots + x_d)/d$. Show that s is Schur convex.

8

Maximal inequalities

8.1 The Hardy–Riesz inequality $(1 < p < \infty)$

In this chapter, we shall again suppose either that (Ω, Σ, μ) is an atom-free measure space, or that $\Omega = \mathbf{N}$ or $\{1, \ldots, n\}$, with counting measure. As its name implies, Muirhead's maximal function enjoys a maximal property:

$$f^\dagger(t) = \sup \left\{ \frac{1}{t} \int_E |f| \, d\mu : \ \mu(E) \leq t \right\} \quad \text{for } t > 0.$$

In this chapter we shall investigate this, and some other maximal functions of greater importance. Many of the results depend upon the following easy but important inequality.

Theorem 8.1.1 *Suppose that h and g are non-negative measurable functions in $M_0(\Omega, \Sigma, \mu)$, satisfying*

$$\alpha \mu(h > \alpha) \leq \int_{(h > \alpha)} g \, d\mu, \quad \text{for each } \alpha > 0.$$

If $1 < p < \infty$ then $\|h\|_p \leq p' \|g\|_p$, and $\|h\|_\infty \leq \|g\|_\infty$.

Proof Suppose first that $1 < p < \infty$. We only need to consider the case where $h \neq 0$ and $\|g\|_p < \infty$. Let

$$\begin{aligned}
h_n(\omega) &= 0 && \text{if } h(\omega) \leq 1/n, \\
&= h(\omega) && \text{if } 1/n < h(\omega) \leq n, \quad \text{and} \\
&= n && \text{if } h(\omega) > n.
\end{aligned}$$

Then $h_n \nearrow h$, and so, by the monotone convergence theorem, it is sufficient to show that $\|h_n\|_p \leq p' \|g\|_p$. Note that $\int h_n^p \, d\mu \leq n^p \mu(h \geq 1/n)$, so that

$h_n \in L^p$. Note also that if $0 < \alpha < 1/n$, then

$$\alpha\mu(h_n > \alpha) \le (1/n)\mu(h > 1/n) \le \int_{(h>1/n)} g \, d\mu = \int_{(h_n>\alpha)} g \, d\mu$$

and so h_n and g also satisfy the conditions of the theorem.

Using Fubini's theorem and Hölder's inequality,

$$\int_\Omega h_n^p \, d\mu = p \int_0^\infty t^{p-1} \mu(h_n > t) \, dt$$

$$\le p \int_0^\infty t^{p-2} \left(\int_{(h_n>t)} g(\omega) \, d\mu(\omega) \right) dt$$

$$= p \int_\Omega g(\omega) \left(\int_0^{h_n(\omega)} t^{p-2} \, dt \right) d\mu(\omega)$$

$$= \frac{p}{p-1} \int_\Omega g(\omega)(h_n(\omega))^{p-1} \, d\mu(\omega)$$

$$\le p' \|g\|_p \left(\int_\Omega (h_n)^{(p-1)p'} \, d\mu \right)^{1/p'} = p' \|g\|_p \|h_n\|_p^{p-1}.$$

We now divide, to get the result.

When $p = \infty$, $\alpha\mu(h > \alpha) \le \int_{(h>\alpha)} g \, d\mu \le \|g\|_\infty \mu(h > \alpha)$, and so $\mu(h > \alpha) = 0$ if $\alpha > \|g\|_\infty$; thus $\|h\|_\infty \le \|g\|_\infty$. □

Corollary 8.1.1 (The Hardy–Riesz inequality) *Suppose that* $1 < p < \infty$.

(i) *If* $f \in L^p(\Omega, \Sigma, \mu)$ *then* $\left\| f^\dagger \right\|_p \le p' \|f\|_p$.

(ii) *If* $f \in L^p[0, \infty)$ *and* $A(f)(t) = (\int_0^t f(s) \, ds)/t$ *then*

$$\|A(f)\|_p \le \left\| f^\dagger \right\|_p \le p' \|f\|_p.$$

(iii) *If* $x \in l_p$ *and* $(A(x))_n = (\sum_{i=1}^n x_i)/n$ *then*

$$\|A(x)\|_p \le \left\| x^\dagger \right\|_p \le p' \|x\|_p.$$

Proof (i) If $\alpha > 0$ and $t = \lambda(f^\dagger > \alpha) > 0$ then

$$\alpha\lambda(f^\dagger > \alpha) = \alpha t \le \int_0^t f^*(s) \, ds = \int_{(f^\dagger>\alpha)} f^*(s) \, ds,$$

so that $\left\| f^\dagger \right\|_p \le p' \|f^*\|_p = p' \|f\|_p$.

(ii) and (iii) follow, since $|A(f)| \le f^\dagger$ and $|A(x)| \le x^\dagger$. □

The constant p' is best possible, in the theorem and in the corollary. Take $\Omega = [0, 1]$, with Lebesgue measure. Suppose that $1 < r < p'$, and let $g(t) = t^{1/r-1}$. Then $g \in L^p$, and $h = g^{\dagger} = rg$, so that $\|g^{\dagger}\|_p \geq r \|g\|_p$. Similar examples show that the constant is also best possible for sequences.

This result was given by Hardy [Har 20], but he acknowledged that the proof that was given was essentially provided by Marcel Riesz. It enables us to give another proof of Hilbert's inequality, in the absolute case, with slightly worse constants.

Theorem 8.1.2 *If $a = (a_n)_{n \geq 0} \in l_p$ and $b = (b_n)_{n \geq 0} \in l_{p'}$, where $1 < p < \infty$, then*

$$\sum_{j=0}^{\infty} \sum_{k=0}^{\infty} \frac{|a_j b_k|}{j + k + 1} \leq (p + p') \|a\|_p \|b\|_{p'} .$$

Proof Using Hölder's inequality,

$$\sum_{k=0}^{\infty} \sum_{j=0}^{k} \frac{|a_j b_k|}{j + k + 1} \leq \sum_{k=0}^{\infty} \left(\sum_{j=0}^{k} \frac{|a_j|}{j + 1} \right) |b_k|$$

$$\leq \|A(|a|)\|_p \|b\|_{p'} \leq p' \|a\|_p \|b\|_{p'} .$$

Similarly,

$$\sum_{j=1}^{\infty} \sum_{k=0}^{j-1} \frac{|a_j b_k|}{j + k + 1} \leq p \|a\|_p \|b\|_p .$$

Adding, we get the result.   □

In exactly the same way, we have a corresponding result for functions on $[0, \infty)$.

Theorem 8.1.3 *If $f \in L^p[0, \infty)$ and $g \in L^{p'}[0, \infty)$, where $1 < p < \infty$, then*

$$\int_0^{\infty} \int_0^{\infty} \frac{|f(x)g(y)|}{x + y} \, dx \, dy \leq (p + p') \|f\|_p \|g\|_{p'} .$$

8.2 The Hardy–Riesz inequality $(p = 1)$

What happens when $p = 1$? If $\mu(\Omega) = \infty$ and f is any non-zero function in L^1 then $f^{\dagger}(t) \geq (f^{\dagger}(1))/t$ for $t \geq 1$, so that $f^{\dagger} \notin L^1$. When $\mu(\Omega) < \infty$, there are functions f in L^1 with $f^{\dagger} \notin L^1$ (consider $f(t) = 1/t(\log(1/t))^2$ on $(0, 1)$).

But in the finite-measure case there is an important and interesting result, due to Hardy and Littlewood [HaL 30], which indicates the importance of the space $L \log L$. We consider the case where $\mu(\Omega) = 1$.

Theorem 8.2.1 *Suppose that $\mu(\Omega) = 1$ and that $f \in L^1$. Then $f^\dagger \in L^1(0,1)$ if and only if $f \in L \log L$. If so, then*

$$\|f\|_{L \log L} \leq \left\|f^\dagger\right\|_1 \leq 6 \|f\|_{L \log L},$$

so that $\left\|f^\dagger\right\|_1$ is a norm on $L \log L$ equivalent to $\|f\|_{L \log L}$.

Proof Suppose first that $f^\dagger \in L^1$ and that $\left\|f^\dagger\right\|_1 = 1$. Then, integrating by parts, if $\epsilon > 0$,

$$1 = \left\|f^\dagger\right\|_1 \geq \int_\epsilon^1 \frac{1}{t} \left(\int_0^t f^*(s)\, ds \right) dt = \left(\epsilon \log \frac{1}{\epsilon} \right) f^\dagger(\epsilon) + \int_\epsilon^1 f^*(t) \log \frac{1}{t}\, dt.$$

Thus $\int_0^1 f^*(t) \log(1/t)\, dt \leq 1$. Also $\|f\|_1 = \|f^*\|_1 \leq \left\|f^\dagger\right\|_1 = 1$, so that $f^*(t) \leq f^\dagger(t) \leq 1/t$. Thus

$$\int |f| \log^+(|f|)\, d\mu = \int_0^1 f^*(t) \log^+ f^*(t)\, dt \leq \int_0^1 f^*(t) \log \frac{1}{t}\, dt \leq 1,$$

and so $f \in L \log L$ and $\|f\|_{L \log L} \leq \left\|f^\dagger\right\|_1$. By scaling, the same result holds for all $f \in L^1$ with $\left\|f^\dagger\right\|_1 < \infty$.

Conversely, suppose that $\int |f| \log^+(|f|) = 1$. Let $B = \{t \in (0,1]: f^*(t) > 1/\sqrt{t}\}$ and let $S = \{t \in (0,1]: f^*(t) \leq 1/\sqrt{t}\}$. If $t \in B$ then $\log^+(f^*(t)) = \log(f^*(t)) > \frac{1}{2} \log(1/t)$, and so

$$\begin{aligned}
\left\|f^\dagger\right\|_1 &= \int_0^1 f^*(t) \log \frac{1}{t}\, dt \\
&\leq 2 \int_B f^*(t) \log^+(f^*(t))\, dt + \int_S \frac{1}{\sqrt{t}} \log \frac{1}{t}\, dt \\
&\leq 2 + \int_0^1 \frac{1}{\sqrt{t}} \log \frac{1}{t}\, dt = 6.
\end{aligned}$$

Thus, by scaling, if $f \in L \log L$ then $f^\dagger \in L^1(0,1)$ and $\left\|f^\dagger\right\|_1 \leq 6 \|f\|_{L \log L}$. $\square$

8.3 Related inequalities

We can obtain similar results under weaker conditions.

Proposition 8.3.1 *Suppose that f and g are non-negative measurable functions in $M_0(\Omega, \Sigma, \mu)$, and that*

$$\alpha\mu(f > \alpha) \le \int_{(f>\alpha)} g \, d\mu, \quad \text{for each } \alpha > 0.$$

Then

$$\alpha\mu(f > \alpha) \le 2 \int_{(g>\alpha/2)} g \, d\mu, \quad \text{for } \alpha > 0.$$

Proof

$$\alpha\mu(f > \alpha) \le \int_{(g>\alpha/2)} g \, d\mu + \int_{(g\le\alpha/2)\cap(f>\alpha)} g \, d\mu$$

$$\le \int_{(g>\alpha/2)} g \, d\mu + \frac{\alpha}{2}\mu(f > \alpha).$$

$\square$

Proposition 8.3.2 *Suppose that f and g are non-negative measurable functions in $M_0(\Omega, \Sigma, \mu)$, and that*

$$\alpha\mu(f > \alpha) \le \int_{(g>\alpha)} g \, d\mu, \quad \text{for each } \alpha > 0.$$

Suppose that ϕ is a non-negative measurable function on $[0, \infty)$ and that $\Phi(t) = \int_0^t \phi(\alpha) \, d\alpha < \infty$ for all $t > 0$. Let $\Psi(t) = \int_0^t (\phi(\alpha)/\alpha) \, d\alpha$. Then

$$\int_X \Phi(f) \, d\mu \le \int_X g\Psi(g) \, d\mu.$$

Proof Using Fubini's theorem,

$$\int_X \Phi(f) \, d\mu = \int_0^\infty \phi(\alpha)\mu(f > \alpha) \, d\alpha \le \int_0^\infty \frac{\phi(\alpha)}{\alpha} \left(\int_{(g>\alpha)} g \, d\mu \right) d\alpha$$

$$= \int_X \left(\int_0^g \frac{\phi(\alpha)}{\alpha} \, d\alpha \right) g \, d\mu = \int_X g\Psi(g) \, d\mu.$$

$\square$

Corollary 8.3.1 *Suppose that f and g are non-negative measurable functions in $M_0(\Omega, \Sigma, \mu)$, and that*

$$\alpha\mu(f > \alpha) \le \int_{(g>\alpha)} g \, d\mu, \quad \text{for each } \alpha > 0.$$

If $1 < p < \infty$ then $\|f\|_p \le (p')^{1/p} \|g\|_p$.

Proof Take $\phi(t) = t^{p-1}$.   □

We also have an L^1 inequality.

Corollary 8.3.2 *Suppose that f and g are non-negative measurable functions in $M_0(\Omega, \Sigma, \mu)$, and that*

$$\alpha\mu(f > \alpha) \leq \int_{(g>\alpha)} g \, d\mu, \quad \text{for each } \alpha > 0.$$

If $\mu(B) < \infty$ then

$$\int_B f \, d\mu \leq \mu(B) + \int_X g \log^+ g \, d\mu.$$

Proof Take $\phi = I_{[1,\infty)}$. Then $\Phi(t) = (t-1)^+$ and $\Psi(t) = \log^+ t$, so that

$$\int_X (f-1)^+ \, d\mu \leq \int_X g \log^+ g \, d\mu.$$

Since $fI_B \leq I_B + (f-1)^+$, the result follows.

Combining this with Proposition 8.3.1, we also obtain the following corollary.

Corollary 8.3.3 *Suppose that f and g are non-negative measurable functions in $M_0(\Omega, \Sigma, \mu)$, and that*

$$\alpha\mu(f > \alpha) \leq \int_{(f>\alpha)} g \, d\mu, \quad \text{for each } \alpha > 0.$$

If $\mu(B) < \infty$ then

$$\int_B f \, d\mu \leq \mu(B) + \int_X 2g \log^+(2g) \, d\mu.$$

8.4 Strong type and weak type

The mapping $f \to f^\dagger$ is sublinear, and so are many other mappings that we shall consider. We need conditions on sublinear mappings comparable to the continuity, or boundedness, of linear mappings. Suppose that E is a normed space, that $0 < q < \infty$ and that $T : E \to M(\Omega, \Sigma, \mu)$ is sublinear. We say that T is of *strong type* (E, q) if there exists $M < \infty$ such that if $f \in E$ then $T(f) \in L^q$ and $\|T(f)\|_q \leq M\|f\|_E$. The least constant M for which the inequality holds for all $f \in E$ is called the *strong type* (E, q) *constant*. When T is linear and $1 \leq q < \infty$, 'strong type (E, q)' and 'bounded from E

to $L^{q'}$ are the same, and the strong type constant is then just the norm of T. When $E = L^p$, we say that T is of *strong type* (p, q).

We also need to consider weaker conditions, and we shall introduce more than one of these. For the first of these, we say that T is of *weak type* (E, q) if there exists $L < \infty$ such that

$$\mu\{w: \ |T(f)(w)| > \alpha\} \le L^q \frac{\|f\|_E^q}{\alpha^q}$$

for all $f \in E$, $\alpha > 0$. Equivalently (see Exercise 7.2), T is of weak type (E, q) if

$$(T(f))^*(t) \le Lt^{-1/q} \|f\|_E \quad \text{for all } f \in E, \ 0 < t < \mu(\Omega).$$

The least constant L for which the inequality holds for all $f \in E$ is called the *weak type* (E, q) *constant*.

When $E = L^p(\Omega', \Sigma', \mu')$, we say that T is of *weak type* (p, q). Since

$$\|g\|_q^q = \int |g|^q d\mu \ge \alpha^q \mu\{x: \ |g(x)| > \alpha\},$$

'strong type (E, q)' implies 'weak type (E, q)'.

For completeness' sake, we say that T is of strong type (E, ∞) or weak type (E, ∞) (strong type (p, ∞) or weak type (p, ∞) when $E = L^p$) if there exists M such that if $f \in E$ then $T(f) \in L^\infty(\mathcal{R}^d)$ and $\|T(f)\|_\infty \le M\|f\|_E$.

Here are some basic properties about strong type and weak type.

Proposition 8.4.1 *Suppose that E is a normed space, that $0 < q < \infty$ and that $S, T: E \to M(\Omega, \Sigma, \mu)$ are sublinear and of weak type (E, q), with constants L_S and L_T. If R is sublinear and $|R(f)| \le |S(f)|$ for all f then R is of weak type (E, q), with constants at most L_S. If $a, b > 0$ then $a|S| + b|T|$ is sublinear and of weak type (E, q), with constants at most $2(a^q L_S^q + b^q L_T^q)^{1/q}$. If S and T are of strong type (E, q), with constants M_S and M_T then R and $a|S| + b|T|$ are of strong type (E, q), with constants at most M_S and $aM_S + bM_T$ respectively.*

Proof The result about R is trivial. Suppose that $\alpha > 0$. Then $(a|S(f)| + b|T(f)| > \alpha) \subseteq (a|S(f)| > \alpha/2) \cup (b|T(f)| > \alpha/2)$, so that

$$\mu(a|S(f)| + b|T(f)| > \alpha) \le \mu(|S(f)| > \alpha/2a) + \mu(|T(f)| > \alpha/2b)$$

$$\le \frac{2^q a^q L_S^q}{\alpha^q} \|f\|_E^q + \frac{2^q b^q L_T^q}{\alpha^q} \|f\|_E^q.$$

The proofs of the strong type results are left as an easy exercise. $\qquad \square$

Weak type is important, when we consider convergence almost every-where. First let us recall an elementary result from functional analysis about convergence in norm.

Theorem 8.4.1 *Suppose that $(T_r)_{r \geq 0}$ is a family of bounded linear mappings from a Banach space $(E, \|.\|_E)$ into a Banach space $(G, \|.\|_G)$, such that*

(i) $\sup_r \|T_r\| = K < \infty$, *and*
(ii) *there is a dense subspace F of E such that $T_r(f) \to T_0(f)$ in norm, for $f \in F$, as $r \to 0$.*

Then if $e \in E$, $T_r(e) \to T_0(e)$ in norm, as $r \to 0$.

Proof Suppose that $\epsilon > 0$. There exists $f \in F$ with $\|f - e\| < \epsilon/3M$, and there exists $r_0 > 0$ such that $\|T_r(f) - T_0(f)\| < \epsilon/3$ for $0 < r \leq r_0$. If $0 < r \leq r_0$ then

$$\|T_r(e) - T_0(e)\| \leq \|T_r(e - f)\| + \|T_r(f) - T_0(f)\| + \|T_0(e - f)\| < \epsilon.$$

$\square$

Here is the corresponding result for convergence almost everywhere.

Theorem 8.4.2 *Suppose that $(T_r)_{r \geq 0}$ is a family of linear mappings from a normed space E into $M(\Omega, \Sigma, \mu)$, and that M is a non-negative sublinear mapping of E into $M(\Omega, \Sigma, \mu)$, of weak type (E, q) for some $0 < q < \infty$, such that*

(i) $|T_r(g)| \leq M(g)$ *for all $g \in E$, $r \geq 0$, and*
(ii) *there is a dense subspace F of E such that $T_r(f) \to T_0(f)$ almost everywhere, for $f \in F$, as $r \to 0$.*

Then if $g \in E$, $T_r(g) \to T_0(g)$ almost everywhere, as $r \to 0$.

Proof We use the first Borel–Cantelli lemma. For each n there exists $f_n \in F$ with $\|g - f_n\| \leq 1/2^n$. Let

$$B_n = (M(g - f_n) > 1/n) \cup (T_r(f_n) \not\to T_0(f_n)).$$

Then

$$\mu(B_n) = \mu(M(g - f_n) > 1/n) \leq \frac{Ln^q}{2^{nq}}.$$

Let $B = \limsup(B_n)$. Then $\mu(B) = 0$, by the first Borel–Cantelli lemma. If $x \notin B$, there exists n_0 such that $x \notin B_n$ for $n \geq n_0$, so that

$$|T_r(g)(x) - T_r(f_n)(x)| \leq M(g - f_n)(x) \leq 1/n, \quad \text{for } r \geq 0,$$

and so

$$|T_r(g)(x) - T_0(g)(x)|$$
$$\leq |T_r(g)(x) - T_r(f)(x)| + |T_r(f_n)(x) - T_0(f_n)(x)| + |T_0(f_n)(x) - T_0(g)(x)|$$
$$\leq 2/n + |T_r(f_n)(x) - T_0(f_n)(x)| \leq 3/n$$

for small enough r. □

We can of course consider other directed sets than $[0, \infty)$; for example $\mathbf{N}$, or the set

$$\{(x, t): t \geq 0, |x| \leq kt\} \subset \mathcal{R}^{d+1} \text{ ordered by } (x, t) \leq (y, u) \text{ if } t \leq u.$$

8.5 Riesz weak type

When $E = L^p(\Omega, \Sigma, \mu)$, a condition slightly less weak than 'weak type' is of considerable interest: we say that T is of *Riesz weak type* (p, q) if there exists $0 < L < \infty$ such that

$$\mu\{x: |T(f)(x)| > \alpha\} \leq \frac{L^q}{\alpha^q} \left(\int_{(|T(f)| > \alpha)} |f|^p \, d\mu \right)^{q/p}.$$

This terminology, which is not standard, is motivated by Theorem 8.1.1, and the Hardy–Riesz inequality. We call the least L for which the inequality holds for all f the *Riesz weak type constant*. Riesz weak type clearly implies weak type, but strong type does not imply Riesz weak type (consider the shift operator $T(f)(x) = f(x - 1)$ on $L^p(\mathbf{R})$, and $T(I_{[0,1]})$.

Proposition 8.5.1 *Suppose that S and T are of Riesz weak type (p, q), with weak Riesz type constants L_S and L_T. Then $\max(|S|, |T|)$ is of Riesz weak type (p, q), with constant at most $(L_S^q + L_T^q)^{1/q}$, and λS is of Riesz weak type (p, q), with constant $|\lambda| L_S$.*

Proof Let $R = \max(|S|, |T|)$. Then $(R(f) > \alpha) = (|S(f)| > \alpha) \cup (|T(f)| > \alpha)$, so that

$$\mu(R > \alpha) \leq \frac{L_S^q}{\alpha^q} \left(\int_{(|S(f)| > \alpha)} |f|^p \, d\mu \right)^{q/p} + \frac{L_T^q}{\alpha^q} \left(\int_{(|T(f)| > \alpha)} |f|^p \, d\mu \right)^{q/p}$$

$$\leq \frac{L_S^q + L_T^q}{\alpha^q} \left(\int_{(R(f) > \alpha)} |f|^p \, d\mu \right)^{q/p}.$$

The proof for λS is left as an exercise. □

We have the following interpolation theorem.

Theorem 8.5.1 *Suppose that T is a sublinear mapping of Riesz weak type (p, p), with Riesz weak type constant L. If $p < q < \infty$ then T is of strong type (q, q), with constant at most $L(q/(q - p))^{1/p}$, and T is of strong type (∞, ∞), with constant L.*

Proof Since T is of Riesz weak type (p, p),

$$\mu(|T(f)|^p > \alpha) \leq \frac{L^p}{\alpha} \int_{(|T(f)|^p > \alpha)} |f|^p \, d\mu.$$

Thus $|T(f)^p|$ and $L^p|f|^p$ satisfy the conditions of Theorem 8.1.1. If $p < q < \infty$, put $r = q/p$ (so that $r' = q/(q - p)$). Then

$$\|T(f)\|_q = \||T(f)|^p\|_r^{1/p} \leq (r')^{1/p} \|L^p|f|^p\|_r^{1/p} = L(r')^{1/p} \|f\|_q.$$

Similarly,

$$\|T(f)\|_\infty = \||T(f)|^p\|_\infty^{1/p} \leq \|L^p|f|^p\|_\infty^{1/p} = L \|f\|_\infty.$$

$\square$

8.6 Hardy, Littlewood, and a batsman's averages

Muirhead's maximal function is concerned only with the values that a function takes, and not with where the values are taken. We now begin to introduce a sequence of maximal functions that relate to the geometry of the underlying space. This is very simple geometry, usually of the real line, or $\mathbf{R}^n$, but to begin with, we consider the integers, where the geometry is given by the order.

The first maximal function that we consider was introduced by Hardy and Littlewood [HaL 30] in the following famous way (their account has been slightly edited and abbreviated here).

The problem is most easily grasped when stated in the language of cricket, or any other game in which a player compiles a series of scores in which an average is recorded ... Suppose that a batsman plays, in a given season, a given 'stock' of innings

$$a_1, a_2, \ldots, a_n$$

(determined in everything except arrangement). Suppose that α_ν is ... his maximum average for any consecutive series of innings ending at the ν-th,

so that

$$\alpha_\nu = \frac{a_{\nu^*} + a_{\nu^*+1} + \cdots + a_\nu}{\nu - \nu^* + 1} = \max_{\mu \leq \nu} \frac{a_\mu + a_{\mu+1} + \cdots + a_\nu}{\nu - \mu + 1};$$

we may agree that, in case of ambiguity, ν^* is to be chosen as small as possible. Let $s(x)$ be a positive function which increases (in the wide sense) with x, and let his 'satisfaction' after the ν-th innings be measures by $s_\nu = s(\alpha_\nu)$. Finally let his total satisfaction for the season be measured by $S = \sum s_\nu = \sum s(\alpha_\nu)$. Theorem 2 ... shows that S is ... a maximum when the innings are played in decreasing order.

Of course, this theorem says that $S \leq \sum_{\nu=1}^n s(a_\nu^\dagger)$.

We shall not give the proof of Hardy and Littlewood, whose arguments, as they say, 'are indeed mostly of the type which are intuitive to a student of cricket averages'. Instead, we give a proof due to F. Riesz [Ri(F) 32]. Riesz's theorem concerns functions on **R**, but first we give a discrete version, which establishes the result of Hardy and Littlewood. We begin with a seemingly trivial lemma.

Lemma 8.6.1 *Suppose that $(f_n)_{n \in \mathbf{N}}$ is a sequence of real numbers for which $f_n \to \infty$ as $n \to \infty$. Let*

$$E = \{n \colon \text{there exists } m < n \text{ such that } f_m > f_n\}.$$

Then we can write $E = \cup_j (c_j, d_j)$ (where $(c_j, d_j) = \{n \colon c_j < n < d_j\}$), with $c_1 < d_1 \leq c_2 < d_2 \leq \cdots$, and $f_n < f_{c_j} \leq f_{d_j}$ for $n \in (c_j, d_j)$.

Proof The union may be empty, finite, or infinite. If (f_n) is increasing then E is empty. Otherwise there exists a least c_1 such that $f_{c_1} > f_{c_1+1}$. Let d_1 be the least integer greater than c_1 such that $f_{d_1} \geq f_{c_1}$. Then $c_1 \notin E$, $d_1 \notin E$, and $n \in E$ for $c_1 < n < d_1$. If (f_n) is increasing for $n \geq d_1$, we are finished. Otherwise we iterate the procedure, starting from d_1. It is then easy to verify that $E = \cup_j (c_j, d_j)$. □

Theorem 8.6.1 (F. Riesz's maximal theorem: discrete version) *If $a = (a_n) \in l_1$, let*

$$\alpha_n = \max_{1 \leq k \leq n} \left(|a_{n-k+1}| + |a_{n-k+2}| + \cdots + |a_n| \right) / k.$$

Then the mapping $a \to \alpha$ is a sublinear mapping of Riesz weak type $(1, 1)$, with Riesz weak type constant 1.

Proof The mapping $a \to \alpha$ is certainly sublinear. Suppose that $\beta > 0$. Then the sequence (f_n) defined by $f_n = \beta n - \sum_{j=1}^{n} |a_j|$ satisfies the conditions of the lemma. Let

$$E_\beta = \{n\colon \text{ there exists } m < n \text{ such that } f_m > f_n\} = \cup_j (c_j, d_j).$$

Now $f_n - f_{n-k} = \beta k - \sum_{j=n-k+1}^{n} |a_j|$, and so $n \in E_\beta$ if and only if $\alpha_n > \beta$. Thus

$$\#\{n\colon \alpha_n > \beta\} = \#(E_\beta) = \sum_j (d_j - c_j - 1).$$

But

$$\beta(d_j - c_j - 1) - \sum_{(c_j < n < d_j)} |a_n| = f_{d_j - 1} - f_{c_j} \leq 0,$$

so that

$$\beta\#\{n\colon \alpha_n > \beta\} \leq \sum_j \left(\sum_{(c_j < n < d_j)} |a_n| \right) = \sum_{\{n:\alpha_n > \beta\}} |a_n|.$$

$\square$

Corollary 8.6.1 $\alpha_n^* \leq a_n^\dagger$.

Proof Suppose that $\gamma < \alpha_n^*$, and let $k = \#\{j\colon \alpha_j > \gamma\}$. Then $k \geq n$ and, by the theorem,

$$\gamma k \leq \sum_{(\alpha_j > \gamma)} |a_j| \leq k a_k^\dagger.$$

Thus $\gamma \leq a_k^\dagger \leq a_n^\dagger$. Since this holds for all $\gamma < \alpha_n^*$, $\alpha_n^* \leq a_n^\dagger$. $\square$

The result of Hardy and Littlewood follows immediately from this, since, with their terminology,

$$S = \sum_\nu s(\alpha_\nu) = \sum_\nu s(\alpha_\nu^*) \leq \sum_\nu s(a_\nu^\dagger).$$

[The fact that the batsman only plays a finite number of innings is resolved by setting $a_n = 0$ for other values of n.]

8.7 Riesz's sunrise lemma

We now turn to the continuous case; as we shall see, the proofs are similar to the discrete case. Here the geometry concerns intervals with a given point as an end-point, a mid-point, or an internal point.

Lemma 8.7.1 (Riesz's sunrise lemma) *Suppose that f is a continuous real-valued function on $\mathbf{R}$ such that $f(x) \to \infty$ as $x \to \infty$ and that $f(x) \to -\infty$ as $x \to -\infty$. Let*

$$E = \{x: \text{ there exists } y < x \text{ with } f(y) > f(x)\}.$$

Then E is an open subset of $\mathbf{R}$, every connected component of E is bounded, and if (a, b) is one of the connected components then $f(a) = f(b)$ and $f(x) < f(a)$ for $a < x < b$.

Proof It is clear that E is an open subset of $\mathbf{R}$. If $x \in \mathbf{R}$, let $m(x) = \sup\{f(t): t < x\}$, and let $L_x = \{y: y \le x, f(y) = m(x)\}$. Since f is continuous and $f(t) \to -\infty$ as $t \to -\infty$, L_x is a closed non-empty subset of $(-\infty, x]$: let $l_x = \sup L_x$. Then $x \in E$ if and only if $f(x) < m(x)$, and if and only if $l_x < x$. If so, $m(x) = f(l_x) > f(t)$ for $l_x < t \le x$.

Similarly, let $R_x = \{z: z \ge x, f(z) = m(x)\}$. Since f is continuous and $f(t) \to \infty$ as $t \to \infty$, R_x is a closed non-empty subset of $[x, \infty)$: let $r_x = \inf R_x$. If $x \in E$ then $m(x) = f(r_x) > f(t)$ for $x \le t < r_x$. Further, $l_x, r_x \notin E$, and so (l_x, r_x) is a maximal connected subset of E and the result follows. □

Why is this the 'sunrise' lemma? The function f represents the profile of a mountain, viewed from the north. The set E is the set of points in shadow, as the sun rises in the east.

This lemma was stated and proved by F. Riesz [Ri(F) 32], but the paper also included a simpler proof given by his brother Marcel.

Theorem 8.7.1 (F. Riesz's maximal theorem: continuous version) *For $g \in L^1(\mathbf{R}, d\lambda)$, let*

$$m^-(g)(x) = \sup_{y < x} \frac{1}{x - y} \int_y^x |g(t)| \, dt,$$

Then m^- is a sublinear operator, and if $\alpha > 0$ then

$$\alpha\lambda(m^-(g) > \alpha) = \int_{(m^-(g) > \alpha)} |g(t)| \, dt,$$

so that m^- is of Riesz weak type $(1, 1)$, with constant 1.

Proof It is clear from the definition that m^- is sublinear. Suppose that $g \in L^1(\mathbf{R}, d\lambda)$ and that $\alpha > 0$. Let $G_\alpha(x) = \alpha x - \int_0^x |g(t)| \, dt$. Then G_α satisfies the conditions of the sunrise lemma. Let

$$E_\alpha = \{x: \text{ there exists } y < x \text{ with } G_\alpha(y) > G_\alpha(x)\} = \cup_j I_j,$$

where the $I_j = (a_j, b_j)$ are the connected components of E_α. Since

$$G_\alpha(x) - G_\alpha(y) = \alpha(x - y) - \int_y^x |g(t)|\, dt,$$

$m^-(g)(x) > \alpha$ if and only if $x \in E_\alpha$. Thus

$$\alpha\lambda(m^-(g) > \alpha) = \alpha\lambda(E_\alpha) = \alpha\sum_j (b_j - a_j).$$

But

$$0 = G_\alpha(b_j) - G_\alpha(a_j) = \alpha(b_j - a_j) - \int_{a_j}^{b_j} |g(t)|\, dt,$$

so that

$$\alpha\lambda(m^-(g) > \alpha) = \sum_j \int_{a_j}^{b_j} |g(t)|\, dt = \int_{(m^-(g) > \alpha)} |g(t)|\, dt.$$

□

In the same way, if

$$m^+(g) = \sup_{y > x} \frac{1}{y - x} \int_x^y |g(t)|\, dt,$$

m^+ is a sublinear operator of Riesz weak type $(1,1)$. By Proposition 8.5.1, the operators

$$m_u(g)(x) = \sup_{y < x < z} \frac{1}{z - y} \int_y^z |g(t)|\, dt = \max(m^-(g)(x), m^+(g)(x)),$$
$$M(g)(x) = \max(m_u(g)(x), |g(x)|)$$

are also sublinear operators of Riesz weak type $(1,1)$.

Traditionally, it has been customary to work with the *Hardy–Littlewood maximal operator*

$$m(g)(x) = \sup_{r > 0} \frac{1}{2r} \int_{x-r}^{x+r} |g(t)|\, dt$$

(although, in practice, m_u is usually more convenient).

Theorem 8.7.2 *The Hardy–Littlewood maximal operator is of Riesz weak type $(1,1)$, with Riesz weak type constant at most 4.*

Proof We keep the same notation as in Theorem 8.7.1, and let $c_j = (a_j + b_j)/2$. Let $F_\alpha = (m(g) > \alpha)$. If $x \in (a_j, c_j)$ then $x \in F_\alpha$ (take $r = x - a_j$), so that

$$\int_{(m(g)>\alpha)} |g|\, dt \geq \sum_j \left(\int_{a_j}^{c_j} |g|\, dt \right)$$

$$= \sum_j \left(\alpha(c_j - a_j) - (G(c_j) - G(a_j)) \right)$$

$$\geq \sum_j \alpha(c_j - a_j) = \alpha\lambda(E_\alpha)/2,$$

since $G(c_j) \leq G(a_j)$ for each j. But

$$(m(g) > \alpha) \subseteq (m_u(g) > \alpha) = (m^-(g) > \alpha) \cup (m^+ > \alpha),$$

so that

$$\lambda(m(g) > \alpha) \leq \lambda(m^-(g) > \alpha) + \lambda(m^+(g) > \alpha) = 2\lambda(E_\alpha),$$

and so the result follows. □

8.8 Differentiation almost everywhere

We are interested in the values that a function takes near a point. We introduce yet another space of functions. We say that a measurable function f on $\mathbf{R}^d$ is *locally integrable* if $\int_B |f|\, d\lambda < \infty$, for each bounded subset B of $\mathbf{R}^d$. We write $L^1_{loc} = L^1_{loc}(\mathbf{R}^d)$ for the space of locally integrable functions on $\mathbf{R}^d$. Note that if $1 < p < \infty$ then $L^p \subseteq L^1 + L^\infty \subseteq L^1_{loc}$.

Here is a consequence of the F. Riesz maximal theorem.

Theorem 8.8.1 *Suppose that $f \in L^1_{loc}(\mathbf{R})$. Let $F(x) = \int_0^x f(t)\, dt$. Then F is differentiable almost everywhere, and the derivative is equal to f almost everywhere. If $f \in L^p$, where $1 < p < \infty$, then*

$$\frac{1}{h} \int_x^{x+h} f(t)\, dt \to f(x) \text{ in } L^p \text{ norm}, \quad \text{as } h \to 0.$$

Proof It is sufficient to prove the differentiability result for $f \in L^1$. For if $f \in L^1_{loc}$ then $fI_{(-R,R)} \in L^1$, for each $R > 0$, and if each $fI_{(-R,R)}$ is differentiable almost everywhere, then so is f. We apply Theorem 8.4.2, using $M(f) = \max(m_u(f), |f|)$, and setting

$$T_h(f)(x) = (1/h) \int_x^{x+h} f(t)\, dt \text{ for } h \neq 0, \quad \text{and } T_0(f)(x) = f(x).$$

Then $|T_h(f)| \leq M(f)$, for all h. If g is a continuous function of compact support, then $T_h(g)(x) \to g(x)$, uniformly in x, as $h \to 0$, and the continuous functions of compact support are dense in $L^1(\mathbf{R})$. Thus $T_h(f) \to f$ almost everywhere as $h \to 0$: but this says that F is differentiable, with derivative f, almost everywhere.

If $f \in L^p$, then, applying Corollary 5.4.2,

$$\|T_h(f)\|_p \leq \left(\int_{-\infty}^{\infty} \left(\frac{1}{|h|} \int_0^h |f(x+t)|\, dt \right)^p dx \right)^{1/p}$$

$$\leq \frac{1}{|h|} \int_0^h \left(\int_{-\infty}^{\infty} |f(x+t)|^p\, dx \right)^{1/p} dt = \|f\|_p.$$

If g is a continuous function of compact support K then $T_h(g) \to g$ uniformly, and $T_h(g) - g$ vanishes outside $K_h = \{x \colon d(x, K) \leq |h|\}$, and so $T_h(g) \to g$ in L^p norm as $h \to 0$. The continuous functions of compact support are dense in $L^p(\mathbf{R})$; convergence in L^p norm therefore follows from Theorem 8.4.1. $\qquad\qquad\square$

8.9 Maximal operators in higher dimensions

Although there are further conclusions that we can draw, the results of the previous section are one-dimensional, and it is natural to ask what happens in higher dimensions. Here we shall obtain similar results. Although the sunrise lemma does not seem to extend to higher dimensions, we can replace it by another beautiful lemma. In higher dimensions, the geometry concerns balls or cubes (which reduce in the one-dimensional case to intervals).

Let us describe the notation that we shall use:

$B_r(x)$ is the closed Euclidean ball $\{y \colon |y - x| \leq r\}$ and $U_r(x)$ is the open Euclidean ball $\{y \colon |y - x| < r\}$. Ω_d is the Lebesgue measure of a unit ball in $\mathcal{R}^d$. $S_r(x)$ is the sphere $\{y \colon |y - x| = r\}$. $Q(x, r) = \{y \colon |x_i - y_i| < r$ for $1 \leq i \leq d\}$ is the cube of side $2r$ centred at x.

We introduce several maximal operators: suppose that $f \in L^1_{\mathrm{loc}}(\mathcal{R}^d)$. We set

$$A_r(f)(x) = \frac{\int_{U_r(x)} f\, d\lambda}{\lambda(U_r(x))} = \frac{1}{r^d \Omega_d} \int_{U_r(x)} f\, d\lambda.$$

$A_r(f)(x)$ is the average value of f over the ball $U_r(x)$.

$$m(f)(x) = \sup_{r>0} A_r(|f|)(x) = \sup_{r>0} \frac{1}{r^d \Omega_d} \int_{U_r(x)} |f| \, d\lambda,$$

$$m_u(f)(x) = \sup_{r>0} \sup_{x \in U_r(y)} \frac{1}{r^d \Omega_d} \int_{U_r(y)} |f| \, d\lambda,$$

$$m^Q(f)(x) = \sup_{r>0} \frac{1}{(2r)^d} \int_{Q_r(x)} |f| \, d\lambda,$$

and

$$m_u^Q(f)(x) = \sup_{r>0} \sup_{x \in Q_r(y)} \frac{1}{(2r)^d} \int_{Q_r(y)} |f| \, d\lambda.$$

As before, m is the *Hardy–Littlewood maximal function*.

The maximal operators are all equivalent, in the sense that if m' and m'' are any two of them then there exist positive constants c and C such that

$$cm'(f)(x) \leq m''(f)(x) \leq Cm'(f)(x)$$

for all f and x.

Proposition 8.9.1 *Each of these maximal operators is sublinear. If m' is any one of them, then $m'(f)$ is a lower semi-continuous function from $\mathcal{R}^d$ to $[0, \infty]$: $E_\alpha = \{x : m'(f)(x) > \alpha\}$ is open in $\mathcal{R}^d$ for each $\alpha \geq 0$.*

Proof It follows from the definition that each of the maximal operators is sublinear. We prove the lower semi-continuity for m: the proof for m^Q is essentially the same, and the proofs for the other maximal operators are easier. If $x \in E_\alpha$, there exists $r > 0$ such that $A_r(|f|)(x) > \alpha$. If $\epsilon > 0$ and $|x - y| < \epsilon$ then $U_{r+\epsilon}(y) \supseteq U_r(x)$, and $\int_{U_{r+\epsilon}} |f| \, d\lambda \geq \int_{U_r} |f| \, d\lambda$, so that

$$m(f)(y) \geq A_{r+\epsilon}(|f|)(y) \geq \left(\frac{r}{r+\epsilon}\right)^d m(f)(x) > \alpha$$

for small enough $\epsilon > 0$. $\qquad\square$

We now come to the d-dimensional version of Riesz's maximal theorem.

Theorem 8.9.1 *The maximal operators m_u and m_u^Q are of Riesz weak type $(1,1)$, each with constant at most 3^d.*

Proof We prove the result for m_u: the proof for m_u^Q is exactly similar. The key result is the following covering lemma.

Lemma 8.9.1 *Suppose that G is a finite set of open balls in $\mathcal{R}^d$, and that λ is Lebesgue measure. Then there is a finite subcollection F of disjoint balls such that*

$$\sum_{U \in F} \lambda(U) = \lambda\left(\bigcup_{U \in F} U\right) \geq \frac{1}{3^d}\lambda\left(\bigcup_{U \in G} U\right).$$

Proof We use a greedy algorithm. If $U = U_r(x)$ is a ball, let $U^* = U_{3r}(x)$ be the ball with the same centre as U, but with three times the radius. Let U_1 be a ball of maximal radius in G. Let U_2 be a ball of maximal radius in G, disjoint from U_1. Continue, choosing U_j of maximal radius, disjoint from $U_1, \ldots, U_{j-1}$, until the process stops, with the choice of U_k.

Let $F = \{U_1, \ldots, U_k\}$. Suppose that $U \in G$. There is a least j such that $U \cap U_j \neq \emptyset$. Then the radius of U is no greater than the radius of U_j (otherwise we would have chosen U to be U_j) and so $U \subseteq U_j^*$. Thus $\bigcup_{U \in G} U \subseteq \bigcup_{U \in F} U^*$ and

$$\lambda(\bigcup_{U \in G} U) \leq \lambda(\bigcup_{U \in F} U^*) \leq \sum_{U \in F} \lambda(U^*) = 3^d \sum_{U \in F} \lambda(U).$$

$\square$

Proof of Theorem 8.9.1 Let $f \in L^1(\mathcal{R}^d)$ and let $E_\alpha = \{x \colon m_u(f)(x) > \alpha\}$. Let K be a compact subset of E_α. For each $x \in K$, there exist $y_x \in \mathbf{R}^d$ and $r_x > 0$ such that $x \in U_{r_x}(y_x)$ and $A_{r_x}(|f|)(y_x) > \alpha$. (Note that it follows from the definition of m_u that $U_{r_x}(y_x) \subseteq E_\alpha$; this is why m_u is easier to work with than m.) The sets $U_{r_x}(y_x)$ cover K, and so there is a finite subcover G. By the lemma, there is a subcollection F of disjoint balls such that

$$\sum_{U \in F} \lambda(U) \geq \frac{1}{3^d}\lambda(\bigcup_{U \in G} U) \geq \frac{\lambda(K)}{3^d}.$$

But if $U \in F$, $\alpha\lambda(U) \leq \int_U |f|\,d\lambda$, so that since $\bigcup_{U \in F} U \subseteq E_\alpha$,

$$\sum_{U \in F} \lambda(U) \leq \frac{1}{\alpha}\sum_{U \in F}\int |f|\,d\lambda \leq \frac{1}{\alpha}\int_{E_\alpha} |f|\,d\lambda.$$

Thus $\lambda(K) \leq 3^d(\int_{E_\alpha} |f|\,d\lambda)/\alpha$, and

$$\lambda(E_\alpha) = \sup\{\lambda(K) \colon K \text{ compact}, K \subseteq E_\alpha\} \leq \frac{3^d}{\alpha}\int_{E_\alpha} |f|\,d\lambda.$$

$\square$

Corollary 8.9.1 *Each of the maximal operators defined above is of weak type* $(1,1)$ *and of strong type* (p,p), *for* $1 < p \leq \infty$.

I do not know if the Hardy–Littlewood maximal operator m is of Riesz weak type $(1,1)$. This is interesting, but not really important; the important thing is that $m \leq m_u$, and m_u is of Riesz weak type $(1,1)$.

8.10 The Lebesgue density theorem

We now have the equivalent of Theorem 8.8.1, with essentially the same proof.

Theorem 8.10.1 *Suppose that* $f \in L^1_{loc}(\mathbf{R}^d)$. *Then* $A_r(f) \to f$ *almost everywhere, as* $r \to 0$, *and* $|f| \leq m(f)$ *almost everywhere. If* $f \in L^p$, *where* $1 < p < \infty$, *then* $A_r(f) \to f$ *in* L^p *norm.*

Corollary 8.10.1 (The Lebesgue density theorem) *If* E *is a measurable subset of* $\mathcal{R}^d$ *then*

$$\frac{1}{r^d \Omega_d} \lambda(U_r(x) \cap E) = \frac{\lambda(U_r \cap E)}{\lambda(U_r)} \to 1 \ \ as \ r \to 0 \ \ \ for \ almost \ all \ x \in E$$

and

$$\frac{1}{r^d \Omega_d} \lambda(U_r(x) \cap E) = \frac{\lambda(U_r \cap E)}{\lambda(U_r)} \to 0 \ \ as \ r \to 0 \ \ \ for \ almost \ all \ x \notin E.$$

Proof Apply the theorem to the indicator function I_E. $\qquad\qquad\square$

8.11 Convolution kernels

We can think of Theorem 8.10.1 as a theorem about convolutions. Let $J_r(x) = I_{U_r(0)}/\lambda(U_r(0))$. Then

$$A_r(f)(x) = \int_{\mathcal{R}^d} J_r(x - y)f(y)\,dy = \int_{\mathcal{R}^d} f(x - y)J_r(y)\,dy = (J_r * f)(x).$$

Then $J_r * f \to f$ almost surely as $r \to 0$, and if $f \in L^p$ then $J_r * f \to f$ in L^p norm.

We can use the Hardy–Littlewood maximal operator to study other convolution kernels. We begin by describing two important examples. The *Poisson kernel* P is defined on the upper half space $H^{d+1} = \{(x,t) \colon x \in \mathbf{R}^d, t > 0\}$ as

$$P(x, t) = P_t(x) = \frac{c_d t}{(|x|^2 + t^2)^{(d+1)/2}}.$$

$P_t \in L^1(\mathcal{R}^d)$, and the constant c_d is chosen so that $\|P_1\|_1 = 1$. A change of variables then shows that $\|P_t\|_1 = \|P_1\|_1 = 1$ for all $t > 0$.

The Poisson kernel is harmonic on H^{d+1} – that is,

$$\frac{\partial^2 P}{\partial t^2} + \sum_{j=1}^{d} \frac{\partial^2 P}{\partial x_j^2} = 0$$

– and is used to solve the Dirichlet problem in H^{d+1}: if f is a bounded continuous function on $\mathbf{R}^d$ and we set

$$u(x, t) = u_t(x) = P_t(f)(x) = (P_t * f)(x)$$

$$= \int_{\mathcal{R}^d} P_t(x - y) f(y) \, dy = \int_{\mathcal{R}^d} f(x - y) P_t(y) \, dy,$$

then u is a harmonic function on H^{d+1} and $u(x, t) \to f(x)$ uniformly on the bounded sets of $\mathbf{R}^d$ as $t \to 0$. We want to obtain convergence results for a larger class of functions f.

Second, let

$$H(x, t) = H_t(x) = \frac{1}{(2\pi t)^{d/2}} e^{-|x|^2/2t}$$

be the *Gaussian kernel.* Then H satisfies the heat equation

$$\frac{\partial H}{\partial t} = \frac{1}{2} \sum_{j=1}^{d} \frac{\partial^2 H}{\partial x_j^2}$$

on H^{d+1}. If f is a bounded continuous function on $\mathbf{R}^d$ and we set

$$v(x, t) = v_t(x) = H_t(f)(x) = (H_t * f)(x)$$

$$= \int_{\mathcal{R}^d} H_t(x - y) f(y) \, dy = \int_{\mathcal{R}^d} f(x - y) H_t(y) \, dy,$$

then v satisfies the heat equation on H^{d+1}, and $v(x, t) \to f(x)$ uniformly on the bounded sets of $\mathbf{R}^d$ as $t \to 0$. Again, we want to obtain convergence results for a larger class of functions f.

The Poisson kernel and the Gaussian kernel are examples of bell-shaped approximate identities. A function $\Phi = \Phi_t(x)$ on $(0, \infty] \times \mathbf{R}^d$ is a *bell-shaped approximate identity* if

(i) $\Phi_t(x) = t^{-d}\Phi_1(x/t)$;
(ii) $\Phi_1 \geq 0$, and $\int_{\mathcal{R}^d} \Phi_1(x) \, dx = 1$;
(iii) $\Phi_1(x) = \phi(|x|)$ where $\phi(r)$ is a strictly decreasing continuous function on $(0, \infty)$, taking values in $[0, \infty]$.

[In fact, the results that we present hold when ϕ is a decreasing function (as for example when we take $\phi = I_{[0,1]}/\lambda(U_1(0)))$, but the extra requirements make the analysis easier, without any essential loss.]

If Φ is a bell-shaped approximate identity, and if $f \in L^1 + L^\infty$, we set

$$\Phi_t(f)(x) = (\Phi_t * f)(x) = \int_{\mathbf{R}^d} \Phi_t(x - y)f(y)\,d\lambda(y).$$

Theorem 8.11.1 *Suppose that Φ is a bell-shaped approximate identity and that $f \in (L^1 + L^\infty)(\mathbf{R}^d)$. Then*

(i) the mapping $(x, t) \to \Phi_t(f)(x)$ is continuous on H^{d+1};

(ii) if $f \in C_b(\mathbf{R}^d)$ then $\Phi_t(f) \to f$ uniformly on the compact sets of $\mathbf{R}^d$;

(iii) if $f \in L^p(\mathbf{R}^d)$, where $1 \le p < \infty$, then $\|\Phi_t(f)\|_p \le \|f\|_p$ and $\Phi_t(f) \to f$ in L^p-norm.

Proof This is a straightforward piece of analysis (using Theorem 8.4.1 and Proposition 7.5.1) which we leave to the reader. □

The convergence in (iii) is convergence in mean. What can we say about convergence almost everywhere? The next theorem enables us to answer this question.

Theorem 8.11.2 *Suppose that Φ is a bell-shaped approximate identity, and that $f \in (L^1 + L^\infty)(\mathcal{R}^d)$. Then $|\Phi_t(f)(x)| \le m(f)(x)$.*

Proof Let $\Phi(x) = \phi(|x|)$, and let us denote the inverse function to ϕ: $(0, \phi(0)] \to [0, \infty)$ by γ. Then, using Fubini's theorem,

$$\Phi_t(f)(x) = \frac{1}{t^d} \int_{\mathbf{R}^d} \Phi_1\left(\frac{x-y}{t}\right) f(y)\,dy$$

$$= \frac{1}{t^d} \int_{\mathbf{R}^d} \left(\int_0^{\Phi_1(\frac{x-y}{t})} du\right) f(y)\,dy$$

$$= \frac{1}{t^d} \int_0^{\phi(0)} \left(\int_{(\Phi_1(\frac{x-y}{t})) > u} f(y)\,dy\right) du$$

$$= \frac{1}{t^d} \int_0^{\phi(0)} \left(\int_{|\frac{x-y}{t}| < \gamma(u)} f(y)dy \right) du$$

$$= \frac{1}{t^d} \int_0^{\phi(0)} \left(\int_{U_{t\gamma(u)}(x)} f(y)dy \right) dy$$

$$= \frac{1}{t^d} \int_0^{\phi(0)} \Omega_d t^d \gamma(u)^d A_{t\gamma(u)}(f)(x)du,$$

so that

$$|\Phi_t(f)(x)| \le m(f)(x) \int_0^{\phi(0)} \Omega_d \gamma(u)^d du$$

$$= m(f)(x) \int_0^{\phi(0)} \lambda(w: \Phi_1(w) > u)du$$

$$= m(f)(x) \int_{\mathbf{R}^d} \Phi_1(w)dw = m(f)(x).$$

$\square$

Corollary 8.11.1 *Let* $\Phi^*(f)(x) = \sup_{t>0} \Phi_t(|f|)$. *Then* Φ^* *is of weak type* $(1,1)$ *and strong type* (p,p), *for* $1 < p \le \infty$.

Corollary 8.11.2 *Suppose that* $f \in L^1(\mathcal{R}^d)$. *Then* $\Phi_t(f)(x) \to f(x)$ *as* $t \to 0$, *for almost all* x.

Proof We apply Theorem 8.9.1, with $M(f) = \Phi^*(f)$. The result holds for continuous functions of compact support; these functions are dense in $L^1(\mathcal{R}^d)$. $\square$

Theorem 8.11.3 *Suppose that* $f \in L^\infty(\mathcal{R}^d)$. *Then* $\Phi_t(f) \to f$ *almost everywhere*.

Proof Let us consider what happens in $\|x\| < R$. Let $g = fI_{\|x\| \le 2R}$, $h = f - g$. Then $g \in L^1(\mathcal{R}^d)$, so $\Phi_t(g) \to g$ almost everywhere. If $|x'| < R$,

$$|\Phi_t(h)(x')| = \left| \int \Phi_t(y - x')h(y)dy \right|$$

$$\le \|h\|_\infty \int_{|z| \ge R} \Phi_t(y)dy \to 0 \quad \text{as} \quad t \to 0.$$

$\square$

Corollary 8.11.3 *If $f \in L^p(\mathcal{R}^d)$ for $1 \leq p \leq \infty$, then $\Phi_t(f) \to f$ almost everywhere.*

Proof $L^p \subseteq L^1 + L^\infty$. □

8.12 Hedberg's inequality

Our next application concerns potential theory. Suppose to begin with that f is a smooth function of compact support on $\mathbf{R}^3$: that is to say, f is infinitely differentiable, and vanishes outside a bounded closed region S. We can think of f as the distribution of matter, or of electric charge. The *Newtonian potential* $I_2(f)$ is defined as

$$I_2(f)(x) = \frac{1}{4\pi} \int_{\mathbf{R}^3} \frac{f(y)}{|x - y|} \, dy = \frac{1}{4\pi} \int_{\mathbf{R}^3} \frac{f(x - u)}{|u|} \, du.$$

This is well-defined, since $1/|x| \in L^1 + L^\infty$.

Since I_2 is a convolution operator, we can expect it to have some continuity properties, and these we now investigate. In fact, we shall do this in a more general setting, which arises naturally from these ideas. We work in $\mathbf{R}^d$, where $d \geq 2$. Suppose that $0 < \alpha < d$. Then $1/|x|^{d-\alpha} \in L^1 + L^\infty$. Thus if $f \in L^1 \cap L^\infty$, we can consider the integrals

$$I_{d,\alpha}(f)(x) = \frac{1}{\gamma_{d,\alpha}} \int_{\mathbf{R}^d} \frac{f(y)}{|x - y|^{d-\alpha}} \, dx = \frac{1}{\gamma_{d,\alpha}} \int_{\mathbf{R}^d} \frac{f(x - u)}{|u|^{d-\alpha}} \, du,$$

where $\gamma = \gamma_{d,\alpha}$ is an appropriate constant. The operator $I_{d,\alpha}$ is called the *Riesz potential operator*, or *fractional integral operator*, of order α.

The function $|x|^{\alpha-d}/\gamma_{d,\alpha}$ is locally integrable, but it is not integrable, and so it is not a scalar multiple of a bell-shaped approximate identity. But as Hedberg [Hed 72] observed, we can split it into two parts, to obtain continuity properties of $I_{d,\alpha}$.

Theorem 8.12.1 (Hedberg's inequality) *Suppose that $0 < \alpha < d$ and that $1 \leq p < d/\alpha$. If $f \in (L^1 + L^\infty)(\mathbf{R}^d)$ and $x \in \mathbf{R}^d$ then*

$$|I_{d,\alpha}(f)(x)| \leq C_{d,\alpha,p} \|f\|_p^{\alpha p/d} \left(m(f)(x)\right)^{1-\alpha p/d},$$

where $m(f)$ is the Hardy–Littlewood maximal function, and $C_{d,\alpha,p}$ is a constant depending only on d, α, and p.

Proof In what follows, $A, B, \ldots$ are constants depending only on d, α and p. Suppose that $R > 0$. Let

$$\Theta_R(x) = \frac{A}{R^d} \left(\frac{|x|}{R}\right)^{\alpha-d} I_{(|x|<R)} = \frac{AI_{(|x|<R)}}{R^\alpha |x|^{d-\alpha}},$$

$$\Psi_R(x) = \frac{A}{R^d} \left(\frac{|x|}{R}\right)^{\alpha-d} I_{(|x|\geq R)} = \frac{AI_{(|x|\geq R)}}{R^\alpha |x|^{d-\alpha}},$$

where A is chosen so that Θ_R is a bell-shaped approximate identity (the lack of continuity at $|x| = R$ is unimportant). Then $\|\Psi_R\|_\infty \leq A/R^d$, and if $1 < p < d/\alpha$ then

$$\|\Psi_R\|_{p'} = \frac{B}{R^\alpha} \left(\int_R^\infty \frac{r^{d-1}}{r^{(d-\alpha)p'}} \, dr\right)^{1/p'} = DR^{-d/p}.$$

Thus, using Theorem 8.11.2, and Hölder's inequality,

$$|I_{d,\alpha}(f)(x)| \leq \frac{R^\alpha}{A\gamma} \left(|\int_{\mathbf{R}^d} f(y)\Theta_\alpha(x-y)\,dy| + |\int_{\mathbf{R}^d} f(y)\Psi_\alpha(x-y)\,dy|\right)$$

$$\leq \frac{R^\alpha}{A\gamma} \left(m(f)(x) + D\|f\|_p R^{-d/p}\right).$$

We now choose $R = R(x)$ so that the two terms are equal: thus $R^{d/p}m(f)(x) = E\|f\|_p$, and so

$$|I_{d,\alpha}(f)(x)| \leq C\|f\|_p^{\alpha p/d} \left(m(f)(x)\right)^{1-\alpha p/d}.$$

$\square$

Applying Corollary 8.9.1, we obtain the following.

Corollary 8.12.1 *Suppose that $0 < \alpha < d$.*
(i) $I_{d,\alpha}$ is of weak type $(1, d/(d-\alpha))$.
(ii) If $1 < p < d/\alpha$ and $q = pd/(d-\alpha p)$ then $\|I_{d,\alpha}(f)\|_q \leq C'_{d,\alpha,p} \|f\|_p$.

Proof (i) Suppose that $\|f\|_1 = 1$ and that $\beta > 0$. Then

$$\lambda(|I_{d,\alpha}(f)| > \beta) \leq \lambda(m(f) > (\beta/C)^{d/(d-\alpha)}) \leq F/\beta^{d/(d-\alpha)}.$$

(ii)

$$\|I_{d,\alpha}(f)\|_q \leq C_{d,\alpha,p} \|f\|_p^{\alpha p/d} \left\|m(f)^{1-\alpha p/d}\right\|_q$$

$$\leq C'_{d,\alpha,p} \|f\|_p^{\alpha p/d} \left\||f|^{1-\alpha p/d}\right\|_q$$

$$= C'_{d,\alpha,p} \|f\|_p.$$

$\square$

Thus in $\mathbf{R}^3$, $\|I_2(f)\|_{3p/(3-2p)} \le C'_{3,2,p} \|f\|_p$, for $1 < p < 3/2$.

Simple scaling arguments show that $q = pd/(d - \alpha)$ is the only index for which the inequality in (ii) holds (Exercise 8.9).

8.13 Martingales

Our final example in this chapter comes from the theory of martingales. This theory was developed as an important part of probability theory, but it is quite as important in analysis. We shall therefore consider martingales defined on a σ-finite measure space (Ω, Σ, μ).

First we describe the setting in which we work. We suppose that there is an increasing sequence $(\Sigma_j)_{j=0}^\infty$ or $(\Sigma_j)_{j=-\infty}^\infty$ of sub-σ-fields of Σ, such that Σ is the smallest σ-field containing $\cup_j \Sigma_j$. We shall also suppose that each of the σ-fields is σ-finite. We can think of this as a system evolving in discrete time. The sets of Σ_j are the events that we can describe at time j. By time $j + 1$, we have learnt more, and so we have a larger σ-field Σ_{j+1}.

As an example, let

$$\mathbf{Z}_j^d = \{a = (a_1, \ldots, a_d): a_i = n_i/2^j, n_i \in \mathbf{Z} \text{ for } 1 \le i \le d\},$$

for $-\infty < j < \infty$. $\mathbf{Z}_j^d$ is a lattice of points in $\mathbf{R}^d$, with mesh size 2^{-j}. If $a \in \mathbf{Z}_j^d$,

$$Q_j(a) = \{x \in \mathbf{R}^d: a_i - 1/2^j < x_i \le a_i, \text{ for } 1 \le i \le d\}$$

is the *dyadic cube* of side 2^{-j} with a in the top right-hand corner. Σ_j is the collection of sets which are unions of dyadic cubes of side 2^{-j}; it is a discrete σ-field whose atoms are the dyadic cubes of side 2^{-j}. We can think of the atoms of Σ_j as pixels; at time $j+1$, a pixel in Σ_j splits into 2^d smaller pixels, and so we have a finer resolution. (Σ_j) is an increasing sequence of σ-fields, and the Borel σ-field is the smallest σ-field containing $\cup_j \Sigma_j$. This is the *dyadic filtration* of $\mathbf{R}^d$.

In general, to avoid unnecessary complication, we shall suppose that each Σ_j is either atom-free, or (as with the dyadic filtration) purely atomic, with each atom of equal measure.

A sequence (f_j) of functions on Ω such that each f_j is Σ_j-measurable is called an *adapted sequence*, or *adapted process*. (Thus, in the case of the dyadic filtration, f_j is constant on the dyadic cubes of side 2^{-j}.) If (f_j) is an adapted sequence of real-valued functions, and if $f_j \in L^1 + L^\infty$, we say that (f_j) is

$$\text{a } \textit{local sub-martingale} \quad \text{if} \quad \int_A f_j \, d\mu \leq \int_A f_{j+1} \, d\mu,$$

$$\text{a } \textit{local super-martingale} \quad \text{if} \quad \int_A f_j \, d\mu \geq \int_A f_{j+1} \, d\mu,$$

$$\text{and a } \textit{local martingale} \quad \text{if} \quad \int_A f_j \, d\mu = \int_A f_{j+1} \, d\mu,$$

whenever A is a set of finite measure in Σ_j. If in addition each $f_j \in L^1$, we say that (f_j) is a *sub-martingale*, *super-martingale* or *martingale*, as the case may be. The definition of local martingale extends to complex-valued functions, and indeed to vector-valued functions, once a suitable theory of vector-valued integration is established.

These ideas are closely related to the idea of a conditional expectation operator, which we now develop.

Theorem 8.13.1 *Suppose that $f \in (L^1 + L^\infty)(\Omega, \Sigma, \mu)$, and that Σ_0 is a σ-finite sub-σ-field of Σ. Then there exists a unique f_0 in $(L^1 + L^\infty)(\Omega, \Sigma_0, \mu)$ such that $\int_A f \, d\mu = \int_A f_0 \, d\mu$ for each $A \in \Sigma_0$ with $\mu(A) < \infty$. Further, if $f \geq 0$ then $f_0 \geq 0$, if $f \in L^1$ than $\|f_0\|_1 \leq \|f\|_1$, and if $f \in L^\infty$ then $\|f_0\|_\infty \leq \|f\|_\infty$.*

Proof We begin with the existence of f_0. Since Σ_0 is σ-finite, by restricting attention to sets of finite measure in Σ_0, it is enough to consider the case where $\mu(\Omega) < \infty$ and $f \in L^1$. By considering f^+ and f^-, we may also suppose that $f \geq 0$. If $B \in \Sigma_0$, let $\nu(B) = \int_B f \, d\mu$. Then ν is a measure on Σ_0, and if $\mu(B) = 0$ then $\nu(B) = 0$. Thus it follows from the Lebesgue decomposition theorem that there exists $f_0 \in L^1(\Omega, \Sigma_0, \mu)$ such that $\int_B f \, d\mu = \nu(B) = \int_B f_0 \, d\mu$ for all $B \in \Sigma_0$. If f_1 is another function with this property then

$$\int_{(f_1 > f_0)} (f_1 - f_0) \, d\mu = \int_{(f_1 < f_0)} (f_1 - f_0) \, d\mu = 0,$$

so that $f_1 = f_0$ almost everywhere.

We now return to the general situation. It follows from the construction that if $f \geq 0$ then $f_0 \geq 0$. If $f \in L^1$, then $f_0 = f_0^+ - f_0^-$, so that

$$\int |f_0| \, d\mu \leq \int |f_0^+| \, d\mu + \int |f_0^-| \, d\mu = \int f^+ \, d\mu + \int f^- \, d\mu = \int |f| \, d\mu.$$

If $f \in L^\infty$ and B is a Σ_0-set of finite measure in $(f_0 > \|f\|_\infty)$, then

$$\int_B (f_0 - \|f\|_\infty) \, d\mu = \int_B (f - \|f\|_\infty) \, d\mu \leq 0,$$

from which it follows that $f_0 \leq \|f\|_\infty$ almost everywhere. Similarly, it follows that $-f_0 \leq \|f\|_\infty$ almost everywhere, and so $\|f_0\|_\infty \leq \|f\|_\infty$. Thus if $f \in (L^1 + L^\infty)(\Omega, \Sigma, \mu)$ then $f_0 \in (L^1 + L^\infty)(\Omega, \Sigma_0, \mu)$. $\square$

The function f_0 is denoted by $\mathbf{E}(f|\Sigma_0)$, and called the *conditional expectation* of f with respect to Σ_0. The conditional expectation operator $f \to \mathbf{E}(f|\Sigma_0)$ is clearly linear. As an example, if Σ_0 is purely atomic, and A is an atom in Σ_0, then $\mathbf{E}(f|\Sigma_0)$ takes the constant value $(\int_A f \, d\mu)/\mu(A)$ on A. The following corollary now follows immediately from Calderón's interpolation theorem.

Corollary 8.13.1 *Suppose that* $(X, \|.\|_X)$ *is a rearrangement invariant Banach function space. If* $f \in X$, *then* $\|\mathbf{E}(f|\Sigma_0)\|_X \leq \|f\|_X$.

In these terms, an adapted process (f_j) in $L^1 + L^\infty$ is a sub-martingale if $f_j \leq \mathbf{E}(f_{j+1}|\Sigma_j)$, for each j, and super-martingales and martingales are characterized in a similar way.

Proposition 8.13.1 *(i) If* (f_j) *is a local martingale, then* $(|f_j|)$ *is a local sub-martingale.*

(ii) If $(X, \|.\|_X)$ *is a rearrangement invariant function space on* (Ω, Σ, μ) *and* (f_n) *is a non-negative local sub-martingale then* $(\|f_j\|_X)$ *is an increasing sequence.*

Proof (i) If $A, B \in \Sigma_j$ then

$$\int_B \mathbf{E}(f_{j+1}|\Sigma_j) I_A \, d\mu = \int_{A \cap B} f_{j+1} \, d\mu = \int_B f_{j+1} I_A \, d\mu$$
$$= \int_B \mathbf{E}(f_{j+1} I_A | \Sigma_j) \, d\mu,$$

so that

$$\mathbf{E}(f_{j+1} I_A | \Sigma_j) = \mathbf{E}(f_{j+1}|\Sigma_j) I_A = f_j I_A.$$

Thus

$$\int_A |f_j| \, d\mu = \int |\mathbf{E}(f_{j+1} I_A | \Sigma_j)| d\mu \leq \int |f_{j+1} I_A| \, d\mu = \int_A |f_{j+1}| \, d\mu.$$

(ii) This follows from Corollary 8.13.1. $\square$

8.14 Doob's inequality

If $f \in (L^1 + L^\infty)(\Sigma)$ then the sequence $\mathbf{E}(f|\Sigma_j)$ is a local martingale. Conversely, if (f_j) is a local martingale and there exists $f \in (L^1 + L^\infty)(\Sigma)$ such that $f_j = \mathbf{E}(f|\Sigma_j)$, for each j, then we say that (f_j) is *closed* by f.

If (f_j) is an adapted process, we set

$$f_k^*(x) = \sup_{j \leq k} |f_j|, \quad f^*(x) = \sup_{j < \infty} |f_j|.$$

Then (f_j^*) is an increasing adapted process, the *maximal process*, and $f_j^* \to f^*$ pointwise.

Theorem 8.14.1 (Doob's inequality) *Suppose that $(g_j)_{j=0}^\infty$ is a nonnegative local submartingale. Then $\alpha\mu(g_k^* > \alpha) \leq \int_{(g_k^* > \alpha)} g_k \, d\mu$.*

Proof Let $\tau(x) = \inf\{j\colon g_j(x) > \alpha\}$. Note that $\tau(x) > k$ if and only if $g_k^*(x) \leq \alpha$, and that $\tau(x) = \infty$ if and only if $g^*(x) \leq \alpha$. Note also that the sets $(\tau = j)$ and $(\tau \leq j)$ are in Σ_j; this says that τ is a *stopping time*. Then

$$\int_{(g_k^* > \alpha)} g_k \, d\mu = \int_{(\tau \leq k)} g_k \, d\mu = \sum_{j=0}^k \int_{(\tau = j)} g_k \, d\mu$$

$$\geq \sum_{j=0}^k \int_{(\tau = j)} g_j \, d\mu \quad \text{(by the local sub-martingale property)}$$

$$\geq \sum_{j=0}^k \alpha\mu(\tau = j) = \alpha\mu(\tau \leq k).$$

$\square$

Although this inequality is always known as Doob's inequality, it was first established by Jean Ville [1937]. It appears in Doob's fundamental paper (Doob [1940]) (where, as elsewhere, he fully acknowledges Ville's priority).

Corollary 8.14.1 *If $1 < p < \infty$ then $\|g_k^*\|_p \leq p' \|g_k\|_p$ and $\|g^*\|_p \leq p' \sup_k \|g_k\|_p$.*

Proof This follows immediately from Theorem 8.1.1. $\square$

8.15 The martingale convergence theorem

We say that a local martingale is *bounded in* L^p if $\sup_j \|f_j\|_p < \infty$.

Theorem 8.15.1 *If $1 < p \leq \infty$ and (f_j) is a local martingale which is bounded in L^p then (f_j) is closed by some f in L^p.*

Proof We use the fact that a bounded sequence in L^p is weakly sequentially compact if $1 < p < \infty$, and is weak* sequentially compact, when $p = \infty$. Thus there exists a subsequence (f_{j_k}) which converges weakly (or weak*, when $p = \infty$) to f in $L^p(\Sigma)$. Then if A is a set of finite measure in Σ_j, $\int_A f_{j_k} d\lambda \to \int_A f d\lambda$. But if $j_k \geq j$, $\int_A f_{j_k} d\lambda = \int_A f_j d\lambda$, and so $\int_A f d\lambda = \int_A f_j d\lambda$. □

We now prove a version of the martingale convergence theorem.

Theorem 8.15.2 *Suppose that (f_j) is a local martingale which is closed by f, for some f in L^p, where $1 \leq p < \infty$. Then $f_j \to f$ in L^p-norm, and almost everywhere.*

Proof Let $F = \mathrm{span}\,(\cup_j L^p(\Sigma_j))$. Then F is dense in $L^p(\Sigma)$, since Σ is the smallest σ-field containing $\cup_j \Sigma_j$. The result is true if $f \in F$, since then $f \in L^p(\Sigma_j)$ for some j, and then $f_k = f$ for $k \geq j$. Let $T_j(f) = \mathbf{E}(f|\Sigma_j)$, let $T_\infty(f) = f$, and let $M(f) = \max(f^*, |f|)$. Then $\|T_j\| = 1$ for all j, and so $f_j \to f$ in norm, for all $f \in L^p$, by Theorem 8.4.1.

In order to prove convergence almost everywhere, we show that the sublinear mapping $f \to M(f) = \max(f^*, |f|)$ is of Riesz weak type $(1,1)$: the result then follows from Theorem 8.4.2. Now $(|f_k|)$ is a local submartingale, and $\int_A |f_k| d\mu \leq \int_A |f| d\mu$ for each A in Σ_k, and so, using Doob's inequality,

$$\alpha\mu(f^* > \alpha) = \lim_{k \to \infty} \alpha\mu(f_k^* > \alpha)$$

$$\leq \lim_{k \to \infty} \int_{(|f_k^*| > \alpha)} |f_k|\, d\mu$$

$$\leq \lim_{k \to \infty} \int_{(|f_k^*| > \alpha)} |f|\, d\mu$$

$$= \int_{(|f^*| > \alpha)} |f|\, d\mu,$$

and so the sublinear mapping $f \to f^*$ is of Riesz weak type: M is therefore also of Riesz weak type $(1,1)$. □

Corollary 8.15.1 *If $1 < p < \infty$, every L^p-bounded local martingale converges in L^p-norm and almost everywhere.*

Although an L^1-bounded martingale need not be closed, nor converge in norm, it converges almost everywhere.

Theorem 8.15.3 *Suppose that $(f_j)_{j=0}^\infty$ is an L^1-bounded martingale. Then f_j converges almost everywhere.*

Proof Since (Ω, Σ_0, μ) is σ-finite, it is enough to show that f_j converges almost everywhere on each set in Σ_0 of finite measure. Now if A is a set of finite measure in Σ_0 then $(f_j I_A)$ is an L^1-bounded martingale. We can therefore suppose that $\mu(\Omega) < \infty$. Let $M = \sup \|f_j\|_1$. Suppose that $N > 0$. Let T be the stopping time $T = \inf\{j \colon |f_j| > N\}$, so that T takes values in $[0, \infty]$. Let $B = (T < \infty)$ and $S = (T = \infty)$. Let

$$g_j(\omega) = f_j(\omega) \qquad \text{if } j \le T(\omega),$$
$$= f_{T(\omega)}(\omega) \quad \text{if } j > T(\omega).$$

If $A \in \Sigma_j$, then

$$\int_A g_{j+1} \, d\mu = \int_{A \cap (j+1 \le T)} f_{j+1} \, d\mu + \int_{A \cap (j+1 > T)} f_T \, d\mu$$

$$= \int_{A \cap (j \le T)} f_{j+1} \, d\mu + \int_{A \cap (j+1=T)} f_{j+1} \, d\mu + \int_{A \cap (j+1 > T)} f_T \, d\mu$$

$$= \int_{A \cap (j \le T)} f_j \, d\mu + \int_{A \cap (j > T)} f_T \, d\mu$$

$$= \int_A g_j \, d\mu,$$

by the martingale property, since $A \cap (j \le T) \in \Sigma_j$. Thus (g_j) is a martingale, the martingale (f_j) *stopped at time T*. Further,

$$\|g_j\|_1 = \sum_{k \le j} \int_{(T=k)} |f_k| \, d\lambda + \int_{(T>j)} |f_j| \, d\lambda \le \|f_j\|_1 \le M,$$

so that g is an L^1-bounded martingale.

Now let $h = |f_T| I_B$. Then $h \le \liminf |g_j|$, so that $\|h\|_1 \le M$, by Fatou's lemma. Thus $h + N I_S \in L^1$, and $|g_j| \le h + N I_S$, for each j. Thus we can write $g_j = m_j(h + N I_S)$, where $\|m_j\|_\infty \le 1$. By weak*-compactness, there exists a subsequence (m_{j_k}) converging weak* in L^∞ to some $m \in L^\infty$. Then (g_{j_k}) converges weakly in L^1 to some $g \in L^1$. We now use the argument of Theorem 8.15.1 to conclude that (g_j) is closed by g, and so g_j converges almost everywhere to g, by Theorem 8.15.2. But $f_j = g_j$ for all j in S, and $\mu(B) = \lim_{k \to \infty} \mu(f_k^* > N) \le M/N$, by Doob's inequality. Thus f_j

converges pointwise except on a set of measure at most M/N. But this holds for all N, and so f_j converges almost everywhere. □

8.16 Notes and remarks

The great mathematical collaboration between Hardy and Littlewood was carried out in great part by correspondence ([Lit 86], pp. 9–11). Reading Hardy's papers of the 1920s and 1930s, it becomes clear that he also corresponded frequently with European mathematicians: often he writes to the effect that the proof that follows is due to Marcel Riesz (or whomsoever), and is simpler, or more general, than his original proof. Mathematical collaboration is a wonderful thing! But it was Hardy who revealed the mathematical power of maximal inequalities.

The term 'Riesz weak type' is introduced here, since it fits very naturally into the development of the theory. Probabilists, with Doob's inequality in mind, might prefer to call it 'Doob weak type'.

The martingale convergence theorem was proved by Doob in a beautiful paper [Doo 40], using Doob's inequality, and an upcrossing argument. The version of the martingale convergence theorem that we present here is as simple as it comes. The theory extends to more general families of σ-fields, to continuous time, and to vector-valued processes. It lies at the heart of the theory of stochastic integration, a theory which has been developed in fine detail, exposed over many years in the Seminar Notes of the University of Strasbourg, and the Notes on the Summer Schools of Probability at Saint-Flour, published in the Springer-Verlag Lecture Notes in Mathematics series. Progress in mathematical analysis, and in probability theory, was handicapped for many years by the failure of analysts to learn what probabilists were doing, and conversely.

Exercises

8.1   Give examples of functions f and g which satisfy the conditions of Theorem 8.1.1, for which $\int f \, d\mu = \infty$ and $\int g \, d\mu = 1$.

8.2   Show that if $f \neq 0$ and $f \geq 0$ then $\int_{\mathbf{R}^d} A(f) \, d\lambda = \infty$.

8.3   Suppose that f is a non-negative decreasing function on $(0, \infty)$. Show that $f^\dagger = m^-(f) = m_u(f)$. What is $m^+(f)$?

8.4   [*The Vitali covering lemma.*] Suppose that E is a bounded measurable subset of $\mathbf{R}^d$. A *Vitali covering* of E is a collection $\mathcal{U}$ of open balls with the property that if $x \in E$ and $\epsilon > 0$ then there exists $U \in \mathcal{U}$ with radius less than ϵ such that $x \in U$. Show that if $\mathcal{U}$ is a Vitali covering of E then there exists a sequence (U_n) of disjoint balls in $\mathcal{U}$ such that $\lambda(E \setminus \cup_n U_n) = 0$.

[Hint: repeated use of Lemma 8.9.1.]

8.5   Suppose that S is a set of open intervals in the line which cover a compact set of measure m. Show that there is a finite disjoint subset T whose union has measure more than $m/2$.

8.6   Give a proof of Theorem 8.11.1.

8.7   Consider the Fejér kernel

$$\sigma_n(t) = \frac{1}{n+1} \left(\frac{\sin(n+1)t/2}{\sin t/2} \right)^2$$

on the unit circle $\mathbf{T}$. Show that if $1 \leq p < \infty$ and $f \in L^p$ then $\sigma_n \star f \to f$ in $L^p(\mathbf{T})$-norm. What about convergence almost everywhere?

8.8   For $t \in \mathbf{R}^d$ let $\Phi(t) = \phi(|t|)$, where ϕ is a continuous strictly decreasing function on $[0, \infty)$ taking values in $[0, \infty]$. Suppose that $\Phi \in L^1 + L^p$, where $1 < p < \infty$. State and prove a theorem about Φ which generalizes Hedberg's inequality, and its corollary.

8.9   Suppose that $f \in (L^1 + L^\infty)(\mathbf{R}^d)$. If $t > 0$ let $\delta_t(f)(x) = f(x/t)$: δ_t is a dilation operator.

   (i) Suppose that $f \in L^p(\mathbf{R}^d)$. Show that $\|\delta_t(f)\|_p = t^{d/p} \|f\|_p$.
   (ii) Show that $\delta_t(I_{d,\alpha}(f)) = t^{-\alpha} I_{d,\alpha}(\delta_t(f))$.
   (iii) Show that if $1 < p < d/\alpha$ then $q = pd/(d - \alpha p)$ is the only index for which $I_{d,\alpha}$ maps $L^p(\mathbf{R}^d)$ continuously into $L^q(\mathbf{R}^d)$.

8.10   Suppose that (Ω, Σ, μ) is a measure space and that Σ_0 is a sub-σ-field of Σ. Suppose that $1 \leq p \leq \infty$, and that J_p is the natural inclusion of $L^p(\Omega, \Sigma_0, \mu)$ into $L^p(\Omega, \Sigma, \mu)$. Suppose that $f \in L^{p'}(\Omega, \Sigma, \mu)$. What is $J_p^*(f)$?

8.11   Let $f_j(t) = 2^j$ for $0 < t \leq 2^{-j}$ and $f_j(t) = 0$ for $2^{-j} < t \leq 1$. Show that (f_j) is an L^1-bounded martingale for the dyadic filtration of $(0, 1]$ which converges everywhere, but is not closed in L^1.

8.12   Let $K = [0, 1]^d$, with its dyadic filtration. Show that if (f_j) is an L^1-bounded martingale then there exists a signed Borel measure ν such that $\nu(A) = \int_A f_j \, d\lambda$ for each $A \in \Sigma_j$. Conversely, suppose that ν is a (non-negative) Borel measure. If A is an atom of Σ_j, let $f_j(x) = 2^{dj} \nu(A)$, for $x \in A$. Show that (f_j) is an L^1-bounded martingale. Let $f = \lim_{j \to \infty} f_j$, and let $\pi = \nu - f \, d\lambda$. Show that π is a non-negative measure which is *singular* with respect to λ: that is, there is a set N such that $\lambda(N) = 0$ and $\nu([0, 1]^d \setminus N) = 0$.

9

Complex interpolation

9.1 Hadamard's three lines inequality

Calderón's interpolation theorem and Theorem 8.5.1 have strong and satisfactory conclusions, but they require correspondingly strong conditions to be satisfied. In many cases, we must start from a weaker position. In this chapter and the next we consider other interpolation theorems; in this chapter, we consider complex interpolation, and all Banach spaces will be assumed to be complex Banach spaces. We shall turn to real interpolation in the next chapter.

We shall be concerned with the Riesz–Thorin Theorem and related results. The original theorem, which concerns linear operators between L^p-spaces, was proved by Marcel Riesz [Ri(M) 26] in 1926; Thorin [Tho 39] gave a different proof in 1939. Littlewood described this in his *Miscellany* [Lit 86] as 'the most impudent in mathematics, and brilliantly successful'. In the 1960s, Thorin's proof was deconstructed, principally by Lions [Lio 61] and Calderón [Cal 63], [Cal 64], [Cal 66], so that the results could be extended to a more general setting. We shall need these more general results, and so we shall follow Lions and Calderón.

The whole theory is concerned with functions, possibly vector-valued, which are bounded and continuous on the closed strip $\bar{S} = \{z = x + iy \in \mathbf{C}: 0 \le x \le 1\}$ and analytic on the open strip $S = \{z = x + iy \in \mathbf{C}: 0 < x < 1\}$, and we shall begin by establishing the first fundamental inequality, from complex analysis, that we shall need.

Proposition 9.1.1 (Hadamard's three lines inequality) *Suppose that f is a non-zero bounded continuous complex-valued function on $\bar{S}$ which is analytic on the open strip S. Let*

$$M_\theta = \sup\{|f(\theta + iy)|: y \in \mathbf{R}\}.$$

135

Then $M_\theta \leq M_0^\theta M_1^{1-\theta}$.

Proof First we simplify the problem. Suppose that $N_0 > M_0$, $N_1 > M_1$. Let

$$g(z) = N_0^{z-1} N_1^{-z} f(z).$$

Then g satisfies the conditions of the proposition, and

$$\sup\{|g(iy)|: y \in \mathbf{R}\} = \sup\{|g(1+iy)|: y \in \mathbf{R}\} < 1.$$

We shall show that $|g(z_0)| \leq 1$ for all $z_0 \in S$; then

$$|f(\theta + iy)| = N_0^{1-\theta} N_1^\theta |g(\theta + iy)| \leq N_0^{1-\theta} N_1^\theta.$$

Since this holds for all $N_0 > M_0$, $N_1 > M_1$, we have the required result.

Let $K = \sup\{|g(z)|: z \in S\}$. We want to apply the maximum modulus principle: the problem is the behaviour of g as $|y| \to \infty$. We deal with this by multiplying by functions that decay at infinity. Suppose that $\epsilon > 0$. Let $h_\epsilon(z) = e^{\epsilon z^2} g(z)$. If $z = x + iy \in S$ then

$$|h_\epsilon(z)| = e^{\epsilon(x^2 - y^2)} |g(z)| \leq e^\epsilon e^{-\epsilon y^2} K,$$

so that $h_\epsilon(z) \to 0$ as $|y| \to \infty$.

Now suppose that $z_0 = x_0 + iy_0 \in S$. Choose $R > 1$ such that $e^{-\epsilon R^2 y_0^2} K \leq 1$. Then z_0 is an interior point of the rectangle with vertices $\pm iRy_0$ and $1 \pm iRy_0$, and $|h(z)| \leq e^\epsilon$ on the sides of the rectangle. Thus, by the maximum modulus principle, $|h_\epsilon(z_0)| \leq e^\epsilon$, and so

$$|g(z_0)| = e^{\epsilon y_0^2} e^{-\epsilon x_0^2} |h(z_0)| \leq e^{\epsilon(1 + y_0^2)}.$$

But ϵ is arbitrary, and so $|g(z_0)| \leq 1$. $\square$

9.2 Compatible couples and intermediate spaces

We now set up the machinery for complex interpolation. Suppose that two Banach spaces $(A_0, \|.\|_{A_0})$ and $(A_1, \|.\|_{A_1})$ are linear subspaces of a Banach space $(V, \|.\|_V)$ (in fact, a Hausdorff topological vector space (V, τ) will do) and that the inclusion mappings $(A_j, \|.\|_{A_j}) \to (V, \|.\|_V)$ are continuous, for $j = 0, 1$. Then the pair $(A_0, \|.\|_{A_0})$, $(A_1, \|.\|_{A_1})$ is called a *compatible couple*. A word about terminology here: the two Banach spaces play a symmetric role, and we shall always use j to denote either 0 or 1, without repeating 'for $j = 0, 1$'.

It is straightforward to show (Exercise 9.1) that the spaces $A_0 \cap A_1$ and $A_0 + A_1$ are then Banach spaces, under the norms

$$\|a\|_{A_0 \cap A_1} = \max(\|a\|_{A_0}, \|a\|_{A_1}).$$
$$\|a\|_{A_0 + A_1} = \inf\{\|a_0\|_{A_0} + \|a_1\|_{A_1} : a = a_0 + a_1, a_j \in A_j\}.$$

A Banach space $(A, \|.\|_A)$ contained in $A_0 + A_1$ and containing $A_0 \cap A_1$ for which the inclusions

$$(A_0 \cap A_1, \|.\|_{A_0 \cap A_1}) \to (A, \|.\|_A) \to (A_0 + A_1, \|.\|_{A_0 + A_1})$$

are continuous is then called an *intermediate space*.

The obvious and most important example is given when $1 \leq p_j \leq \infty$. Then $(L^{p_0}, \|.\|_{p_0})$, $(L^{p_1}, \|.\|_{p_1})$ form a compatible couple, and if p is between p_0 and p_1 then $(L^p, \|.\|_p)$ is an intermediate space (Theorem 5.5.1).

With Hadamard's three lines inequality in mind, we now proceed as follows. Suppose that $(A_0, \|.\|_{A_0})$, $(A_1, \|.\|_{A_1})$ is a compatible couple. Let $L_0 = \{iy : y \in \mathbf{R}\}$ and $L_1 = \{1 + iy : y \in \mathbf{R}\}$ be the two components of the boundary of S. We set $\mathcal{F}(A_0, A_1)$ to be the vector space of all functions F on the closed strip $\bar{S}$ taking values in $A_0 + A_1$ for which

- F is continuous and bounded on $\bar{S}$;
- F is analytic on S (in the sense that $\phi(F)$ is analytic for each continuous linear functional ϕ on $A_0 + A_1$);
- $F(L_j) \subset A_j$, and F is a bounded continuous map from L_j to A_j.

We give $\mathcal{F}(A_0, A_1)$ the norm

$$\|F\|_{\mathcal{F}} = \max_{j=0,1}(\sup\{\|F(z)\|_{A_j} : z \in L_j\}).$$

Proposition 9.2.1 *If $F \in \mathcal{F}(A_0, A_1)$ and $z \in S$ then $\|F(z)\|_{A_0 + A_1} \leq \|F\|_{\mathcal{F}}$.*

Proof There exists $\phi \in (A_0 + A_1)^*$ with $\|\phi\|^* = 1$ and $\phi(F(z)) = \|F(z)\|_{A_0 + A_1}$. Then $\phi(F)$ satisfies the conditions of Proposition 9.1.1, and so $|\phi(F(z))| \leq \|F\|_{\mathcal{F}}$. $\square$

If (F_n) is an $\mathcal{F}$-Cauchy sequence, then it follows that $F_n(z)$ converges uniformly, to $F(z)$ say, on $\bar{S}$; then $F \in \mathcal{F}(A_0, A_1)$ and $F_n \to F$ in $\mathcal{F}$-norm. Thus $(\mathcal{F}(A_0, A_1), \|.\|_{\mathcal{F}})$ is a Banach space.

Now suppose that $0 < \theta < 1$. The mapping $F \to F(\theta)$ is a continuous linear mapping from $\mathcal{F}(A_0, A_1)$ into $A_0 + A_1$. We denote the image by $(A_0, A_1)_{[\theta]} = A_{[\theta]}$, and give it the quotient norm:

$$\|a\|_{[\theta]} = \inf\{\|F\|_{\mathcal{F}} : F_{[\theta]} = a\}.$$

Then $(A_{[\theta]}, \|.\|_{[\theta]})$ is an intermediate space.

With all this in place, the next fundamental theorem follows easily.

Theorem 9.2.1 *Suppose that (A_0, A_1) and (B_0, B_1) are compatible couples and that T is a linear mapping from $A_0 + A_1$ into $B_0 + B_1$, mapping A_j into B_j, with $\|T(a)\|_{B_j} \leq M_j \|a\|_{A_j}$ for $a \in A_j$, for $j = 0, 1$. Suppose that $0 < \theta < 1$. Then $T(A_{[\theta]}) \subseteq B_{[\theta]}$, and $\|T(a)\|_{[\theta]} \leq M_0^{1-\theta} M_1^\theta \|a\|_{[\theta]}$ for $a \in A_{[\theta]}$.*

Proof Suppose that a is a non-zero element of $A_{[\theta]}$ and that $\epsilon > 0$. Then there exists $F \in \mathcal{F}(A_0, A_1)$ such that $F(\theta) = a$ and $\|F\|_{\mathcal{F}} \leq (1+\epsilon) \|a\|_{[\theta]}$. Then the function $T(F(z))$ is in $\mathcal{F}(B_0, B_1)$, and

$$\|T(F(z))\|_{B_j} \leq (1+\epsilon) M_j \|F(z)\|_{A_j} \text{ for } z \in L_j.$$

Thus $T(a) = T(F(\theta)) \in B_{[\theta]}$. Set $G(z) = M_0^{z-1} M_1^{-z} T(F)(z)$. Then $G \in \mathcal{F}(B_0, B_1)$, and $\|G(z)\|_{B_j} \leq (1+\epsilon) \|F(z)\|_{A_j}$ for $z \in L_j$. Thus

$$\|G(\theta)\|_{[\theta]} = M_0^{\theta-1} M_1^{-\theta} \|T(a)\|_{[\theta]} \leq (1+\epsilon) \|a\|_{[\theta]},$$

so that $\|T(a)\|_{[\theta]} \leq (1+\epsilon) M_0^{1-\theta} M_1^\theta \|a\|_{[\theta]}$. Since ϵ is arbitrary, the result follows. $\qquad\square$

9.3 The Riesz–Thorin interpolation theorem

Theorem 9.2.1 is the first ingredient of the Riesz–Thorin interpolation theorem. Here is the second.

Theorem 9.3.1 *Suppose that $1 \leq p_0, p_1 \leq \infty$ and that $0 < \theta < 1$. Let $1/p = (1-\theta)/p_0 + \theta/p_1$. If (A_0, A_1) is the compatible couple $(L^{p_0}(\Omega, \Sigma, \mu), L^{p_1}(\Omega, \Sigma, \mu))$ then $A_{[\theta]} = L^p(\Omega, \Sigma, \mu)$, and $\|f\|_{[\theta]} = \|f\|_p$ for $f \in L^p(\Omega, \Sigma, \mu)$.*

Proof The result is trivially true if $p_0 = p_1$. Suppose that $p_0 \neq p_1$. Let us set $u(z) = (1-z)/p_0 + z/p_1$, for $z \in \bar{S}$; note that $u(\theta) = 1/p$ and that $\Re(u(z)) = 1/p_j$ for $z \in L_j$. First, let us consider a simple function $f = \sum_{k=1}^K r_k e^{i\alpha_k} I_{E_k}$ with $\|f\|_p = 1$. Set $F(z) = \sum_{k=1}^K r_k^{pu(z)} e^{i\alpha_k} I_{E_k}$, so that $F(\theta) = f$. If $z \in L_j$ then $|F(z)| = \sum_{k=1}^K r_k^{p/p_j} I_{E_k}$, and so $\|F(z)\|_{p_j} = \|f\|_p^{p/p_j} = 1$. Thus F is continuous on $\bar{S}$, analytic on S, and bounded in $A_0 + A_1$ on $\bar{S}$. Consequently $\|f\|_{[\theta]} \leq 1$. By scaling, $\|f\|_{[\theta]} \leq \|f\|_p$ for all simple f.

Now suppose that $f \in L^p$. Then there exists a sequence (f_n) of simple functions which converge in L^p-norm and almost everywhere to f. Then (f_n) is Cauchy in $\|.\|_{[\theta]}$, and so converges to an element g of $(A_0, A_1)_{[\theta]}$. But then a subsequence converges almost everywhere to g, and so $g = f$. Thus $L^p(\Omega, \Sigma, \mu) \subseteq (A_0, A_1)_{[\theta]}$, and $\|f\|_{[\theta]} \le \|f\|_p$ for $f \in L^p(\Omega, \Sigma, \mu)$.

To prove the converse, we use a duality argument. Suppose that f is a non-zero function in $(A_0, A_1)_{[\theta]}$. Suppose that $\epsilon > 0$. Then there exists $F \in \mathcal{F}(A_0, A_1)$ with $F(\theta) = f$ and $\|F\|_{\mathcal{F}} \le (1+\epsilon)\|f\|_{[\theta]}$. Now let us set $B_j = L^{p'_j}$, so that (B_0, B_1) is a compatible couple, $L^{p'}(\Omega, \Sigma, \mu) \subseteq (B_0, B_1)_{[\theta]}$, and $\|g\|_{[\theta]} \le \|g\|_{p'}$ for $g \in L^{p'}(\Omega, \Sigma, \mu)$. Thus if g is a non-zero simple function, there exists $G \in \mathcal{F}(B_0, B_1)$ with $G(\theta) = g$ and $\|G\|_{\mathcal{F}} \le (1+\epsilon)\|g\|_{p'}$. Let us now set $I(z) = \int F(z)G(z)\,d\mu$. Then I is a bounded continuous function on $\bar{S}$, and is analytic on S. Further, if $z \in L_j$ then, using Hölder's inequality,

$$|I(z)| \le \int |F(z)||G(z)|\,d\mu \le \|F(z)\|_{p_j} \cdot \|G(z)\|_{p'_j}$$
$$\le (1+\epsilon)^2 \|f\|_{[\theta]} \|g\|_{[\theta]} \le (1+\epsilon)^2 \|f\|_{[\theta]} \|g\|_{p'} .$$

We now apply Hadamard's three lines inequality to conclude that

$$|I(\theta)| = \left| \int fg\,d\mu \right| \le (1+\epsilon)^2 \|f\|_{[\theta]} \|g\|_{p'} .$$

Since this holds for all simple g and all $\epsilon > 0$, it follows that $f \in L^p$ and $\|f\|_p \le \|f\|_{[\theta]}$. □

There is also a vector-valued version of this theorem.

Theorem 9.3.2 *Suppose that E is a Banach space. Suppose that $1 \le p_0, p_1 \le \infty$ and that $0 < \theta < 1$. Let $1/p = (1 - \theta)/p_0 + \theta/p_1$. If (A_0, A_1) is the compatible couple $(L^{p_0}(\Omega; E), L^{p_1}(\Omega; E))$ then $A_{[\theta]} = L^p(\Omega; E)$, and $\|f\|_{[\theta]} = \|f\|_p$ for $f \in L^p(\Omega; E)$.*

Proof The proof is exactly the same, making obvious changes. (Consider a simple function $f = \sum_{k=1}^K r_k x_k I_{E_k}$ with $r_k \in \mathbf{R}$, $x_k \in E$ and $\|x_k\| = 1$, and with $\|f\|_p = 1$. Set $F(z) = \sum_{k=1}^K r_k^{pu(z)} x_k I_{E_k}$, so that $F(\theta) = f$.) □

Combining Theorems 9.2.1 and 9.3.1, we obtain the Riesz–Thorin interpolation theorem.

Theorem 9.3.3 (The Riesz–Thorin interpolation theorem) *Suppose that (Ω, Σ, μ) and (Φ, T, ν) are measure spaces. Suppose that $1 \le p_0, p_1 \le \infty$ and that $1 \le q_0, q_1 \le \infty$, and that T is a linear mapping from $L^{p_0}(\Omega, \Sigma, \mu) + L^{p_1}(\Omega, \Sigma, \mu)$ into $L^{q_0}(\Phi, T, \nu) + L^{q_1}(\Phi, T, \nu)$ and that T maps $L^{p_j}(\Omega, \Sigma, \mu)$ continuously into $L^{q_j}(\Phi, T, \nu)$ with norm M_j, for $j = 0, 1$. Suppose that*

$0 < \theta < 1$, *and define* p_θ *and* q_θ *by*

$$\frac{1}{p_\theta} = \frac{1-\theta}{p_0} + \frac{\theta}{p_1}, \quad \frac{1}{q_\theta} = \frac{1-\theta}{q_0} + \frac{\theta}{q_1},$$

(with the obvious conventions if any of the indices are infinite). Then T *maps* $L^p(\Omega, \Sigma, \mu)$ *continuously into* $L^q(\Phi, T, \nu)$ *with norm at most* $M_0^{1-\theta} M_1^\theta$.

There is also a vector-valued version of the Riesz–Thorin theorem, which we leave the reader to formulate.

9.4 Young's inequality

We now turn to applications. These involve harmonic analysis on locally compact abelian groups. Let us describe what we need to know about this – an excellent account is given in Rudin [Rud 79]. Suppose that G is a locally compact abelian group. Since we are restricting our attention to σ-finite measure spaces, we shall suppose that G is σ-compact (a countable union of compact sets). Since we want the dual group (defined in the next section) to have the same property, we shall also suppose that G is metrizable. In fact, neither condition is really necessary, but both are satisfied by the examples that we shall consider. There exists a measure μ, *Haar measure*, on the Borel sets of G for which (if the group operation is addition) $\mu(A) = \mu(-A) = \mu(A + g)$ for each Borel set A and each $g \in G$; further μ is unique up to scaling. If G is compact, we usually normalize μ so that $\mu(G) = 1$. In fact, we shall only consider the following examples:

- **R**, under addition, with Lebesgue measure, and finite products $\mathbf{R}^d$, with product measure;
- $\mathbf{T} = \{z \in \mathbf{C}: |z| = 1\} = \{e^{i\theta}: 0 \le \theta < 2\pi\}$, under multiplication, and with measure $d\theta/2\pi$, and finite products $\mathbf{T}^d$, with product measure;
- **Z**, under addition, with counting measure #, and finite products $\mathbf{Z}^d$, with counting measure;
- $\mathbf{D}_2 = \{1, -1\}$, under multiplication, with probability measure $\mu(\{1\}) = \mu(\{-1\}) = 1/2$, finite products $\mathbf{D}_2^d = \{\omega = (\omega_1, \dots, \omega_d): \omega_i = \pm 1\}$, with product measure, under which each point has measure $1/2^d$, and the countable product $D_2^{\mathbf{N}}$, with product measure.
- $\mathbf{Z}_2 = \{0, 1\}$, under addition mod 2, with counting measure $\#(\{0\}) = \#(\{1\}) = 1$, finite products $\mathbf{Z}_2^d = \{v = (v_1, \dots, v_d): v_i = 0 \text{ or } 1\}$, with counting measure, and the countable sum $\mathbf{Z}_2^{(\mathbf{N})}$, consisting of all $\mathbf{Z}_2$ valued sequences with only finitely many non-zero terms, again with counting measure. Let $\mathbf{P}_d$ denote the set of subsets of $\{1, \dots, d\}$. If $A \in \mathbf{P}_d$, then

we can consider I_A as an element of $\mathbf{Z}_2^d$; thus we can identify $\mathbf{Z}_2^d$ with $\mathbf{P}_d$. Under this identification, the group composition of two sets A and B is the symmetric difference $A \Delta B$.

Note that although $\mathbf{D}_2^d$ and $\mathbf{Z}_2^d$ are isomorphic as groups, we have given then different measures.

Our first application concerns convolution. Suppose that G is a locally compact abelian group and that $1 < p < \infty$. It follows from Proposition 7.5.1 that if $f \in L^1(G)$ and $g \in L^p(G)$ then $f \star g \in L^p(G)$ and $\|f \star g\|_p \le \|f\|_1 \|g\|_p$. On the other hand, if $h \in L^{p'}(G)$ then

$$\left| \int h(x-y)g(y)\,d\mu(y) \right| \le \|g\|_p \|h\|_{p'},$$

by Hölder's inequality, so that $h \star g$ is defined as an element of L^∞ and $\|h \star g\|_\infty \le \|h\|_{p'} \|f\|_p$. If now $k \in L^q(G)$, where $1 < q < p'$, then $k \in L^1 + L^{p'}$, and so we can define the convolution $k \star g$. What can we say about $k \star g$?

Theorem 9.4.1 (Young's inequality) *Suppose that G is a σ-compact locally compact metrizable abelian group, that $1 < p, q < \infty$ and that $1/p + 1/q = 1 + 1/r > 1$. If $g \in L^p(G)$ and $k \in L^q(G)$ then $k \star g \in L^r(G)$, and $\|k \star g\|_r \le \|k\|_p \|g\|_q$.*

Proof If $f \in L^1(G) + L^{p'}(G)$, let $T_g(f) = f \star g$. Then $T \in L(L^1, L^p)$, and $\left\| T \colon L^1 \to L^p \right\| \le \|g\|_p$. Similarly, $T \in L(L^{p'}, L^\infty)$, and $\left\| T \colon L^{p'} \to L^\infty \right\| \le \|g\|_p$. We take $p_0 = 1$, $p_1 = p'$ and $q_0 = p$, $q_1 = \infty$. If we set $\theta = p/q' = q/r$ we find that

$$\frac{1-\theta}{1} + \frac{\theta}{p'} = \frac{1}{q}, \quad \frac{1-\theta}{p} + \frac{\theta}{\infty} = \frac{1}{r};$$

the result therefore follows from the Riesz–Thorin interpolation theorem. $\square$

In fact, it is not difficult to prove Young's inequality without using interpolation (Exercise 9.3).

9.5 The Hausdorff–Young inequality

For our second application, we consider group duality, and the Fourier transform. A *character* on a σ-compact locally compact metrizable abelian group

G is a continuous homomorphism of G into $\mathbf{T}$. Under pointwise multiplication, the characters form a group, the *dual group* G', and G' becomes a σ-compact locally compact metrizable abelian group when it is given the topology of uniform convergence on the compact subsets of G. If G is compact, then G' is discrete, and if G is discrete, then G' is compact. The dual of a finite product is (naturally isomorphic to) the product of the duals. The dual G'' of G' is naturally isomorphic to G. For the examples above, we have the following duals:

- $\mathbf{R}' = \mathbf{R}$; if $x \in \mathbf{R}$ and $\phi \in \mathbf{R}'$ then $\phi(x) = e^{2\pi i \phi x}$.
- $(\mathbf{R}^d)' = \mathbf{R}^d$; if $x \in \mathbf{R}^d$ and $\phi \in (\mathbf{R}^d)'$ then $\phi(x) = e^{2\pi i \langle \phi, x \rangle}$.
- $\mathbf{T}' = \mathbf{Z}$ and $\mathbf{Z}' = \mathbf{T}$; if $n \in \mathbf{Z}$ and $e^{i\theta} \in \mathbf{T}$ then $n(e^{i\theta}) = e^{in\theta}$.
- $(\mathbf{D}_2^d)' = \mathbf{Z}_2^d$ and $(\mathbf{Z}_2^d)' = \mathbf{D}_2^d$. If $\omega \in \mathbf{D}_2^d$ and $A \in \mathbf{P}_d$, let $w_A(\omega) = \prod_{i \in A} \omega_i$.
  The function w_A is a character on $\mathbf{D}_2^d$, and is called a *Walsh function*. If $A = \{i\}$, we write ϵ_i for $w_{\{i\}}$; the functions $\epsilon_1, \ldots, \epsilon_d$ are called *Bernoulli random variables*. $\epsilon_i(\omega) = \omega_i$, and $w_A = \prod_{i \in A} \epsilon_i$.

  $(\mathbf{D}_2^{\mathbf{N}})' = \mathbf{Z}_2^{(\mathbf{N})}$ and $(\mathbf{Z}_2^{(\mathbf{N})})' = \mathbf{D}_2^{\mathbf{N}}$. Again, the Walsh functions are the characters on $\mathbf{D}_2^{\mathbf{N}}$.

If $f \in L^1(G)$, we define the *Fourier transform* $\mathcal{F}(f) = \hat{f}$ as

$$\mathcal{F}(f)(\gamma) = \int_G f(g)\overline{\gamma(g)}\, d\mu(g) \quad (\gamma \in G').$$

It follows from the theorem of dominated convergence that $\mathcal{F}(f)$ is a bounded continuous function on G', and the mapping $\mathcal{F}$ is a norm-decreasing linear mapping of $L^1(G)$ into $C_b(G')$. We also have the Plancherel theorem.

Theorem 9.5.1 (The Plancherel theorem) *Suppose that G is a σ-compact locally compact metrizable abelian group. If $f \in L^1(G) \cap L^2(G)$, then $\mathcal{F}(f) \in L^2(G', \mu')$ (where μ' is Haar measure on G'), and we can scale the measure μ' so that $\|\mathcal{F}(f)\|_2 = \|f\|_2$. We can then extend $\mathcal{F}$ by continuity to a linear isometry of $L^2(G)$ onto $L^2(G')$; the inverse mapping is given by*

$$f(g) = \int_{G'} \mathcal{F}(f)(\gamma)\gamma(g)\, d\mu'(\gamma).$$

Proof We give an outline of the proof in the case where G is a compact group, and Haar measure has been normalized so that $\mu(G) = 1$. First, the characters form an orthonormal set in $L^2(G)$. For if $\gamma \in G'$ then

$$\langle \gamma, \gamma \rangle = \int_G \gamma\bar{\gamma}\, d\mu = \int_G 1\, d\mu = 1,$$

while if γ_1 and γ_2 are distinct elements of G', and $\gamma_1(h) \neq \gamma_2(h)$, then, using the invariance of Haar measure,

$$
\begin{aligned}
\langle \gamma_1, \gamma_2 \rangle = \int_G \gamma_1 \bar{\gamma}_2 \, d\mu &= \int_G (\gamma_1 \gamma_2^{-1})(g) \, d\mu(g) \\
&= \int_G (\gamma_1 \gamma_2^{-1})(g+h) \, d\mu(g) = (\gamma_1 \gamma_2^{-1})(h) \int_G (\gamma_1 \gamma_2^{-1})(g) \, d\mu(g) \\
&= \gamma_1(h) \gamma_2^{-1}(h) \, \langle \gamma_1, \gamma_2 \rangle .
\end{aligned}
$$

Thus $\langle \gamma_1, \gamma_2 \rangle = 0$. Finite linear combinations of characters are called *trigonometric polynomials*. The trigonometric polynomials form an algebra of functions, closed under conjugation (since $\bar{\gamma} = \gamma^{-1}$). The next step is to show that the characters separate the points of G; we shall not prove this, though it is clear when $G = \mathbf{T}^d$ or $\mathbf{D}_2^{\mathbf{N}}$. It then follows from the complex Stone–Weierstrass theorem that the trigonometric polynomials are dense in $C(G)$. Further, $C(G)$ is dense in $L^2(G)$: this is a standard result from measure theory, but again is clear if $G = \mathbf{T}^d$ or $\mathbf{D}_2^{\mathbf{N}}$. Thus the characters form an orthonormal basis for $L^2(G)$. Thus if $f \in L^2(G)$ we can write f uniquely as $\sum_{\gamma \in G'} a_\gamma \gamma$, and then $\|f\|_2^2 = \sum_{\gamma \in G'} |a_\gamma|^2$. But then $\mathcal{F}(f)(\gamma) = a_\gamma$ and $f(g) = \sum_\gamma \mathcal{F}(f)(\gamma)\gamma(g)$.

The proof for locally compact groups is harder: the Plancherel theorem for $\mathbf{R}$, and so for $\mathbf{R}^d$, comes as an exercise later (Exercise 13.1). □

After all this, the next result may seem to be an anti-climax.

Theorem 9.5.2 (The Hausdorff–Young inequality) *Suppose that $f \in L^r(G)$, where G is a σ-compact locally compact metrizable abelian group and $1 < r < 2$. Then the Fourier transform $\mathcal{F}(f)$ is in $L^{r'}(G')$, and $\|\mathcal{F}(f)\|_{r'} \leq \|f\|_p$.*

Proof The Fourier transform is an isometry on L^2, and is norm-decreasing from L^1 to L^∞. We therefore apply the Riesz–Thorin interpolation theorem, taking $p_0 = 1$, $p_1 = 2$, $q_0 = \infty$ and $q_1 = 2$, and taking $\theta = 2/r$. □

9.6 Fourier type

We now turn to the Fourier transform of vector-valued functions. If $f \in L^1(G; E)$, where E is a Banach space, we can define the Fourier transform $\mathcal{F}(f)$ by setting $\mathcal{F}(f)(\gamma) = \int_G f(g)\overline{\gamma(g)} \, d\mu(g)$. Then $\mathcal{F}(f) \in C_b(G', E)$, and $\|\mathcal{F}(f)\|_\infty \leq \|f\|_1$. In general though, neither the Plancherel theorem nor the Hausdorff–Young inequalities extend to this setting, as the following example shows. Let us take $G = \mathbf{T}$, $E = c_0$, and $f(\theta) = (\lambda_n e^{in\theta})$, where $\lambda = (\lambda_n) \in c_0$. Then $\left\| f(e^{i\theta}) \right\|_\infty = \|\lambda\|_\infty$ for all θ, so that $\|f\|_{L^p(c_0)} = \|\lambda\|_\infty$, for $1 \le p \le \infty$. On the other hand $(\mathcal{F}(f))_k = \lambda_k e_k$, where e_k is the kth unit vector in c_0, and so

$$\sum_k \|(\mathcal{F}(f))_k\|^p = \sum_k |\lambda_k|^p.$$

Thus if we choose λ in c_0, but not in l^p, for any $1 \le p < \infty$, it follows that $\mathcal{F}(f)$ is not in l^p, for any $1 \le p < \infty$.

On the other hand, there are cases where things work well. For example, if H is a Hilbert space with orthonormal basis (e_n), and $f = \sum_n f_n e_n \in L^2(G; H)$, then $f_n \in L^2(G)$ for each n, and $\|f\|_2^2 = \sum_n \|f_n\|_2^2$. We can apply the Plancherel theorem to each f_n. Then $\mathcal{F}(f) = \sum_n \mathcal{F}(f_n) e_n$, and $\mathcal{F}$ is an isometry of $L^2(G; H)$ onto $L^2(G'; H)$; we have a vector-valued Plancherel theorem. Using the vector-valued Riesz–Thorin interpolation theorem, we also obtain a vector-valued Hausdorff–Young inequality.

This suggests a way of classifying Banach spaces. Suppose that E is a Banach space, that G is a σ-compact locally compact metrizable abelian group and that $1 \le p \le 2$. Then we say that E is of *Fourier type p with respect to G* if $\mathcal{F}(f) \in L^{p'}(G'; E)$ for all $f \in L^p(G; E) \cap L^1(G; E)$ and the mapping $f \to \mathcal{F}(f)$ extends to a continuous linear mapping from $L^p(G; E)$ into $L^{p'}(G', E)$. It is not known whether this condition depends on G, for infinite G, though Fourier type p with respect to $\mathbf{R}$, $\mathbf{T}$ and $\mathbf{Z}$ are known to be the same. If the condition holds for all G, we say that E is of *Fourier type p*. Every Banach space is of Fourier type 1. We have seen that c_0 is not of Fourier type p with respect to T for any $1 < p \le 2$, and that Hilbert space is of Fourier type 2.

Proposition 9.6.1 *If E is of Fourier type p with respect to G then E is of Fourier type r with respect to G, for $1 < r < p$.*

Proof The result follows from the vector-valued Riesz–Thorin theorem, since

$$L^r(G; E) = (L^1(G; E), L^p(G; E))_{[\theta]} \text{ and } L^{r'}(G; E) = (L^\infty(G; E), L^{p'}(G; E))_{[\theta]},$$

where $\theta = p'/r'$.   □

This shows that 'Fourier type p' forms a scale of conditions, the condition becoming more stringent as p increases. Kwapień [Kwa 72] has shown that a Banach space is of Fourier type 2 if and only if it is isomorphic to a Hilbert space.

Fourier type extends to subspaces. We also have the following straight-forward duality result.

Proposition 9.6.2 *A Banach space E is of Fourier type p with respect to G if and only if its dual E^* is of Fourier type p with respect to G'.*

Proof Suppose that E is of Fourier type p with respect to G, and that $\left\|\mathcal{F}: L^p(G) \to L^{p'}(G')\right\| = K$. Suppose that $h \in L^p(G'; E^*) \cap L^1(G'; E^*)$. If f is a simple E-valued function on G then, by Fubini's theorem

$$\int_G f(g)\mathcal{F}(h)(g)\,d\mu(g) = \int_G f(g)\left(\int_{G'} h(\gamma)\overline{\gamma(g)}\,d\mu'(\gamma)\right)d\mu(g)$$

$$= \int_{G'} h(\gamma)\left(\int_G f(g)\overline{\gamma(g)}\,d\mu(g)\right)d\mu'(\gamma)$$

$$= \int_{G'} h(\gamma)\mathcal{F}(f)(\gamma)\,d\mu'(\gamma).$$

Thus

$$\|\mathcal{F}(h)\|_{p'} = \sup\left\{\left|\int_G f\mathcal{F}(h)\,d\mu\right|: f \text{ simple}, \|f\|_p \le 1\right\}$$

$$= \sup\left\{\left|\int_{G'} \mathcal{F}(f)h\,d\mu'\right|: f \text{ simple}, \|f\|_p \le 1\right\}$$

$$\le \sup\left\{\|\mathcal{F}(f)\|_{p'}\|h\|_p: f \text{ simple}, \|f\|_p \le 1\right\} \le K\|h\|_p.$$

Thus E^* is of Fourier type p with respect to G'. Conversely, if E^* is of Fourier type p with respect to G', then E^{**} is of Fourier type p with respect to $G'' = G$, and so E is of Fourier type p with respect to G, since E is isometrically isomorphic to a subspace of E^{**}. $\square$

Thus if L^1 is infinite-dimensional, then L^1 does not have Fourier type p with respect to Z, for any $p > 1$, since $(L^1)^*$ has a subspace isomorphic to c_0.

9.7 The generalized Clarkson inequalities

What about the L^p spaces?

Theorem 9.7.1 *Suppose that $1 < p < \infty$. Then $L^p(\Omega, \Sigma, \nu)$ is of Fourier type r for $1 < r \le \min(p, p')$, and if $f \in L^r(G; L^p)$ then*

$$\|\mathcal{F}(f)\|_{L^{r'}(G'; L^p)} \le \|f\|_{L^r(G; L^p)}.$$

Proof We use Corollary 5.4.2 twice.

$$\left(\int_{G'} \|\mathcal{F}(f)\|_{L^p(\Omega)}^{r'} \, d\mu' \right)^{1/r'} = \left(\int_{G'} \left(\int_{\Omega} |\mathcal{F}(f)(\gamma,\omega)|^p \, d\nu(\omega) \right)^{r'/p} d\mu'(\gamma) \right)^{1/r'}$$

$$\leq \left(\int_{\Omega} \left(\int_{G'} |\mathcal{F}(f)(\gamma,\omega)|^{r'} \, d\mu'(\gamma) \right)^{p/r'} d\nu(\omega) \right)^{1/p}$$

by Corollary 5.4.2, and

$$\left(\int_{\Omega} \left(\int_{G'} |\mathcal{F}(f)(\gamma,\omega)|^{r'} \, d\mu'(\gamma) \right)^{p/r'} d\nu(\omega) \right)^{1/p}$$

$$\leq \left(\int_{\Omega} \left(\int_{G} |f(g,\omega)|^r \, d\mu(g) \right)^{p/r} d\nu(\omega) \right)^{1/p},$$

by the Hausdorff–Young inequality. Finally

$$\left(\int_{\Omega} \left(\int_{G} |f(g,\omega)|^r \, d\mu(g) \right)^{p/r} d\nu(\omega) \right)^{1/p}$$

$$\leq \left(\int_{G} \left(\int_{\Omega} |f(g,\omega)|^p \, d\nu(\omega) \right)^{r/p} d\mu(g) \right)^{1/r}$$

$$= \left(\int_{G} \|f\|_{L^p(\Omega)}^r \, d\mu(g) \right)^{1/r},$$

by Corollary 5.4.2, again.   □

This enables us to prove the following classical inequalities concerning L^p spaces.

Theorem 9.7.2 (Generalized Clarkson inequalities) *Suppose that $f, g \in L^p(\Omega, \Sigma, \nu)$, where $1 < p < \infty$, and suppose that $1 < r \leq \min(p, p')$.*

(i) $\|f+g\|_p^{r'} + \|f-g\|_p^{r'} \leq 2(\|f\|_p^r + \|g\|_p^r)^{r'-1}.$

(ii) $2(\|f\|_p^{r'} + \|g\|_p^{r'})^{r-1} \leq \|f+g\|_p^r + \|f-g\|_p^r.$

(iii) $2(\|f\|_p^{r'} + \|g\|_p^{r'}) \leq \|f+g\|_p^{r'} + \|f-g\|_p^{r'} \leq 2^{r'-1}(\|f\|_p^{r'} + \|g\|_p^{r'}).$

(iv) $2^{r-1}(\|f\|_p^r + \|g\|_p^r) \leq \|f+g\|_p^r + \|f-g\|_p^r \leq 2(\|f\|_p^r + \|g\|_p^r).$

Proof (i) Define $h \in L^r(D_2; L^p)$ by setting $h(1) = f$, $h(-1) = g$. Then $h = ((f+g)/2)1 + ((f-g)/2)\epsilon$,   so   that $\mathcal{F}(h)(0) = (f + g)/2$ and

$\mathcal{F}(h)(1) = (f - g)/2$. Thus, applying the Hausdorff–Young inequality,

$$\|\mathcal{F}(h)\|_{L^{r'}(Z_2;L^p)} = \tfrac{1}{2}(\|f + g\|_p^{r'} + \|f - g\|_p^{r'})^{1/r'}$$

$$\le \|h\|_{L^r(D_2;L^p)} = (\tfrac{1}{2}(\|f\|_p^r + \|g\|_p^r))^{1/r}$$

$$= \frac{1}{2^{1/r}}(\|f\|_p^r + \|g\|_p^r)^{1/r}.$$

Multiplying by 2, and raising to the r'-th power, we obtain (i).

(ii) Apply (i) to $u = f + g$ and $v = f - g$:

$$\|2f\|_p^{r'} + \|2g\|_p^{r'} \le 2(\|f + g\|_p^r + \|f - g\|_p^r)^{r'-1}.$$

Dividing by 2, and raising to the $(r-1)$-st power, we obtain (ii).

(iii) Since $\|h\|_{L^r(D_2,L^p)} \le \|h\|_{L^{r'}(D_2,L^p)}$,

$$2^{-1/r}(\|f\|_p^r + \|g\|_p^r)^{1/r} \le 2^{-1/r'}(\|f\|_p^{r'} + \|g\|_p^{r'})^{1/r'}.$$

Substituting this in (i), and simplifying, we obtain the right-hand inequality. Also,

$$2^{-1/r}(\|f + g\|_p^r + \|f - g\|_p^r)^{1/r} \le 2^{-1/r'}(\|f + g\|_p^{r'} + \|f - g\|_p^{r'})^{1/r'}.$$

Substituting this in (ii), and simplifying, we obtain the left-hand inequality.

(iv) These are proved in the same way as (iii); the details are left to the reader. □

In fact, Clarkson [Cla 36] proved these inequalities in the case where $r = \min(p, p')$ (see Exercise 9.5).

9.8 Uniform convexity

Clarkson's inequalities give strong geometric information about the unit ball of the L^p spaces, for $1 < p < \infty$. The unit ball of a Banach space $(E, \|.\|_E)$ is convex, but its unit sphere $S_E = \{x\colon \|x\| = 1\}$ can contain large flat spots. For example, in L^1, the set $S_{L^1}^+ = \{f \in S_{L^1}\colon f \ge 0\} = \{f \in S_{L^1}\colon \int f \, d\mu = 1\}$ is a convex set, so that if $f_1, f_2 \in S_{L^1}^+$ then $\|(f_1 + f_2)/2\| = 1$. By contrast, a Banach space $(E, \|.\|_E)$ is said to be *uniformly convex* if, given $\epsilon > 0$, there exists $\delta > 0$ such that if $x, y \in S_E$ and $\|(x + y)/2\| > 1 - \delta$ then $\|x - y\| < \epsilon$. In particular, $(E, \|.\|_E)$ is *p-uniformly convex*, where $2 \le p < \infty$, if there exists $C > 0$ such that if $x, y \in S_E$ then

$$\left\|\frac{x + y}{2}\right\| \le 1 - C \|x - y\|^p.$$

Theorem 9.8.1 *If $2 \le p < \infty$ then $L^p(\Omega, \Sigma, \mu)$ is p-uniformly convex. If $1 < p \le 2$ then $L^p(\Omega, \Sigma, \mu)$ is 2-uniformly convex.*

Proof When $p \ge 2$, the result follows from the first of the generalized Clarkson inequalities, since if $\|f\|_p = \|g\|_p = 1$ then

$$\left\| \frac{f+g}{2} \right\|^p \le 1 - \left\| \frac{f-g}{2} \right\|^p, \quad \text{so that} \quad \left\| \frac{f+g}{2} \right\| \le 1 - \frac{1}{p2^p} \|f - g\|^p.$$

When $1 < p < 2$, a similar argument shows that L^p is p'-uniformly convex. To show that it is 2-uniformly convex, we need to work harder. We need the following inequality.

Lemma 9.8.1 *If $1 < p < \infty$ and $s, t \in \mathbf{R}$ then there exists $C_p > 0$ such that*

$$\left(\frac{|s|^p + |t|^p}{2} \right)^{2/p} \ge \left(\frac{s+t}{2} \right)^2 + C_p(s-t)^2.$$

Proof By homogeneity, it is sufficient to prove the result for $s = 1$ and $|t| \le 1$. For $0 \le t \le 1$, let $f_p(t) = ((1 + |t|^p)/2)^{1/p}$. Then by Taylor's theorem with remainder, if $0 \le t < 1$ there exists $t < r < 1$ such that

$$f_p(t) = f_p(1) + (t-1)f'_p(t) + \frac{(t-1)^2}{2} f''_p(r).$$

Now

$$f'_p(t) = \frac{t^{p-1}}{2}(f_p(t))^{1-p} \quad \text{and} \quad f''_p(t) = \frac{(p-1)t^{p-2}}{4}(f_p(t))^{1-2p}$$

so that $f_p(1) = 1$, $f'_p(1) = 1/2$ and $f''_p(t) \ge (p-1)/2^p$ for $1/2 \le t \le 1$. Thus

$$((1 + t^p)/2)^{1/p} - (1+t)/2 \ge \frac{p-1}{2^{p+1}}(1-t)^2$$

for $1/2 \le t \le 1$. On the other hand, $f_p(t) - (1+t)/2 > 0$ on $[-1, 1/2]$, by Hölder's inequality, so that $(((1 + |t|^p)/2)^{1/p} - (1+t)/2)/(1-t)^2 > 0$ on $[-1, 1/2]$, and is therefore bounded below by a positive constant. Thus there exists $B_p > 0$ such that

$$((1 + |t|^p)/2)^{1/p} - (1+t)/2 \ge B_p(1-t)^2 \text{ for } t \in [-1, 1].$$

On the other hand,

$$((1 + |t|^p)/2)^{1/p} + (1+t)/2 \ge ((1 + |t|^p)/2)^{1/p} \ge 2^{-1/p} \text{ for } t \in [-1, 1];$$

the result follows by multiplying these inequalities. $\qquad \square$

Now suppose that $f, g \in S_{L^p}$. By the lemma,

$$\frac{|f|^p + |g|^p}{2} \geq \left(\left|\frac{f+g}{2}\right|^2 + C_p|f-g|^2\right)^{p/2},$$

so that, integrating and using the reverse Minkowski inequality for $L^{p/2}$,

$$1 \geq \left(\int_\Omega \left(\left|\frac{f+g}{2}\right|^2 + C_p|f-g|^2\right)^{p/2} d\mu\right)^{1/p}$$

$$\geq \left(\left(\int_\Omega \left|\frac{f+g}{2}\right|^p d\mu\right)^{2/p} + C_p\left(\int_\Omega |f-g|^p d\mu\right)^{2/p}\right)^{1/2}$$

$$= \left(\left\|\frac{f+g}{2}\right\|_p^2 + C_p \|f-g\|_p^2\right)^{1/2},$$

and the result follows from this. $\qquad \square$

Uniformly convex spaces have strong properties. Among them is the following, which provides a geometrical proof that L^p spaces are reflexive, for $1 < p < \infty$.

Theorem 9.8.2 *A uniformly convex Banach space is reflexive.*

Proof We consider the uniformly convex space $(E, \|.\|_E)$ as a subspace of its bidual E^{**}. We use the fact, implied by the Hahn–Banach theorem, that the unit sphere S_E is weak*-dense in $S_{E^{**}}$. Suppose that $\Phi \in S_{E^{**}}$. We shall show that for each $n \in \mathbf{N}$ there exists $x_n \in S_E$ with $\|x_n - \Phi\| \leq 1/n$. Thus $x_n \to \Phi$ in norm, so that $\Phi \in S_E$, since S_E is a closed subset of the complete space E.

Suppose that $n \in \mathbf{N}$. By uniform convexity, there exists $\eta > 0$ such that if $x, y \in S_E$ and $\|(x+y)/2\| > 1 - \eta$ then $\|x - y\| < 1/3n$. There exists $\phi \in S_{E^*}$ such that $|\Phi(\phi)| > 1 - \eta/2$. Let M be the non-empty set $\{x \in S_E : |\phi(x) - \Phi(\phi)| < \eta/2\}$. If $x, y \in M$ then $|\phi((x+y)/2) - \Phi(\phi)| < \eta/2$, so that $|\phi((x+y)/2)| > 1 - \eta$; thus $\|(x+y)/2\| > 1 - \eta$ and so $\|x - y\| < 1/3n$. Now pick $x_n \in M$. There exists $\psi \in S_{E^*}$ such that $|\psi(x_n) - \Phi(\psi)| > \|x_n - \Phi\| - 1/3n$. Let N be the non-empty set

$$\{x \in S_E : |\phi(x) - \Phi(\phi)| < \eta/2, |\psi(x) - \Phi(\psi)| < 1/3n\}.$$

Note that $N \subseteq M$. Pick $y_n \in N$. Then

$$\|x_n - \Phi\| \leq |\psi(x_n) - \Phi(\psi)| + 1/3n$$
$$\leq |\psi(x_n - y_n)| + |\psi(y_n) - \Phi(\psi)| + 1/3n$$
$$\leq 1/3n + 1/3n + 1/3n = 1/n.$$

$\square$

9.9 Notes and remarks

Fourier type was introduced by Peetre [Pee 69]. The introduction of Fourier type gives the first example of a general programme of classifying Banach spaces, according to various criteria. We begin with a result which holds for the scalars (in this case, the Hausdorff–Young inequality) and find that it holds for some, but not all, Banach spaces. The extent to which it holds for a particular space then provides a classification (in this case, Fourier type). Results of Kwapień [Kwa 72] show that a Banach space has Fourier type 2 if and only if it is isomorphic to a Hilbert space.

Uniform convexity provides another way of classifying Banach spaces. The uniform convexity of a Banach space $(E, \|.\|_E)$ is related to the behaviour of martingales taking values in E. Theorem 9.8.2 can be extended in an important way. We say that a Banach space $(E, \|.\|_E)$ is *finitely represented* in $(F, \|.\|_F)$ if the finite-dimensional subspaces of F look like finite-dimensional subspaces of E: if G is a finite-dimensional subspace of F and $\epsilon > 0$ then there is a linear mapping $T : G \to E$ such that

$$\|T(g)\| \leq \|g\| \leq (1 + \epsilon) \|T(g)\| \quad \text{for all } g \in G.$$

A Banach space $(E, \|.\|_E)$ is *super-reflexive* if every Banach space which is finitely represented in E. It is an easy exercise (Exercise 9.9) to show that a uniformly convex space is super-reflexive. A remarkable converse holds: if $(E, \|.\|_E)$ is super-reflexive, then E is linearly isomorphic to a uniformly convex Banach space, and indeed to a p-uniformly convex space, for some $2 \leq p < \infty$ ([Enf 73], [Pis 75]). More information about uniform convexity, and the dual notion of uniform smoothness, is given in [LiT 79].

Exercises

9.1   Suppose that $(A_0, \|.\|_{A_0})$ and $A_1, \|.\|_{A_1})$ form a compatible couple.
         (i) Show that if (x_n) is a sequence in $A_0 \cap A_1$ and that $x_n \to l_0$ in $(A_0, \|.\|_{A_0})$ and $x_n \to l_1$ in $(A_1, \|.\|_{A_1})$ then $l_0 = l_1$.
         (ii) Show that $(A_0 \cap A_1, \|.\|_{A_0 \cap A_1})$ is a Banach space.

(iii) Show that $\{(a, -a): a \in A_0 \cap A_1\}$ is a closed linear subspace of $(A_0, \|.\|_{A_0}) \times (A_1, \|.\|_{A_1})$.

(iv) Show that $(A_0 + A_1, \|.\|_{A_0+A_1})$ is a Banach space.

9.2　Suppose that f is a non-zero bounded continuous complex-valued function on the closed strip $\bar{S} = \{z = x + iy: 0 \leq x \leq 1\}$ which is analytic on the open strip $S = \{z = x + iy: 0 < x < 1\}$, and which satisfies $|f(iy)| \leq 1$ and $|f(1 + iy)| \leq 1$ for $y \in \mathbf{R}$. Show that

$$\phi(w) = \frac{1}{i\pi} \log\left(i\frac{1 - z}{1 + z}\right)$$

maps the unit disc D conformally onto S. What happens to the boundary of D?

Let $g(w) = f(\phi(w))$. Show that if $w \in D$ then

$$g(w) = \frac{1}{2\pi} \int_0^{2\pi} \frac{g(e^{i\theta})e^{i\theta}}{e^{i\theta} - w} d\theta.$$

Deduce that $|f(z)| \leq 1$ for $z \in S$.

9.3　Suppose that $1 < p, q < \infty$ and that $1/p + 1/q = 1 + 1/r > 1$. Let $\alpha = r'/p'$, $\beta = r'/q'$. Show that $\alpha + \beta = 1$, and that if $h \in L^{r'}$ and $\|h\|_{r'} = 1$ then $|h|^\alpha \in L^{p'}$, with $\||h|^\alpha\|_{p'} = 1$ and $|h|^\beta \in L^{q'}$, with $\||h|^\beta\|_{q'} = 1$. Use this to give a direct proof of Young's inequality.

9.4　Suppose that $a = (a_n) \in l_2(\mathbf{Z})$.

(i) Use the Cauchy–Schwarz inequality to show that

$$\sum_{n \neq m} \left|\frac{a_n}{m - n}\right| \leq \left(2\sum_{n=1}^\infty \frac{1}{n^2}\right)^{1/2} \|a\|_2.$$

(ii) Let T be the the saw-tooth function

$$T(e^{i\theta}) = \pi - \theta \quad \text{for } 0 < t < \pi,$$
$$= -\pi - \theta \quad \text{for } -\pi \leq t < 0,$$
$$= 0 \quad \text{for } t = 0.$$

Show that $\hat{T}_0 = 0$ and that $\hat{T}_n = -i/n$ for $n \neq 0$.

(iii) Calculate $\|T\|_2$, and use the Plancherel theorem to show that $\sum_{n=1}^\infty (1/n)^2 = \pi^2/6$.

(iv) Let $A(e^{i\theta}) = \sum_{m=-\infty}^\infty ia_n e^{in\theta}$, so that $A \in L^2(\mathbf{T})$ and $\hat{A}_n = ia_n$. Let $C = AT$. Show that $\|C\|_2 \leq \pi \|A\|_2$.

(v) What is $\hat{c}_n$? Show that

$$\sum_{m=-\infty}^{\infty} \left| \sum_{n \neq m} \frac{a_n}{m-n} \right|^2 \leq \pi^2 \|a\|_2^2 .$$

(vi) *(Hilbert's inequality for $l_2(\mathbf{Z})$).* Suppose that $b = (b_m) \in l_2(\mathbf{Z})$. Show that

$$\sum_{m=-\infty}^{\infty} \left| \sum_{n \neq m} \frac{a_n b_m}{m-n} \right| \leq \pi \|a\|_2 \|b\|_2 .$$

9.5   Verify that the generalized Clarkson inequalities establish Clarkson's original inequalities, in the following form. Suppose that $f, g \in L^p(\Omega, \Sigma, \nu)$. If $2 \leq p < \infty$ then

   (a) $2(\|f\|_p^p + \|g\|_p^p) \leq \|f+g\|_p^p + \|f-g\|_p^p \leq 2^{p-1}(\|f\|_p^p + \|g\|_p^p)$.

   (b) $2(\|f\|_p^p + \|g\|_p^p)^{p'-1} \leq \|f+g\|_p^{p'} + \|f-g\|_p^{p'}$.

   (c) $\|f+g\|_p^p + \|f-g\|_p^p \leq 2(\|f\|_p^{p'} + \|g\|_p^{p'})^{p-1}$.

   If $1 < p < 2$ then the inequalities are reversed.

9.6   Show that the restrictions of the norm topology and the weak topology to the unit sphere S_E of a uniformly convex space are the same. Does a weak Cauchy sequence in S_E converge in norm?

9.7   Say that a Banach space is of *strict Fourier type p* if it is of Fourier type p and $\|\mathcal{F}(f)\|_{L^{p'}(G',E)} \leq \|f\|_{L^p(G,E)}$ for all $f \in L^p(G, E)$, and all G. Show that a Banach space of strict Fourier type p is p'-uniformly convex.

9.8   Suppose that $f_1, \ldots, f_d \in L^p(\Omega, \Sigma, \nu)$ and that $\epsilon_1, \ldots, \epsilon_d$ are Bernoulli functions on D_2^d.

   (i) Show that if $1 < p < 2$ then

$$\left(\frac{1}{2^d} \sum_{w \in D_2^d} \left\| \sum_{j=1}^{d} \epsilon_j(w) f_j \right\|_p^{p'} \right)^{1/p'} \leq \left(\sum_{j=1}^{d} \|f_j\|_p^p \right)^{1/p} .$$

   (ii) Use a duality argument to show that if $2 < p < \infty$ then

$$\left(\frac{1}{2^d} \sum_{w \in D_2^d} \left\| \sum_{j=1}^{d} \epsilon_j(w) f_j \right\|_p^{p'} \right)^{1/p'} \geq \left(\sum_{j=1}^{d} \|f_j\|_p^p \right)^{1/p} .$$

9.9   Suppose that a Banach space $(E, \|.\|_E)$ is finitely represented in a uniformly convex Banach space $(F, \|.\|_F)$. Show that $(E, \|.\|_E)$ is uniformly convex. Show that a uniformly convex space is super-reflexive.

10

Real interpolation

10.1 The Marcinkiewicz interpolation theorem: I

We now turn to real interpolation, and in particular to the Marcinkiewicz theorem, stated by Marcinkiewicz in 1939 [Mar 39]. Marcinkiewicz was killed in the Second World War, and did not publish a proof; this was done by Zygmund in 1956 [Zyg 56]. The theorem differs from the Riesz–Thorin theorem in several respects: it applies to sublinear mappings as well as to linear mappings; the conditions at the end points of the range are weak type ones and the conclusions can apply to a larger class of spaces than the L^p spaces. But the constants in the inequalities are worse than those that occur in the Riesz–Thorin theorem.

We begin by giving a proof in the simplest case. This is sufficient for many purposes; the proof is similar to the proof of the more sophisticated result that we shall prove later, and introduces techniques that we shall use there.

Theorem 10.1.1 (The Marcinkiewicz interpolation theorem: I) *Suppose that $0 < p_0 < p < p_1 \leq \infty$, and that $T : L^{p_0}(\Omega, \Sigma, \mu) + L^{p_1}(\Omega, \Sigma, \mu) \to L^0(\Phi, T, \nu)$ is sublinear. If T is of weak type (p_0, p_0), with constant c_0, and weak type (p_1, p_1), with constant c_1, then T is of strong type (p, p), with a constant depending only on c_0, c_1, p_0, p_1 and p.*

Proof First we consider the case when $p_1 < \infty$. Suppose that $f \in L^p$. The idea of the proof is to decompose f into two parts, one in L^{p_0}, and one in L^{p_1}, and to let this decomposition vary. For $\alpha > 0$, let $E_\alpha = \{x : |f(x)| > \alpha\}$, let $g_\alpha = fI_{E_\alpha}$ and let $h_\alpha = f - g_\alpha$. Then $g_\alpha \in L^{p_0}$, since $\|g_\alpha\|_{p_0} \leq \mu(E_\alpha)^{1/p - 1/p_0} \|f\|_p$, by Hölder's inequality, and $h_\alpha \in L^{p_1}$, since $\int (|h_\alpha|/\alpha)^{p_1} d\mu \leq \int (|h_\alpha|/\alpha)^p d\mu$. Since $f = g_\alpha + h_\alpha$,

$$|T(f)| \leq |T(g_\alpha)| + |T(h_\alpha)|,$$

so that

$$(|T(f)| > \alpha) \subseteq (|T(g_\alpha)| > \alpha/2) \cup (|T(h_\alpha)| > \alpha/2)$$

and

$$\nu(|T(f)| > \alpha) \leq \nu(|T(g_\alpha)| > \alpha/2) + \nu(|T(h_\alpha)| > \alpha/2).$$

Thus

$$\int |T(f)|^p d\nu = p \int_0^\infty \alpha^{p-1} \nu(|T(f)| > \alpha)\, d\alpha$$

$$\leq p \int_0^\infty \alpha^{p-1} \nu(|T(g_\alpha)| > \alpha/2)\, d\alpha$$

$$+ p \int_0^\infty \alpha^{p-1} \nu(|T(h_\alpha)| > \alpha/2)\, d\alpha$$

$$= I_0 + I_1, \quad \text{say.}$$

Since T is of weak type (p_0, p_0),

$$I_0 \leq c_0 p \int_0^\infty \alpha^{p-1} \left(\int |g_\alpha(x)|^{p_0} d\mu(x) \right) / (\alpha/2)^{p_0}\, d\alpha$$

$$= 2^{p_0} c_0 p \int_0^\infty \alpha^{p-p_0-1} \left(\int_{(|f|>\alpha)} |f(x)|^{p_0} d\mu(x) \right) d\alpha$$

$$= 2^{p_0} c_0 p \int_\Omega |f(x)|^{p_0} \left(\int_0^{|f(x)|} \alpha^{p-p_0-1} d\alpha \right) d\mu(x)$$

$$= \frac{2^{p_0} c_0 p}{p - p_0} \int_\Omega |f(x)|^{p_0} |f(x)|^{p-p_0} d\mu(x) = \frac{2^{p_0} c_0 p}{p - p_0} \|f\|_p^p.$$

Similarly, since T is of weak type (p_1, p_1),

$$I_1 \leq c_1 p \int_0^\infty \alpha^{p-1} \left(\int |h_\alpha(x)|^{p_1} d\mu(x) \right) / (\alpha/2)^{p_1}\, d\alpha$$

$$= 2^{p_1} c_1 p \int_0^\infty \alpha^{p-p_1-1} \left(\int_{(|f|\leq\alpha)} |f(x)|^{p_1} d\mu(x) \right) d\alpha$$

$$= 2^{p_1} c_1 p \int_\Omega |f(x)|^{p_1} \left(\int_{|f(x)|}^\infty \alpha^{p-p_1-1} d\alpha \right) d\mu(x)$$

$$= \frac{2^{p_1} c_1 p}{p_1 - p} \int_\Omega |f(x)|^{p_1} |f(x)|^{p-p_1} d\mu(x) = \frac{2^{p_1} c_0 p}{p_1 - p} \|f\|_p^p.$$

Combining these two, we have the desired result.

Secondly, suppose that $p_1 = \infty$, and that $f \in L^p$. Write $f = g_\alpha + h_\alpha$, as before. Then $\|T(h_\alpha)\|_\infty \le c_1 \alpha$, so that if $|T(f)(x)| > 2c_1 \alpha$ then $|T(g_\alpha)(x)| > c_1 \alpha$. Thus, arguing as for I_0 above,

$$\int |T(f)|^p d\nu = p \int_0^\infty t^{p-1} \nu(|T(f)| > t) dt$$

$$= p(2c_1)^p \int_0^\infty \alpha^{p-1} \nu(|T(f)| > 2c_1 \alpha) \, d\alpha$$

$$\le p(2c_1)^p \int_0^\infty \alpha^{p-1} \nu(|T(g_\alpha)| > c_1 \alpha) \, d\alpha$$

$$\le c_1 p(2c_1)^p c_0 \int_0^\infty \alpha^{p-1} \left(\int_\Omega |g_\alpha|^{p_0} d\mu \right) \Big/ (c_1 \alpha)^{p_0} \, d\alpha$$

$$= 2^p p c_1^{p-p_0} c_0 \int_0^\infty \alpha^{p-p_0-1} \left(\int_{(|f|>\alpha)} |f|^{p_0} d\mu \right) d\alpha$$

$$= \frac{2^p p c_1^{p-p_0} c_0}{p - p_0} \|f\|_p^p.$$

$\square$

10.2 Lorentz spaces

In order to obtain stronger results, we need to spend some time introducing a new class of function spaces, the *Lorentz* spaces, and to prove a key inequality due to Hardy. The Lorentz spaces are a refinement of the L^p spaces, involving a second parameter; they fit well with the proof of the Marcinkiewicz theorem. The Muirhead maximal function $f^\dagger$ is an important ingredient in their study; for this reason we shall assume either that (Ω, Σ, μ) is atom-free or that it is discrete, with counting measure.

We begin with *weak-L^p*. If $0 < p < \infty$, the *weak-L^p* space $L_w^p = L_w^p(\Omega, \Sigma, \mu)$, or *Lorentz space* $L_{p,\infty} = L_{p,\infty}(\Omega, \Sigma, \mu)$, is defined as

$$L_{p,\infty} = \{ f \in L^1 + L^\infty : \|f\|_{p,\infty}^* = \sup_{\alpha > 0} \alpha (\mu(|f| > \alpha))^{1/p} < \infty \}.$$

Note that $\|f\|_{p,\infty}^* = \sup\{t^{1/p} f^*(t) : 0 < t < \mu(\Omega)\}$. This relates to weak type: a sublinear mapping T of a Banach space E into $M(\Omega, \Sigma, \mu)$ is of weak type (E, p) if and only if $T(E) \subseteq L_{p,\infty}$ and there exists a constant c such that $\|T(f)\|_{p,\infty}^* \le c \|f\|_E$. Note that, in spite of the notation, $\|.\|_{p,\infty}^*$ is not a norm (and in fact if $p \le 1$, there is no norm on $L_{p,\infty}$ equivalent to $\|.\|_{p,\infty}^*$). When $1 < p < \infty$ we can do better.

Proposition 10.2.1 *Suppose that $1 < p < \infty$. Then $f \in L_{p,\infty}$ if and only if*

$$\|f\|_{p,\infty}^\dagger = \sup\{t^{1/p} f^\dagger(t) \colon 0 < t < \mu(\Omega)\} < \infty.$$

Further $\|.\|_{p,\infty}^\dagger$ is a norm on $L_{p,\infty}$, and

$$\|f\|_{p,\infty}^* \leq \|f\|_{p,\infty}^\dagger \leq p' \|f\|_{p,\infty}^* .$$

$(L_{p,\infty}, \|.\|_{p,\infty}^\dagger)$ *is a rearrangement-invariant function space.*

Proof If $\|f\|_{p,\infty}^\dagger < \infty$, then since $f^* \leq f^\dagger$, $\|f\|_{p,\infty}^* \leq \|f\|_{p,\infty}^\dagger$ and $f \in L_{p,\infty}$. On the other hand, if $f \in L_{p,\infty}$ then

$$\int_0^t f^*(s)\, ds \leq \|f\|_{p,\infty}^* \int_0^t s^{-1/p}\, ds = p' \|f\|_{p,\infty}^* t^{1-1/p}$$

so that $t^{1/p} f^\dagger(t) \leq p' \|f\|_{p,\infty}^*$, and $\|f\|_{p,\infty}^\dagger \leq p' \|f\|_{p,\infty}^*$. Since the mapping $f \to f^\dagger$ is sublinear, $\|.\|_{p,\infty}^\dagger$ is a norm, and finally all the conditions for $(L_{p,\infty}, \|.\|_{p,\infty}^\dagger)$ to be a rearrangement invariant function space are readily verified. $\qquad\square$

The form of the weak-L^p spaces $L_{p,\infty}$ suggests a whole spectrum of rearrangement-invariant function spaces. We define the *Lorentz space* $L_{p,q}$ for $0 < p < \infty$ and $0 < q < \infty$ as

$$L_{p,q} = \left\{ f \colon \|f\|_{p,q}^* = \left(\frac{q}{p} \int_0^{\mu(\Omega)} t^{q/p} f^*(t)^q \frac{dt}{t} \right)^{1/q} < \infty \right\}.$$

Note that $\|f\|_{p,q}^*$ is the L^q norm of f^* with respect to the measure

$$(q/p)t^{q/p-1}\, dt = d(t^{q/p}).$$

Note also that $L_{p,p} = L^p$, with equality of norms. In general, however, $\|.\|_{p,q}$ is not a norm, and if $p < 1$ or $q < 1$ there is no equivalent norm. But if $1 < p < \infty$ and $1 \leq q < \infty$ then, as in Proposition 10.2.1, there is an equivalent norm. In order to prove this, we need Hardy's inequality, which is also at the heart of the general Marcinkiewicz interpolation theorem which we shall prove.

10.3 Hardy's inequality

Theorem 10.3.1 (Hardy's inequality) *Suppose that f is a non-negative measurable function on $[0, \infty)$. Let*

$$A_{\theta,\beta}(f)(t) = t^{-\beta} \int_0^t s^\theta f(s) \frac{ds}{s},$$

$$B_{\theta,\beta}(f)(t) = t^\beta \int_t^\infty s^\theta f(s) \frac{ds}{s},$$

for $-\infty < \theta < \infty$ and $\beta > 0$. If $1 \le q < \infty$ then

$$(i) \qquad \int_0^\infty (A_{\theta,\beta}(f)(t))^q \frac{dt}{t} \le \frac{1}{\beta^q} \int_0^\infty (t^{\theta-\beta} f(t))^q \frac{dt}{t},$$

and

$$(ii) \qquad \int_0^\infty (B_{\theta,\beta}(f)(t))^q \frac{dt}{t} \le \frac{1}{\beta^q} \int_0^\infty (t^{\theta+\beta} f(t))^q \frac{dt}{t}.$$

Proof We shall first prove this in the case where $\theta = 1$ and $q = 1$. Then

$$\int_0^\infty A_{1,\beta}(f)(t) \frac{dt}{t} = \int_0^\infty t^{-1-\beta} \left(\int_0^t f(u)\, du \right) dt$$

$$= \int_0^\infty \left(\int_u^\infty t^{-1-\beta}\, dt \right) f(u)\, du$$

$$= \frac{1}{\beta} \int_0^\infty u^{-\beta} f(u)\, du,$$

and so in this case we have equality.

Next, suppose that $\theta = 1$ and $1 < q < \infty$. We write $f(s) = s^{(\beta-1)/q'} s^{(1-\beta)/q'} f(s)$, and apply Hölder's inequality:

$$\int_0^t f(s)\, ds \le \left(\int_0^t s^{\beta-1}\, ds \right)^{1/q'} \left(\int_0^t s^{(1-\beta)q/q'} f(s)^q\, ds \right)^{1/q}$$

$$= \left(\frac{t^\beta}{\beta} \right)^{1/q'} \left(\int_0^t s^{(1-\beta)q/q'} f(s)^q\, ds \right)^{1/q},$$

so that, since $q/q' = q - 1$,

$$(A_{1,\beta}(f)(t))^q \le \frac{1}{\beta^{q-1}} t^{-\beta} \int_0^t s^{(1-\beta)(q-1)} f(s)^q\, ds.$$

Thus

$$
\int_0^\infty (A_{1,\beta}(f)(t))^q \, \frac{dt}{t} \leq \frac{1}{\beta^{q-1}} \int_0^\infty t^{-\beta-1} \left(\int_0^t s^{(1-\beta)(q-1)} f(s)^q \, ds \right) dt
$$

$$
= \frac{1}{\beta^{q-1}} \int_0^\infty \left(\int_s^\infty t^{-\beta-1} \, dt \right) s^{(1-\beta)(q-1)} f(s)^q \, ds
$$

$$
= \frac{1}{\beta^q} \int_0^\infty s^{-\beta+(1-\beta)(q-1)} f(s)^q \, ds
$$

$$
= \frac{1}{\beta^q} \int_0^\infty (s^{(1-\beta)} f(s))^q \, \frac{ds}{s}.
$$

The general form of (i) now follows by applying this to the function $s^{\theta-1} f(s)$.
To prove (ii), we set $g(u) = f(1/u)$ and $u = 1/s$. Then

$$
B_{\theta,\beta}(f)(t) = t^\beta \int_t^\infty s^\theta f(s) \, \frac{ds}{s}
$$

$$
= t^\beta \int_0^{1/t} u^{-\theta} g(u) \, \frac{du}{u}
$$

$$
= A_{-\theta,\beta}(g)(1/t),
$$

so that

$$
\int_0^\infty (B_{\theta,\beta}(f)(t))^q \, \frac{dt}{t} = \int_0^\infty (A_{-\theta,\beta}(g)(1/t))^q \, \frac{dt}{t} = \int_0^\infty (A_{-\theta,\beta}(g)(t))^q \, \frac{dt}{t}
$$

$$
\leq \frac{1}{\beta^q} \int_0^\infty (t^{-\theta-\beta} g(t))^q \, \frac{dt}{t} = \frac{1}{\beta^q} \int_0^\infty (t^{\theta+\beta} f(t))^q \, \frac{dt}{t}.
$$

$\square$

If we set $\theta = 1$ and apply the result to f^*, we obtain the following:

Corollary 10.3.1 *If $f \in (L^1 + L^\infty)(\Omega, \Sigma, \mu)$ then*

$$
\int t^{(1-\beta)q} f^\dagger(t)^q \, \frac{dt}{t} \leq \frac{1}{\beta^q} \int t^{(1-\beta)q} f^*(t)^q \, \frac{dt}{t}.
$$

Note that if we set $\theta = 1$ and $\beta = 1/q'$, we obtain the Hardy–Riesz inequality.

10.4 The scale of Lorentz spaces

We now have the following result, which complements Proposition 10.2.1.

Theorem 10.4.1 *Suppose that $1 < p < \infty$, $1 \leq q < \infty$. Then $f \in L_{p,q}$ if and only if*

$$\|f\|_{p,q}^{\dagger} = \left(\frac{q}{p} \int_0^{\mu(\Omega)} t^{q/p} f^{\dagger}(t)^q \frac{dt}{t} \right)^{1/q} < \infty.$$

Further $\|.\|_{p,q}^{\dagger}$ is a norm on $L_{p,q}$, and

$$\|f\|_{p,q}^* \leq \|f\|_{p,q}^{\dagger} \leq p' \|f\|_{p,q}^*.$$

$(L_{p,q}, \|.\|_{p,q}^{\dagger})$ is a rearrangement-invariant function space.

Proof The result follows from the corollary to Hardy's inequality, setting $\beta = 1/p'$. $\|f\|_{p,q}^{\dagger}$ is a norm, since $f^{\dagger}$ is sublinear, and the rest follows as in Proposition 10.2.1. $\qquad\square$

What is the relation between the various $L_{p,q}$ spaces, as the indices vary? First, let us keep p fixed, and let q vary.

Theorem 10.4.2 *If $0 < p < \infty$ and $1 \leq q < r \leq \infty$ then $L_{p,q} \subseteq L_{p,r}$, $\|f\|_{p,r}^* \leq \|f\|_{p,q}^*$ and $\|f\|_{p,r}^{\dagger} \leq \|f\|_{p,q}^{\dagger}$,*

Proof If $f \in L_{p,q}^{\dagger}$ and $0 < t < \mu(\Omega)$ then

$$t^{1/p} f^{\dagger}(t) = \left(\frac{q}{p} \int_0^t (s^{1/p} f^{\dagger}(t))^q \frac{ds}{s} \right)^{1/q} \leq \left(\frac{q}{p} \int_0^{\mu(\Omega)} (s^{1/p} f^{\dagger}(s))^q \frac{ds}{s} \right)^{1/q}$$

$$= \|f\|_{p,q}^{\dagger},$$

so that $L_{p,q} \subseteq L_{p,\infty}$, and the inclusion is norm decreasing. The same argument works for the norms $\|f\|_{p,q}^*$ and $\|f\|_{p,\infty}^*$.

Suppose that $1 \leq q < r < \infty$. Since

$$\frac{q}{p} \int_0^{\infty} (t^{1/p} h(t))^q \frac{dt}{t} = q \int_0^{\infty} (th(t^p))^q \frac{dt}{t},$$

for h a non-negative measurable function, we need only show that if g is a decreasing function on $[0, \infty)$ then

$$\left(q \int_0^{\infty} t^q g(t)^q \frac{dt}{t} \right)^{1/q}$$

is a decreasing function of q.

We first consider the case where $1 = q < r$. We can approximate g from below by an increasing sequence of decreasing step functions, and so it is enough to consider such functions. We take g of the form

$$g = \sum_{j=1}^{J} a_j I_{[0,t_j]}, \quad \text{where } a_j > 0 \text{ and } t_j > 0 \text{ for } 1 \le j \le J.$$

Then, applying Minkowski's inequality,

$$\left(r \int_0^\infty t^r (g(t))^r \frac{dt}{t} \right)^{1/r} \le \sum_{j=1}^{J} \left(r a_j^r \int_0^{t_j} t^{r-1}\, dt \right)^{1/r}$$

$$= \sum_{j=1}^{J} a_j t_j = \int_0^\infty t g(t) \frac{dt}{t}.$$

Next, suppose that $1 < q < r$. Let $\lambda = r/q$, let $h(t) = (g(t^{1/q}))^q$ and let $u = t^q$, so that $h(u) = (g(t))^q$. Then changing variables, and using the result above,

$$\left(r \int_0^\infty t^r (g(t))^r \frac{dt}{t} \right)^{1/r} = \left(\lambda \int_0^\infty u^\lambda (h(u))^\lambda \frac{du}{u} \right)^{1/q\lambda}$$

$$\le \left(\lambda \int_0^\infty u h(u) \frac{du}{u} \right)^{1/q}$$

$$= \left(q \int_0^\infty t^q (g(t))^q \frac{dt}{t} \right)^{1/q}. \qquad \square$$

What happens as p varies? If (Ω, Σ, μ) is non-atomic and $\mu(\Omega) = \infty$, we can expect no patterns of inclusions, since there is none for the spaces $L^p = L_{p,p}$. When (Ω, Σ, μ) is non-atomic and of finite measure, we have the following.

Proposition 10.4.1 *Suppose that (Ω, Σ, μ) is non-atomic and that $\mu(\Omega) < \infty$. Then if $0 < p_1 < p_2 \le \infty$, $L_{p_2,q_2} \subseteq L_{p_1,q_1}$ for any q_1, q_2, with continuous inclusion.*

Proof Because of Theorem 10.4.2, it is enough to show that $L_{p_2,\infty} \subseteq L_{p_1,1}$, with continuous inclusion. But if $f \in L_{p_2,\infty}$ then

$$\frac{1}{p_1} \int_0^{\mu(\Omega)} t^{1/p_1} f^*(t) \frac{dt}{t} \le \left(\frac{1}{p_1} \int_0^{\mu(\Omega)} t^{1/p_1 - 1/p_2} \frac{dt}{t} \right) \|f\|_{p_2,\infty}^*$$

$$= \frac{p_2}{p_2 - p_1} (\mu(\Omega))^{1/p_1 - 1/p_2} \|f\|_{p_2,\infty}^*. \qquad \square$$

When (Ω, Σ, μ) is atomic, we can take $\Omega = \mathbf{N}$. We then denote the Lorentz space by $l_{p,q}$. In this case, as you might expect, the inclusions go the other way.

Proposition 10.4.2 *If* $0 < p_1 < p_2 \leq \infty$, *then* $l_{p_1,q_1} \subseteq l_{p_2,q_2}$ *for any* q_1, q_2, *with continuous inclusion.*

Proof Again it is enough to show that $l_{p_1,\infty} \subseteq l_{p_2,1}$, with continuous inclusion. But if $x \in l_{p_1,\infty}$ then

$$\frac{1}{p_2} \sum_{n=1}^{\infty} n^{1/p_2-1} x_n^* \leq \left(\frac{1}{p_2} \sum_{n=1}^{\infty} n^{1/p_2-1/p_1-1} \right) \|x\|_{p_1,\infty}^* .$$

□

10.5 The Marcinkiewicz interpolation theorem: II

We now come to a more general version of the Marcinkiewicz interpolation theorem: we weaken the conditions, and obtain a stronger result. The proof that we give is due to Hunt [Hun 64].

Theorem 10.5.1 (The Markinkiewicz interpolation theorem: II)
Suppose that $1 \leq p_0 < p_1 < \infty$ *and* $1 \leq q_0, q_1 \leq \infty$, *with* $q_0 \neq q_1$, *and that* T *is a sublinear operator from* $L_{p_0,1}(\Omega', \Sigma', \mu') + L_{p_1,1}(\Omega', \Sigma', \mu')$ *to* $M_1(\Omega, \Sigma, \mu)$ *which is of weak types* $(L_{p_0,1}, q_0)$ *and* $(L_{p_1,1}, q_1)$. *Suppose that* $0 < \theta < 1$, *and set*

$$\frac{1}{p} = \frac{1-\theta}{p_0} + \frac{\theta}{p_1}, \quad \frac{1}{q} = \frac{1-\theta}{q_0} + \frac{\theta}{q_1}.$$

Then if $1 \leq r \leq \infty$ *there exists a constant* B, *depending only on* p_0, p_1, q_0, q_1, θ, r *and the weak type constants, such that* $\|T(f)\|_{q,r}^* \leq B \|f\|_{p,r}^*$, *for* $f \in L_{p,r}$.

Corollary 10.5.1 *If* $q \geq p$ *then there exists a constant* B *such that* $\|T(f)\|_q \leq B \|f\|_p$.

Hunt [Hun 64] has shown that the result is false if $q < p$.

Proof Before beginning the proof, some comments are in order. First, it is easy to check that $L_{p,r} \subseteq L_{p_0,1} + L_{p_1,1}$ for $p_0 < p < p_1$ and $1 \leq r \leq \infty$. Second, we shall only give the proof when all of the indices are finite; a separate proof is needed when one or more index is infinite, but the proofs are easier. Thirdly, we shall not keep a close account of the constants that accrue, but will introduce constants C_i without comment.

We set

$$\gamma = \frac{1/q_0 - 1/q_1}{1/p_0 - 1/p_1} \left[= \frac{1/q_0 - 1/q}{1/p_0 - 1/p} = \frac{1/q - 1/q_1}{1/p - 1/p_1} \right].$$

Note that γ can be positive or negative.

Suppose that $f \in L_{p,r}$. We split f in much the same way as in Theorem 10.1.1. We set

$$g_\alpha(x) = f(x) \quad \text{if } |f(x)| > f^*(\alpha^\gamma),$$
$$= 0 \qquad \text{otherwise,}$$

and set $h_\alpha = f - g_\alpha$.

Since T is sublinear, $|T(f)| \leq |T(g_\alpha) + |T(h_\alpha)|$, and so $(T(f))^*(\alpha) \leq T(g_\alpha)^*(\alpha/2) + T(h_\alpha)^*(\alpha/2)$. Thus

$$\|T(f)\|_{q,r}^* \leq \left(\frac{r}{q} \int_0^\infty (\alpha^{1/q}(T(g_\alpha)^*(\alpha/2) + T(h_\alpha)^*(\alpha/2))^r \frac{d\alpha}{\alpha} \right)^{1/r}$$

$$\leq \left(\frac{r}{q} \int_0^\infty (\alpha^{1/q}T(g_\alpha)^*(\alpha/2))^r \frac{d\alpha}{\alpha} \right)^{1/r}$$

$$+ \left(\frac{r}{q} \int_0^\infty (\alpha^{1/q}T(h_\alpha)^*(\alpha/2))^r \frac{d\alpha}{\alpha} \right)^{1/r} = J_0 + J_1, \text{ say.}$$

We consider each term separately.

Since T is of weak type $(L_{p_0,1}, q_0)$,

$$T(g_\alpha)^*(\alpha/2) \leq C_0 \left(\frac{2}{\alpha} \right)^{q_0} \|g_\alpha\|_{p_0,1}^*.$$

But $g_\alpha^* \leq f^*.I_{[0,\alpha^\gamma)}$, so that

$$\|g_\alpha\|_{p_0,1}^* \leq \frac{1}{p_0} \int_0^{\alpha^\gamma} s^{1/p_0} f^*(s) \frac{ds}{s}.$$

Thus

$$J_0^r \le C_1 \int_0^\infty \left(\alpha^{1/q-1/q_0} \int_0^{\alpha^\gamma} s^{1/p_0} f^*(s) \frac{ds}{s} \right)^r \frac{d\alpha}{\alpha}$$

$$= C_2 \int_0^\infty \left(u^{1/p-1/p_0} \int_0^u s^{1/p_0} f^*(s) \frac{ds}{s} \right)^r \frac{du}{u}$$

$$\text{(where } u = \alpha^\gamma\text{)}$$

$$= C_2 \int_0^\infty \left(A_{1/p-1/p_0,1/p_0}(f^*)(u) \right)^r \frac{du}{u}$$

$$\le C_3 \int_0^\infty \left(u^{1/p} f^*(u) \right)^r \frac{du}{u} \quad \text{(using Hardy's inequality)}$$

$$= C_4 (\|f\|_{p,r}^*)^r.$$

Similarly, since T is of weak type $(L_{p_1,1}, q_1)$,

$$T(h_\alpha)^*(\alpha/2) \le C_5 \left(\frac{2}{\alpha} \right)^{q_1} \|h_\alpha\|_{p_1,1}^*.$$

But

$h_\alpha^* \le f^*(\alpha^\gamma)$ and $h_\alpha^* \le f^*$, so that

$$\|h_\alpha\|_{p_1,1} \le \alpha^{\gamma/p_1} f^*(\alpha^\gamma) + \frac{1}{p_1} \int_{\alpha^\gamma}^\infty s^{1/p_1} f^*(s) \frac{ds}{s}.$$

Thus

$$J_1^r \le C_6 \int_0^\infty \left(\alpha^{1/q-1/q_1} \left(\alpha^{1/\gamma p_1} f^*(\alpha^\gamma) + \frac{1}{p_1} \int_{\alpha^\gamma}^\infty s^{1/p_1} f^*(s) \frac{ds}{s} \right) \right)^r \frac{d\alpha}{\alpha},$$

so that $J_1 \le C_7(K_1 + K_2)$, where

$$K_1 = \left(\int_0^\infty (\alpha^{1/q-1/q_1+\gamma/p_1} f^*(\alpha^\gamma))^r \frac{d\alpha}{\alpha} \right)^{1/r}$$

$$= \left(\int_0^\infty (u^{1/p} f^*(u))^r \frac{du}{u} \right)^{1/r} \quad \text{(where } u = \alpha^\gamma\text{)}$$

$$\le C_8 \|f\|_{p,r}^*,$$

and

$$
\begin{aligned}
K_2^r &= \int_0^\infty \left(\alpha^{1/q-1/q_1} \int_{\alpha^\gamma}^\infty s^{1/p_1} f^*(s) \, \frac{ds}{s} \right)^r \frac{d\alpha}{\alpha} \\
&= \frac{1}{|\gamma|^{1/r}} \int_0^\infty \left(u^{1/p-1/p_1} \int_u^\infty s^{1/p_1} f^*(s) \, \frac{ds}{s} \right)^r \frac{du}{u} \quad \text{(where } u = \alpha^\gamma) \\
&= \frac{1}{|\gamma|^{1/r}} \int_0^\infty \left(B_{1/p-1/p_1, 1/p_1}(u) \right)^r \frac{du}{u} \\
&\le C_9 (\|f\|_{p,r}^*)^r,
\end{aligned}
$$

using Hardy's inequality again. This completes the proof. □

We have the following extension of the Hausdorff–Young inequality.

Corollary 10.5.2 (Paley's inequality) *If G is a locally compact abelian group then the Fourier transform is a continuous linear mapping from $L^p(G)$ to the Lorentz space $L_{p',p}(G')$, for $1 < p < 2$.*

In detail, when $G = \mathbf{R}^d$ this says that there are constants C_p and K_p such that

$$
\left(\int_{\mathbf{R}^d} |\hat{f}(u)|^p u^{d(p-2)} \, du \right)^{1/p} \le K_p \left(\int_0^\infty |(\hat{f})^\dagger(t)|^p t^{p-2} \, dt \right)^{1/p} \le K_p C_p \|f\|_p .
$$

(Paley's proof was different!)

10.6 Notes and remarks

The Marcinkiewicz theorem has inspired a whole theory of interpolation spaces. This theory is developed in detail in the books by Bergh and Löfström [BeL 76] and Bennett and Sharpley [BeS 88].

The Lorentz spaces were introduced by Lorentz [Lor 50]. More details can be found in [Hun 66], [StW 71] and [BeS 88].

Exercises

10.1 Show that the simple functions are dense in $L_{p,q}$ when p and q are finite.

10.2 Suppose that $(E, \|.\|_E)$ is a Banach function space, and that $1 \le p < \infty$. Suppose that $\|I_A\| \le \mu(A)^{1/p}$ for all sets A of finite measure. Show that $L_{p,1} \subseteq E$ and that the inclusion mapping is continuous.

10.3 Suppose that $(E, \|.\|_E)$ is a Banach function space in which the simple functions are dense, and that $1 \leq p < \infty$. Suppose that $\|I_A\| \geq \mu(A)^{1/p}$ for all sets A of finite measure. Show that $E \subseteq L_{p,\infty}$ and that the inclusion mapping is continuous.

10.4 Prove Theorem 10.5.1 when $r = \infty$, and when q_0 or q_1 is infinite.

11

The Hilbert transform, and Hilbert's inequalities

11.1 The conjugate Poisson kernel

We now consider the Hilbert transform, one of the fundamental operators of harmonic analysis. We begin by studying the Hilbert transform on the real line **R**, and show how the results that we have established in earlier chapters are used to establish its properties. We then more briefly discuss the Hilbert transform on the circle **T**. Finally we show how the techniques that we have developed can be applied to singular integral operators on $\mathbf{R}^d$.

Suppose that $f \in L^p(\mathbf{R})$, where $1 \leq p < \infty$. Recall that in Section 8.11 we used the Poisson kernel

$$P(x,t) = P_t(x) = \frac{t}{\pi(x^2 + t^2)}$$

to construct a harmonic function $u(x,t) = u_t(x) = (P_t \star f)(x)$ on the upper half space $H^2 = \{(x,t) : t > 0\}$ such that $u_t \in L^p$, and $u_t \to f$ in L^p norm and almost everywhere (Theorem 8.11.1 and Corollary 8.11.3). We can however think of H^2 as the upper half-plane $C^+ = \{z = x + it : t > 0\}$ in the complex plane, and then u is the real part of an analytic function $u + iv$ on C^+, unique up to a constant. We now turn to the study of this function. We start with the Poisson kernel. If $z = x + it$ then

$$\frac{i}{\pi z} = \frac{t}{\pi(x^2 + t^2)} + \frac{ix}{\pi(x^2 + t^2)}$$
$$= P(x,t) + iQ(x,t) = P_t(x) + iQ_t(x).$$

P is the Poisson kernel, and Q is the *conjugate Poisson kernel*. Since $(P + iQ)(x + it)$ is analytic in $x + it$, Q is harmonic. Note that (Q_t) is *not* an approximate identity: it is an odd function and is not integrable. On the other hand, $Q_t \in L^p(\mathcal{R})$ for $1 < p \leq \infty$, and for each such p there exists k_p such that $\|Q_t\|_p \leq k_p/t^{1/p'}$. This is easy to see when $p = \infty$ since

$|Q_t(x)| \le Q_t(t) = 1/2\pi t$. If $1 < p < \infty$,

$$\int_{-\infty}^{\infty} |Q_t(x)|^p \, dx = \frac{2}{\pi} \left(\int_0^t \frac{x^p}{(x^2 + t^2)^p} \, dx + \int_t^{\infty} \frac{x^p}{(x^2 + t^2)^p} \, dx \right)$$

$$\le \frac{2}{\pi} \left(\int_0^t \frac{dx}{t^p} + \int_t^{\infty} \frac{dx}{x^p} \right) = \frac{2}{\pi} \left(1 + \frac{1}{p-1} \right) \frac{1}{t^{p-1}}$$

$$= \frac{2p}{\pi(p-1)t^{p-1}} = k_p^p / t^{p/p'}.$$

If $f \in L^p(\mathbf{R})$, where $1 \le p < \infty$, we can therefore define

$$Q_t(v) = v_t(x) = v(x,t) = Q_t \star f = \frac{1}{\pi} \int_{-\infty}^{\infty} \frac{y f(x-y)}{y^2 + t^2} \, dy,$$

and then $u + iv$ is analytic. Thus v is harmonic in (x,t). Further,

$$|v(x,t)| \le \|Q_t\|_{p'} \|f\|_p \le k_{p'} \|f\|_p / t^{1/p},$$

and v is well-behaved at infinity. But what happens when $t \to 0$?

11.2 The Hilbert transform on $L^2(\mathbf{R})$

We first consider the simplest case, when $p = 2$. Since each Q_t is a convolution operator, it is sensible to consider Fourier transforms. Simple calculations, using the calculus of residues, and Jordan's lemma (Exercise 11.1), show that

$$\mathcal{F}(P_t)(\xi) = \hat{P}_t(\xi) = e^{-2\pi t |\xi|} \quad \text{and} \quad \mathcal{F}(Q_t)(\xi) = \hat{Q}_t(\xi) = -i\,\mathrm{sgn}(\xi)e^{-2\pi t |\xi|}.$$

Here, an essential feature is that the Fourier transforms of Q_t are uniformly bounded. Then

$$\hat{v}_t(\xi) = \hat{Q}_t(\xi)\hat{f}(\xi) = -i\,\mathrm{sgn}(\xi)e^{-2\pi t |\xi|}\hat{f}(\xi),$$

so that

$$\|v_t\|_2 = \|\hat{v}_t\|_2 \le \left\| \hat{f} \right\|_2 = \|f\|_2,$$

by Plancherel's theorem. Let

$$w(\xi) = -i\,\mathrm{sgn}(\xi)\hat{f}(\xi).$$

Then $w \in L^2$ and $\|w\|_2 = \left\| \hat{f} \right\|_2 = \|f\|_2$. Further,

$$|\hat{v}_t(\xi) - w(\xi)|^2 \le 4|w(\xi)|^2,$$

so that by the theorem of dominated convergence, $\hat{v}_t \to w$ in L^2-norm. We define the *Hilbert transform* $H(f)$ to be the inverse Fourier transform of w. Then by Plancherel's theorem again, $\|H(f)\|_2 = \|f\|_2$ and $v_t \to H(f)$ in L^2-norm. Further $\hat{v}_t = \hat{P}_t \widehat{H(f)}$, and so $v_t = P_t(H(f))$. Thus $v_t \to H(f)$ in L^2 norm and almost everywhere, by Theorem 8.11.1 and Corollary 8.11.3. Finally,

$$\widehat{H^2(f)}(\xi) = -i\operatorname{sgn}(\xi)\widehat{H(f)}(\xi) = -\hat{f}(\xi),$$

so that H is an isometry of $L^2(\mathbf{R})$ onto $L^2(\mathbf{R})$. Let us sum up what we have shown. Let

$$Q^*(f)(x) = \sup_{t>0} |Q_t(f)(x)| = \sup_{t>0} |v_t(x)|.$$

Q^* is sublinear.

Theorem 11.2.1 *The Hilbert transform H is an isometry of $L^2(\mathbf{R})$ onto $L^2(\mathbf{R})$, and $H^2(f) = -f$, for $f \in L^2(\mathbf{R})$. $Q_t(f) = P_t(H(f))$, so that $Q_t(f) \to H(f)$ in norm, and almost everywhere, and $\|Q^*(f)\|_2 \leq 2\|f\|_2$.*

We have defined the Hilbert transform in terms of Fourier transforms. Can we proceed more directly? As $t \to 0$, $Q_t(x) \to 1/\pi x$ and $\hat{Q}_t(\xi) \to -i\operatorname{sgn}(\xi)$. This suggests that we should define $H(f)$ as $h \star f$, where $h(x) = 1/\pi x$. But h has a singularity at the origin, which we must deal with. Let us set $h_\epsilon(x) = h(x)$ if $|x| \geq \epsilon$ and $h_\epsilon(x) = 0$ if $|x| < \epsilon$. Then h_ϵ is not integrable, but it is in L^p for $1 < p \leq \infty$. Thus if $f \in L^2$ we can define

$$H_\epsilon(f)(x) = (h_\epsilon \star f)(x) = \frac{1}{\pi} \int_{|y|>\epsilon} \frac{f(y)}{x-y}\, dy,$$

and $|H_\epsilon(f)(x)| \leq \|h_\epsilon\|_2 \|f\|_2$.

Although neither Q_1 nor H_1 is integrable, their difference is, and it can be dominated by a bell-shaped function. This allows us to transfer results from $Q_t(f)$ to $H_\epsilon(f)$. Let $H^*(f)(x) = \sup_{\epsilon>0} |H_\epsilon(f)(x)|$. H^* is sublinear; it is called the *maximal Hilbert transform*.

Proposition 11.2.1 (Cotlar's inequality: $p = 2$) *Suppose that $f \in L^2(\mathbf{R})$. Then $H^*(f) \leq m(H(f)) + 2m(f)$, and H^* is of strong type $(2,2)$.*

Proof Let $\eta = \log(e/2)$, and let

$$L(x) = \tfrac{1}{2} + \eta(1 - |x|) \quad \text{for } |x| \leq 1,$$

$$= \left| \frac{1}{x} - \frac{x}{x^2+1} \right| \quad \text{for } |x| > 1.$$

Then L is a continuous even integrable function on $\mathbf{R}$, and it is strictly decreasing on $[0, \infty)$. $\|L\|_1 = 1 + \eta + \log 2 = 2$. Let $\Phi = L/2$. Then Φ is a bell-shaped approximate identity, and $|h_\epsilon - Q_\epsilon| \le 2\Phi_\epsilon$. Thus if $f \in L^2$, $|H_\epsilon(f)| \le |Q_\epsilon(f)| + 2m(f)$, by Theorem 8.11.2. But $|Q_\epsilon(f)| = |P_\epsilon(H(f))| \le m(H(f))$, again by Theorem 8.11.2. Thus $H^*(f) \le m(H(f)) + 2m(f)$. By Theorem 8.5.1, H^* is of strong type $(2,2)$. $\square$

Theorem 11.2.2 *Suppose that $f \in L^2(\mathbf{R})$. Then $H_\epsilon(f) \to H(f)$ in L^2 norm, and almost everywhere.*

The limit

$$\lim_{\epsilon \to 0} \frac{1}{\pi} \int_{|y| > \epsilon} \frac{f(y)}{x - y} \, dy$$

is the *Cauchy principal value* of $\int f(y)/(x - y) \, dy$.

Proof If f is a step function, $H_\epsilon(f) - Q_\epsilon(f) \to 0$ except at the points of discontinuity of f. Thus it follows from Theorem 8.4.2 that if $f \in L^2$ then $H_\epsilon(f) - Q_\epsilon(f) \to 0$ almost everywhere, and so $H_\epsilon(f) \to f$ almost everywhere. Since $|H_\epsilon(f) - Q_\epsilon(f)|^2 \le 4(m(f))^2$, it follows from the theorem of dominated convergence that $\|H_\epsilon(f) - Q_\epsilon(f)\|_2 \to 0$, and so $H_\epsilon(f) \to f$ in L^2 norm. $\square$

11.3 The Hilbert transform on $L^p(\mathbf{R})$ for $1 < p < \infty$

What about other values of p? The key step is to establish a weak type $(1, 1)$ inequality: we can then use Marcinkiewicz interpolation and duality to deal with other values of p. Kolmogoroff [Kol 25] showed that the mapping $f \to H(f)$ is of weak type $(1,1)$, giving a proof which is a *tour de force* of argument by contradiction. Subsequent proofs have been given, using the harmonicity of the kernels, and the analyticity of $P + iQ$. We shall however introduce techniques due to Calderón and Zygmund [CaZ 52], applying them to the Hilbert transform. These techniques provide a powerful tool for studying other more general singular integral operators, and we shall describe these at the end of the chapter.

Theorem 11.3.1 *The mapping $f \to Q^*(f)$ is of weak type (1,1).*

Proof By Theorem 11.2.1, Q^* is of strong type $(2, 2)$. Suppose that $f \in L^1$. Without loss of generality we need only consider $f \geq 0$. We consider the dyadic filtration $(\mathcal{F}_j)$, and set $f_j = \mathcal{E}(f|\mathcal{F}_j)$.

Suppose that $\alpha > 0$. Let τ be the stopping time $\tau = \inf\{j: f_j > \alpha\}$, as in Doob's lemma. Since $f_j \leq 2^j \|f\|_1$, $\tau > -\infty$. We set $M_j = (\tau = j)$, $M = \cup_j(M_j) = (\tau < \infty)$ and $L = (\tau = \infty)$. We define

$$g(x) = f(x) \text{ if } x \in L,$$
$$= f_j(x) \text{ if } x \in M_j.$$

The function g is the *good* part of f; note that $\|g\|_1 = \|f\|_1$. The function $b = f - g$ is the *bad* part of f; $\|b\|_1 \leq 2\|f\|_1$. Since

$$(|Q^*(f)| > \alpha) \subseteq (|Q^*(g)| > \alpha/2) \cup (|Q^*(b)| > \alpha/2),$$

we can consider the two parts separately.

We begin with the good part. If $x \in M_j$, then $f_{j-1}(x) \leq \alpha$, so that, since $f \geq 0$, $f_j(x) \leq 2\alpha$. If $x \in L$, $f_j(x) \leq \alpha$ for all j, so that by the martingale convergence theorem, $f(x) \leq \alpha$ for almost all $x \in L$. Consequently $\|g\|_\infty \leq 2\alpha$.

Applying Doob's lemma, $\lambda(M) \leq \|f\|_1/\alpha$, and so

$$\int g^2 \, d\lambda = \int_L g^2 \, d\lambda + \int_M g^2 \, d\lambda$$
$$\leq \alpha \int_L g \, d\lambda + \left(\frac{\|f\|_1}{\alpha}\right) 4\alpha^2$$
$$\leq 5\alpha \|f\|_1,$$

so that $\|Q^*(g)\|_2^2 \leq 4\|g\|_2^2 \leq 20\alpha\|f\|_1$. Thus, by Markov's inequality,

$$\lambda(|Q^*(g)| > \alpha/2) \leq (20\alpha\|f\|_1)(2/\alpha)^2$$
$$= 80\|f\|_1/\alpha.$$

We now turn to the bad part b. M is the union of a disjoint sequence (E_k) of dyadic intervals, for each of which $\int_{E_k} b \, d\lambda = 0$. Let F_k be the interval with the same mid-point as E_k, but two times as long, and let $N = \cup_k F_k$. Then

$$\lambda(N) \leq \sum_k \lambda(F_k) = 2 \sum_k \lambda(E_k) = 2\lambda(M) \leq 2\|f\|_1/\alpha.$$

It is therefore sufficient to show that

$$\lambda((|Q^*(b)| > \alpha/2) \cap C(N)) \leq 8\|f\|_1/\alpha,$$

and this of course follows if we show that

$$\int_{C(N)} |Q^*(b)| \, d\lambda \le 4 \, \|f\|_1 \, .$$

Let $b_k = b.I_{E_k}$. Then $b = \sum_k b_k$ and $v_t(b)(x) = \sum_k v_t(b_k)(x)$ for each x. Consequently,

$$Q^*(b) = \sup_{t>0} |v_t(b)| \le \sup_{t>0} \sum_k |v_t(b_k)| \le \sum_k Q^*(b_k).$$

Thus

$$\int_{C(N)} Q^*(b) \, d\lambda \le \int_{C(N)} \sum_k Q^*(b_k) \, d\lambda = \sum_k \int_{C(N)} Q^*(b_k) \, d\lambda$$

$$\le \sum_k \int_{C(F_k)} Q^*(b_k) \, d\lambda.$$

We now need to consider $\int_{C(F_k)} Q^*(b_k) \, d\lambda$ in detail. Let $E_k = (x_0 - l, x_0 + l]$, so that $F_k = (x_0 - 2l, x_0 + 2l]$. If $x_0 + y \in C(F_k)$ then

$$v_t(b_k)(x_0 + y) = \int_{-l}^{l} b_k(x_0 + u) Q_t(y - u) \, d\lambda(u)$$

$$= \int_{-l}^{l} b_k(x_0 + u)(Q_t(y - u) - Q_t(y)) \, d\lambda(u),$$

since $\int_{-l}^{l} b_k(x_0 + u) \, d\lambda(u) = 0$. Thus

$$|v_t(b_k)(x_0 + y)| \le \|b_k\|_1 \sup_{-l \le u \le l} |Q_t(y - u) - Q_t(y)|.$$

Now if $|u| \le l$ and $|y| > 2l$ then $|y| \le 2|y - u| < 3|y|$, and so

$$|Q_t(y - u) - Q_t(y)| = \frac{1}{\pi} \left| \frac{y - u}{(y - u)^2 + t^2} - \frac{y}{y^2 + t^2} \right|$$

$$= \frac{1}{\pi} \left| \frac{u(y(y - u) - t^2)}{((y - u)^2 + t^2)(y^2 + t^2)} \right|$$

$$\le \frac{4l}{\pi y^2} \left| \frac{y(y - u) + t^2}{y^2 + t^2} \right| \le \frac{6l}{\pi y^2}.$$

Thus

$$Q^*(b_k)(x_0 + y) = \sup_{t>0} |v_t(b_k)(x_0 + y)| \le \frac{6l \, \|b_k\|_1}{\pi y^2},$$

and so

$$\int_{C(F_k)} Q^*(b_k) \, d\lambda \le \frac{6 \, \|b_k\|_1}{\pi}.$$

Consequently

$$\int_{C(N)} |Q^*(b)| \, d\lambda \le \frac{6}{\pi} \sum_k \|b_k\|_1 = \frac{6}{\pi} \|b\|_1 \le \frac{12}{\pi} \|f\|_1.$$

Corollary 11.3.1 *Suppose that $1 < p < \infty$. Then Q^* is of strong type (p,p). If $f \in L^p(\mathbf{R})$ then $Q_t(f)$ is convergent, in L^p norm and almost everywhere, to a function $H(f)$, say. $H(f) \in L^p(\mathbf{R})$, and the linear mapping $f \to H(f) : L^p(\mathbf{R}) \to L^p(\mathbf{R})$ is bounded.*

Proof Suppose first that $1 < p \le 2$. It follows from the Marcinkiewicz interpolation theorem that Q^* is of strong type (p,p) for $1 < p < 2$. If $f \in L^p \cap L^2$ then $Q_t(f) - H(f) \to 0$ almost everywhere, as $t \to 0$, and $|Q_t(f) - Q_s(f)| \le 2Q^*(f)$, so that $|Q_t(f) - H(f)| \le 2Q^*(f)$. Thus $Q_t(f) \to H(f)$ in L^p-norm. Since $L^2 \cap L^p$ is dense in L^p, the remaining results of the corollary now follow.

Suppose now that $2 < p < \infty$. If $f \in L^p(\mathbf{R})$ and $g \in L^{p'}(\mathbf{R})$ then

$$\int g Q_t(f) \, d\lambda = \int Q_t(g) f \, d\lambda,$$

and from this it follows that $Q_t(f) \in L^p(\mathbf{R})$, and that the mappings $f \to Q_t(f) : L^p(\mathbf{R}) \to L^p(\mathbf{R})$ are uniformly bounded; there exists K such that $\|Q_t(f)\|_p \le K \|f\|_p$ for all $f \in L^p(\mathbf{R})$ and $t > 0$.

Suppose that $f \in L^2(\mathbf{R}) \cap L^p(\mathbf{R})$. Then $Q_t(f) \to H(f)$ in $L^2(\mathbf{R})$, $Q_t(f) = P_t(H(f))$ and $Q_t(f) \to H(f)$ almost everywhere. Now $\{Q_t(f) : t > 0\}$ is bounded in L^p, and so by Fatou's lemma, $\|H(f)\|_p < K \|f\|_p$. But then $\|Q^*(f)\|_p = \|P^*(H(f))\|_p \le K \|f\|_p$. Since $L^2(\mathbf{R}) \cap L^p(\mathbf{R})$ is dense in $L^p(\mathbf{R})$, this inequality extends to all $f \in L^p(\mathbf{R})$. The remaining results now follow easily from this. $\square$

Corollary 11.3.2 (Hilbert's inequality) *If $1 < p < \infty$ there exists a constant K_p such that if $f \in L^p(\mathbf{R})$ and $g \in L^{p'}(\mathbf{R})$ then*

$$\left| \int_{\mathbf{R}} \left(\int_{\mathbf{R}} \frac{f(x)}{x - y} \, dx \right) g(y) \, dy \right| \le K_p \|f\|_p \|g\|_{p'}.$$

[Here the inner integral is the principal value integral.]

With these results, we can mimic the proof of Proposition 11.2.1 to obtain the following.

Proposition 11.3.1 (Cotlar's inequality) *Suppose that* $1 < p < \infty$ *and that* $f \in L^p(\mathbf{R})$. *Then* $H^*(f) \leq m(H(f)) + 2m(f)$, *and* H^* *is of strong type* (p, p).

Similarly we have the following.

Theorem 11.3.2 *If* $f \in L^p(\mathbf{R})$, *where* $1 < p < \infty$, *then* $H_\epsilon(f) \to H(f)$ *in* L^p-*norm and almost everywhere.*

11.4 Hilbert's inequality for sequences

We can easily derive a discrete version of Hilbert's inequality.

Theorem 11.4.1 (Hilbert's inequality for sequences) *If* $1 < p < \infty$ *there exists a constant* K_p *such that if* $a = (a_n) \in l_p(\mathbf{Z})$ *then*

$$\sum_{m=-\infty}^{\infty} \left| \sum_{n \neq m} \frac{a_n}{m - n} \right|^p \leq K_p \|a\|_p^p.$$

Thus if $b \in l_{p'}$ *then*

$$\left| \sum_{m=-\infty}^{\infty} b_m \left(\sum_{n \neq m} \frac{a_n}{m - n} \right) \right| \leq K_p \|a\|_p \|b\|_{p'}.$$

Proof Let $h_0 = 0$, $h_n = 1/n$ for $n \neq 0$. Then $h \in l_{p'}$ for $1 < p < \infty$, and so the sum $\sum_{n \neq m} a_n/(m - n)$ converges absolutely. For $0 < \epsilon < 1/2$ let $J_\epsilon = (2\epsilon)^{-1/p} I_{(-\epsilon, \epsilon)}$ and let $K_\epsilon = (2\epsilon)^{-1/p'} I_{(-\epsilon, \epsilon)}$, so that J_ϵ and K_ϵ are unit vectors in $L^p(\mathbf{R})$ and $L^{p'}(\mathbf{R})$ respectively. Then the principal value

$$\int J_\epsilon(x)\, dx = \lim_{\eta \to 0} \int_{|x| > \eta} J_\epsilon(x)\, dx$$

is zero, while

$$\frac{1}{|m - n| + 2\epsilon} \leq \int J_\epsilon(x - n) K_\epsilon(y - m)\, dx \leq \frac{1}{|m - n| - 2\epsilon},$$

for $m \neq n$. If (a_n) and (b_m) are sequences each with finitely many non-zero terms, let

$$A_\epsilon(x) = \sum_n a_n J_\epsilon(x - n) \quad \text{and} \quad B_\epsilon(y) = \sum_m b_m K_\epsilon(y - m).$$

Then by Hilbert's inequality, $|\int_{\mathbf{R}} H(A_\epsilon)(y) B_\epsilon(y)\, dy| \le K_p \|A_\epsilon\|_p \|B_\epsilon\|_{p'}$. But $\|A_\epsilon\|_p = \|a\|_p$ and $\|B_\epsilon\|_{p'} = \|b\|_{p'}$, and

$$\int_{\mathbf{R}} H(A_\epsilon)(y) B_\epsilon(y)\, dy \to \sum_m \left(\sum_{n \ne m} \frac{a_n b_m}{m - n} \right) \quad \text{as } \epsilon \to 0.$$

Thus

$$\left| \sum_m b_m \left(\sum_{n \ne m} \frac{a_n}{m - n} \right) \right| \le K_p \|a\|_p \|b\|_{p'};$$

letting b vary,

$$\sum_m \left| \sum_{n \ne m} \frac{a_n}{m - n} \right|^p \le K_p^p \|a\|_p^p.$$

The usual approximation arguments then show that the result holds for general $a \in l_p(\mathbf{Z})$ and $b \in l_{p'}(\mathbf{Z})$. $\qquad\square$

11.5 The Hilbert transform on T

Let us now consider what happens on the circle **T**, equipped with Haar measure $\mathbf{P} = d\theta/2\pi$. If $f \in L^1(\mathbf{T})$, then we write $\mathbf{E}(f)$ for $\int_{\mathbf{T}} f\, d\mathbf{P}$, and set $P_0(f) = f - \mathbf{E}(f)$. For $1 \le p \le \infty$, P_0 is a continuous projection of $L^p(\mathbf{T})$ onto $L_0^p(\mathbf{T}) = \{f \in L^p(\mathbf{T}): \mathbf{E}(f) = 0\}$.

Let $c(z) = (1 + z)/(1 - z)$. If $z = re^{i\theta}$ and $r < 1$ then

$$c(z) = 1 + 2 \sum_{k=1}^{\infty} z^k = 1 + 2 \sum_{k=1}^{\infty} r^k e^{ik\theta}$$

$$= \sum_{k=-\infty}^{\infty} r^{|k|} e^{ik\theta} + \sum_{k=-\infty}^{\infty} \operatorname{sgn}(k) r^{|k|} e^{ik\theta}$$

$$= P_r(e^{i\theta}) + iQ_r(e^{i\theta}) = \left(\frac{1 - r^2}{1 - 2r \cos \theta + r^2} \right) + i \left(\frac{2r \sin \theta}{1 - 2r \cos \theta + r^2} \right).$$

$P(re^{i\theta}) = P_r(e^{i\theta})$ and $Q(re^{i\theta}) = Q_r(e^{i\theta})$ are the *Poisson kernel* and *conjugate Poisson kernel*, respectively. If $f \in L^1(\mathbf{T})$, we define $P_r(f) = P_r \star f$ and $Q_r(f) = Q_r \star f$. $P_r \ge 0$ and $\|P_r\|_1 = \mathbf{E}(P_r) = 1$, and so $\|P_r(f)\|_p \le \|f\|_p$ for $f \in L^p(\mathbf{T})$, for $1 \le p \le \infty$. We define the maximal function

$$m(f)(e^{i\theta}) = \sup_{0 < t \le \pi} \frac{1}{2t} \int_{-t}^{t} |f(e^{i(\theta + \phi)})|\, d\phi.$$

$(P_r)_{0<r<1}$ is an approximate identity, and, arguing as in Theorem 8.11.2, $P^*(f) = \sup_{0<r<1} |P_r(f)| \le m(f)$. From this it follows that if $f \in L^p(\mathbf{T})$, then $P_r(f) \to f$ in L^p norm and almost everywhere, for $1 \le p < \infty$.

Now let us consider the case when $p = 2$. If $f \in L^2(\mathbf{T})$, let

$$H(f) = -i \sum_{k=-\infty}^{\infty} \operatorname{sgn}(k) \hat{f}_k e^{ik\theta};$$

the sum converges in L^2 norm, and $\|H(f)\|_2 = \|f - \mathbf{E}(f)\|_2 = \|P_0(f)\|_2 \le \|f\|_2$. $H(f)$ is the *Hilbert transform* of f. $H^2(f) = P_0(f)$, so that H maps $L_0^2(\mathbf{T})$ isometrically onto itself.

If $f \in L^2(\mathbf{T})$, then $Q_r(f) = P_r(H(f))$, so that $Q^*(f) \le P^*(H(f))$, $Q_r(f) \to f$ in L^2 norm, and almost everywhere, and $\|Q^*(f)\|_2 \le 2\|H(f)\|_2 \le 2\|f\|_2$. Further $Q_r(e^{i\theta}) \to \cot(\theta/2)$ as $r \nearrow 1$. Let us set, for $0 < \epsilon < \pi$,

$$H_\epsilon(e^{i\theta}) = \cot(\theta/2) \text{ for } \epsilon < \theta \le \pi,$$
$$= 0 \text{ for } 0 < \theta \le \epsilon.$$

Then H_{1-r} and Q_r are sufficiently close to show that $H_\epsilon(f) \to H(f)$ in L^2 norm, and almost everywhere, as $\epsilon \to 0$.

What happens when $1 < p < \infty$? It is fairly straightforward to use the Calderón–Zygmund technique, the Marcinkiewicz intepolation theorem, and duality to obtain results that correspond exactly to those for $L^p(\mathbf{R})$. It is however possible to proceed more directly, using complex analysis, and this we shall do.

First we have the following standard result.

Proposition 11.5.1 *Suppose that $1 < p < \infty$ and that u is a harmonic function on $\mathbf{D} = \{z : |z| < 1\}$ with the property that*

$$\sup_{0<r<1} \left(\frac{1}{2\pi} \int_0^{2\pi} |u(re^{i\theta})|^p \, d\theta \right) < \infty.$$

Then there exists $f \in L^p(\mathbf{T})$ such that $u(re^{i\theta}) = P_r(f)(e^{i\theta})$ for all $re^{i\theta} \in \mathbf{D}$.

Proof Let $u_r(e^{i\theta}) = u(re^{i\theta})$. Then $\{u_r : 0 < r < 1\}$ is bounded in $L^p(\mathbf{T})$, and so there exist $r_n \nearrow 1$ and $f \in L^p(\mathbf{T})$ such that $u_{r_n} \to f$ weakly as $n \to \infty$. Thus if $0 < r < 1$ and $0 \le \theta < 2\pi$ then $P_r(u_{r_n})(e^{i\theta}) \to P_r(f)(e^{i\theta})$. But $P_r(u_{r_n}) = u_{rr_n}$, and so $u_r(e^{i\theta}) = P_r(f)(e^{i\theta})$. $\square$

We begin with the weak type $(1, 1)$ result.

Theorem 11.5.1 *Suppose that $f \in L^1(\mathbf{T})$. Then $Q_r(f)$ converges pointwise almost everywhere to a function $H(f)$ on* **T** *as $r \nearrow 1$, and if $\alpha > 0$ then* $\mathbf{P}(|H(f)| > \alpha) \leq 4\,\|f\|_1 \,/(2 + \alpha)$.

Proof By considering positive and negative parts, it is enough to consider $f \geq 0$ with $\|f\|_1 = 1$, and to show that $\mathbf{P}(|H(f)| > \alpha) \leq 2/(1 + \alpha)$. For $z = re^{i\theta}$ set

$$F(z) = P_r(f)(e^{i\theta}) + iQ_r(f)(e^{i\theta}).$$

F is an analytic function on **D** taking values in the right half-plane $H_r = \{x + iy : x > 0\}$, and $F(0) = 1$. First we show that $\mathbf{P}(|Q_r(f)| > \alpha) \leq 2/(1 + \alpha)$ for $0 < r < 1$. Let $w_\alpha(z) = 1 + (z - \alpha)/(z + \alpha)$: w_α ia a Möbius transformation mapping H_r conformally onto $\{z : |z - 1| < 1\}$. Note also that if $z \in H_r$ and $|z| > \alpha$ then $\Re(w_\alpha(z)) > 1$.

Now let $G_\alpha(z) = w_\alpha(F(z)) = J_\alpha(z) + iK_\alpha(z)$. Then $J_\alpha(z) > 0$, and if $|Q_r(f)(z)| > \alpha$ then $J_\alpha(z) > 1$. Further, $J_\alpha(0) = w_\alpha(1) = 2/(1 + \alpha)$. Thus

$$\mathbf{P}(|Q_r(f)| > \alpha) \leq \frac{1}{2\pi} \int_0^{2\pi} J_\alpha(re^{i\theta}) \, d\theta = J_\alpha(0) = \frac{2}{1 + \alpha}.$$

Now let $S(z) = 1/(1 + F(z))$. Then S is a bounded analytic function on **D**, and so by Proposition 11.5.1, there exists $s \in L^2(\mathbf{T})$ such that $S(re^{i\theta}) = P_r(s)(e^{i\theta})$. Thus $S(re^{i\theta}) \to s(e^{i\theta})$ almost everywhere as $r \nearrow 1$. Consequently, F, and so $Q_r(f)$, have radial limits, finite or infinite, almost everywhere. But, since $\mathbf{P}(|Q_r(f)| > \alpha) \leq 2/(1 + \alpha)$ for $0 < r < 1$, the limit $H(f)$ must be finite almost everywhere, and then $\mathbf{P}(|H(f)| > \alpha) \leq 2/(1+\alpha)$.
□

If $f \in L^1(\mathbf{T})$, let $Q^*(f) = \sup_{0 < r < 1} Q_r(f)$.

Theorem 11.5.2 *If $1 < p < \infty$ then Q^* is of strong type (p, p).*

Proof It is enough to show that there exists a constant K_p such that $\|Q_r(f)\| \leq K_p \|f\|_p$ for all $f \in L^p(\mathbf{T})$. For then, by Proposition 11.5.1, there exists $g \in L^p(\mathbf{T})$ such that $Q_r(f) = P_r(g)$, and then $Q^*(f) = P^*(g)$, so that $\|Q^*(f)\|_p \leq p' \|g\|_p \leq p'K_p \|f\|_p$. If $f \in L^p(\mathbf{T})$, $h \in L^{p'}(\mathbf{T})$, then $\mathbf{E}(Q_r(f)h) = \mathbf{E}(fQ_r(\check{h}))$, where $\check{h}(e^{i\theta}) = h(e^{-i\theta})$, and so a standard duality argument shows that we need only prove this for $1 < p < 2$. Finally, we need only prove the result for $f \geq 0$.

Suppose then that $f \in L^p(\mathbf{T})$, that $f \geq 0$ and that $0 < r < 1$. Let $\gamma = \pi/(p + 1)$, so that $0 < \gamma < \pi/2$ and $\pi/2 < p\gamma < p\pi/2 < \pi$. Note that

$\cos p\gamma = -\cos\gamma$. As before, for $z = re^{i\theta}$ set

$$F(z) = P_r(f)(e^{i\theta}) + iQ_r(f)(e^{i\theta}).$$

F is an analytic function on $\mathbf{D}$ taking values in the right half-plane H_r, and so we can define the analytic function $G(z) = (F(z))^p = J(z) + iK(z)$. Then

$$\|Q_r(f)\|_p^p \le \frac{1}{2\pi} \int_0^{2\pi} |G(re^{i\theta})|\, d\theta.$$

We divide the unit circle into two parts: let

$$S = \{e^{i\theta}: 0 \le |\arg F(re^{i\theta})| \le \gamma\},$$
$$L = \{e^{i\theta}: \gamma < |\arg F(re^{i\theta})| < \pi/2\}.$$

If $e^{i\theta} \in S$ then $|F(re^{i\theta})| \le P_r(f)(e^{i\theta})/\cos\gamma$, so that

$$\frac{1}{2\pi}\int_S |G(re^{i\theta})|\, d\theta \le \frac{1}{2\pi(\cos\gamma)^p}\int_S (P_r(f)(e^{i\theta}))^p\, d\theta$$
$$\le (\|P_r(f)\|_p/\cos\gamma)^p \le (\|f\|_p/\cos\gamma)^p.$$

On the other hand, if $e^{i\theta} \in L$ then $\pi\gamma < \arg G(re^{i\theta}) < 2\pi$, so that $J(re^{i\theta}) < 0$ and $|G(re^{i\theta})| \le -J(re^{i\theta})/\cos\gamma$. But

$$\frac{1}{2\pi}\int_L J(re^{i\theta})\, d\theta + \frac{1}{2\pi}\int_S J(re^{i\theta})\, d\theta = J(0) = (\mathbf{E}(f))^p \ge 0,$$

and so

$$\frac{1}{2\pi}\int_L |G(re^{i\theta})|\, d\theta \le \frac{-1}{2\pi\cos\gamma}\int_L J(re^{i\theta})\, d\theta \le \frac{1}{2\pi\cos\gamma}\int_S J(re^{i\theta})\, d\theta$$
$$\le \frac{1}{2\pi\cos\gamma}\int_S |G(re^{i\theta})|\, d\theta \le \|f\|_p^p/(\cos\gamma)^{p+1}.$$

Consequently $\|Q_r(f)\|_p^p \le (2/(\cos\gamma)^{p+1})\, \|f\|_p^p$.   □

The following corollaries now follow, as in Section 11.3.

Corollary 11.5.1 *Suppose that $1 < p < \infty$. If $f \in L^p(\mathbf{T})$ then $Q_r(f)$ is convergent, in L^p norm and almost everywhere, to a function $H(f)$, say, as $r \nearrow 1$. $H(f) \in L^p(\mathbf{R})$, and the linear mapping $f \to H(f): L^p(\mathbf{R}) \to L^p(\mathbf{R})$ is bounded.*

Corollary 11.5.2 *If $f \in L^p(\mathbf{T})$, where $1 < p < \infty$, then $H_\epsilon(f) \to H(f)$ in L^p-norm and almost everywhere, as $\epsilon \to 0$.*

11.6 Multipliers

We now explore how the ideas of Section 11.3 extend to higher dimensions. We shall see that there are corresponding results for *singular integral operators*. These are operators which reflect the algebraic structure of $\mathbf{R}^d$, as we shall describe in the next two sections. We consider bounded linear operators on $L^2(\mathbf{R}^d)$. If $y \in \mathbf{R}^d$, the *translation operator* τ_y is defined as $\tau_y(f)(x) = f(x - y)$. This is an isometry of $L^2(\mathbf{R}^d)$ onto itself; first, we consider operators which commute with all translation operators. (This idea clearly extends to $L^2(G)$, where G is a locally compact abelian group, and is the starting point for commutative harmonic analysis.) Operators which commute with all translation operators are characterized as follows.

Theorem 11.6.1 *Suppose that $T \in L(L^2(\mathbf{R}^d))$. The following are equivalent.*

(i) T commutes with all translation operators.

(ii) If $g \in L^1(\mathbf{R}^d)$ and $f \in L^2(\mathbf{R}^d)$ then $T(g \star f) = g \star T(f)$.

(iii) There exists $h \in L^\infty(\mathbf{R}^d)$ such that $\widehat{T(f)} = h\hat{f}$ for all $f \in L^2(\mathbf{R}^d)$.

If these conditions are satisfied, then $\|T\| = \|h\|_\infty$.

If so, then we write $T = M_h$, and call T a *multiplier*.

Proof Suppose that (i) holds. If $g \in L^1(\mathbf{R}^d)$ and $f, k \in L^2(\mathbf{R}^d)$ then

$$
\begin{aligned}
\langle g \star T(f), k \rangle &= \left\langle \int \tau_y(T(f))g(y)\, dy, k \right\rangle \\
&= \left\langle \int T(\tau_y(f))g(y)\, dy, k \right\rangle \\
&= \left\langle T\!\left(\int \tau_y(f)g(y)\, dy\right), k \right\rangle \\
&= \langle T(g \star f), k \rangle.
\end{aligned}
$$

Thus (ii) holds.

On the other hand, if (ii) holds and if $f \in L^2(\mathbf{R}^d)$ then

$$
\begin{aligned}
T(\tau_y(f)) &= \lim_{t \to 0} T(\tau_y(P_t(f))) = \lim_{t \to 0} T(\tau_y(P_t) \star f) \\
&= \lim_{t \to 0} \tau_y(P_t) \star T(f) = \tau_y(T(f)),
\end{aligned}
$$

where $(P_t)_{t>0}$ is the Poisson kernel on $\mathbf{R}^d$ and convergence is in L^2 norm. Thus (i) holds.

If (iii) holds then

$$(T\widehat{(\tau_y(f))})(\xi) = h(\xi)\widehat{(\tau_y(f))}(\xi) = h(\xi)e^{-2\pi i\langle y,\xi\rangle}\hat{f}(\xi) = (\tau_y\widehat{(T(f))})(\xi),$$

so that $T\tau_y = \tau_y T$, and (i) holds. Further,

$$\|T(f)\|_2 = \left\|\widehat{T(f)}\right\|_2 \le \|h\|_\infty \left\|\hat{f}\right\|_2 = \|h\|_\infty \|f\|_2.$$

Finally, if (i) and (ii) hold, and $f \in L^2(\mathbf{R}^d)$, let

$$\phi(f) = (P_1 \star T(f))(0) = c_d \int \frac{T(f)(x)}{(|x|^2+1)^{d/2}}\, dx.$$

Then $|\phi(f)| \le \|P_1\|_2 \|T(f)\|_2 \le \|P_1\|_2 \|T\| \|f\|_2$, so that ϕ is a continuous linear functional on $L^2(\mathbf{R}^d)$. Thus there exists $k \in L^2(\mathbf{R}^d)$ such that $\phi(f) = \langle f, k\rangle$. Let $j(y) = \overline{k(-y)}$. Then

$$(f \star j)(x) = \int f(y)\overline{k(y-x)}\, dy = \int f(y+x)\overline{k(y)}\, dy$$

$$= \phi(\tau_{-x}(f)) = (P_1 \star T(\tau_{-x}(f)))(0)$$

$$= (P_1 \star \tau_{-x}(T(f)))(0) = \int P_1(-y)T(f)(y+x)\, dy$$

$$= \int P_1(x-y)T(f)(y)\, dy = (P_1 \star T(f))(x).$$

Thus $P_1 \star T(f) = f \star j$. Taking Fourier transforms, $e^{-2\pi|\xi|}\widehat{T(f)}(\xi) = \hat{f}(\xi)\hat{j}(\xi)$, so that $\widehat{T(f)}(\xi) = h(\xi)\hat{f}(\xi)$, where $h(\xi) = e^{2\pi|\xi|}\hat{j}(\xi)$. Suppose that $\lambda(|h| > \|T\|) > 0$. Then there exists B of positive finite measure on which $|h| > \|T\|$. But then there exists $g \in L^2(\mathbf{R}^d)$ for which $\hat{g} = \overline{\mathrm{sgn}\, h}I_B$. Then

$$\|T(g)\|_2^2 = \int_B |h(\xi)|^2\, d\xi > \|T\|^2 \|\hat{g}\|_2^2 = \|T\|^2 \|g\|_2^2,$$

giving a contradiction. Thus $h \in L^\infty(\mathbf{R}^d)$, and $\|h\|_\infty \le \|T\|$. $\qquad\square$

11.7 Singular integral operators

$\mathbf{R}^d$ is not only a locally compact abelian group under addition, but is also a vector space. We therefore consider multipliers on $L^2(\mathbf{R}^d)$ which respect scalar multiplication. If $\lambda > 0$ the *dilation operator* δ_λ is defined as $\delta_\lambda(f)(x) = f(x/\lambda)$. If $f \in L^p(\mathbf{R}^d)$ then $\|\delta_\lambda(f)\|_p = \lambda^{1/p}\|f\|_p$, so that dilation introduces a scaling factor which varies with p.

We consider multipliers on $L^2(\mathbf{R}^d)$ which commute with all dilation operators.

If $f \in L^2(\mathbf{R}^d)$ then $\widehat{\delta_\lambda(f)}(\xi) = \lambda^d \hat{f}(\lambda\xi)$. Thus if M_h commutes with dilations then

$$(\widehat{M_h \delta_\lambda(f)})(\xi) = \lambda^d h(\xi) \hat{f}(\lambda\xi) = (\widehat{\delta_\lambda M_h(f)})(\xi) = \lambda^d h(\lambda\xi) \hat{f}(\lambda\xi),$$

so that $h(\lambda\xi) = h(\xi)$; h is constant on rays from the origin, and $h(\xi) = h(\xi/|\xi|)$. If we now proceed formally, and let K be the inverse Fourier transform of h, then a change of variables shows that $K(\lambda x) = K(x)/\lambda^d$; K is homogeneous of degree $-d$, and if $x \neq 0$ then $K(x) = (1/|x|^d)K(x/|x|)$. Such functions have a singularity at the origin; we need to impose some regularity on K. There are various possibilities here, but we shall suppose that K satisfies a Lipschitz condition on S^{d-1}: there exists $C < \infty$ such that $|K(x) - K(y)| \leq C|x - y|$ for $|x| = |y| = 1$. In particular, K is bounded on S^{d-1}; let $A = \sup\{|K(x)|: |x| = 1\}$.

Thus we are led to consider a formal convolution $K \star f$, where K is homogeneous of degree $-d$, and satisfies this regularity condition. K is not integrable, but if we set $K_\epsilon(x) = K(x)$ for $|x| \geq \epsilon$ and $K(x) = 0$ for $|x| < \epsilon$ then $K_\epsilon \in L^p(\mathbf{R}^d)$ for all $1 < p \leq \infty$. Following the example of the Hilbert transform, we form the convolution $K_\epsilon(f) = K_\epsilon \star f$, and see what happens as $\epsilon \to 0$.

Let us see what happens if f is very well behaved. Suppose that f is a smooth function of compact support, and that $f(x) = 1$ for $|x| \leq 2$. If $|x| \leq 1$ and $0 < \epsilon < \eta \leq 1$ then

$$(K_\eta \star f)(x) - (K_\epsilon \star f)(x) = \left(\int_{S^{d-1}} K(\omega) \, ds(\omega) \right) \log(\eta/\epsilon),$$

so that if the integral is to converge, we require that $(\int_{S^{d-1}} K(\omega) \, ds(\omega)) = 0$. We are therefore led to the following definition.

A function K defined on $\mathbf{R}^d \setminus \{0\}$ is a *regular Calderón–Zygmund kernel* if

(i) K is homogeneous of degree $-d$;

(ii) K satisfies a Lipschitz condition on the unit sphere S^{d-1};

(iii) $\int_{S^{d-1}} K(\omega) \, ds(\omega) = 0$.

The Hilbert transform kernel $K(x) = 1/x$ is, up to scaling, the only regular Calderón–Zygmund kernel on $\mathbf{R}$. On $\mathbf{R}^d$, the *Riesz kernels* $c_d x_j/|x|^{d+1}$ $(1 \leq j \leq d)$ (where c_d is a normalizing constant) are important examples of regular Calderón–Zygmund kernels (see Exercise 11.3).

The regularity conditions lead to the following consequences.

Theorem 11.7.1 *Suppose that K is a regular Calderón–Zygmund kernel.*

(i) There exists a constant D such that $|K(x-y) - K(x)| \leq D|y|/|x|^{d+1}$ for $|x| > 2|y|$.

(ii) (Hörmander's condition) There exists a constant B such that

$$\int_{|x|>2|y|} |K(x-y) - K(x)|\, dx \leq B \quad and \quad \int_{|x|>2|y|} |K_\epsilon(x-y) - K_\epsilon(x)|\, dx \leq B$$

for all $\epsilon > 0$.

(iii) There exists a constant C such that $\left\| \hat{K}_\epsilon \right\|_\infty \leq C$ for all $\epsilon > 0$.

Proof We leave (i) and (ii) as exercises for the reader (Exercise 11.2); (i) is easy, and, for K, (ii) follows by integrating (i). The argument for K_ϵ is elementary, but more complicated, since there are two parameters $|y|$ and ϵ. The fact that the constant does not depend on ϵ follows from homogeneity.

(iii) $\hat{K}_\epsilon(\xi) = \lim_{R \to \infty} I_{\epsilon,R}$, where

$$I_{\epsilon,R} = \int_{\epsilon \leq |x| \leq R} e^{-i\langle x,\xi\rangle} K(x)\, dx.$$

Thus $\hat{K}_\epsilon(0) = 0$, by condition (iii). For $\xi \neq 0$ let $r = \pi/|\xi|$. If $\epsilon < 2r$ then $I_{\epsilon,R} = I_{\epsilon,2r} + I_{2r,R}$ and

$$|I_{\epsilon,2r}| = \left| \int_{\epsilon \leq |x| \leq 2r} (e^{-i\langle x,\xi\rangle} - 1) K(x)\, dx \right|$$

$$\leq |\xi| \int_{\epsilon \leq |x| \leq 2r} |x|(A/|x|^d)\, dx \leq C_d 2r|\xi|A = 2\pi C_d A.$$

We must therefore show that $I_{a,R}$ is bounded, for $a \geq 2r$. Let $z = \pi\xi/|\xi|^2$, so that $|z| = r$ and $e^{i\langle z,\xi\rangle} = e^{i\pi} = -1$. Now

$$I_{a,R} = \int_{a \leq |x-z| \leq R} e^{-i\langle x-z,\xi\rangle} K(x-z)\, dx = -\int_{a \leq |x-z| \leq R} e^{-i\langle x,\xi\rangle} K(x-z)\, dx,$$

so that

$$I_{a,R} = \tfrac{1}{2}\left(I_{a,R} - \int_{a \leq |x-z| \leq R} e^{-i\langle x,\xi\rangle} K(x-z)\, dx \right)$$

$$= F + \tfrac{1}{2}\int_{a+r \leq |x| \leq R-r} e^{-i\langle x,\xi\rangle}(K(x) - K(x-z))\, dx + G,$$

where the fringe function F is of the form $\int_{a-r \leq |x| \leq a+r} f(x)\, dx$, where $|f(x)| \leq A/(a-r)^d$, so that $|F| \leq \Omega_d A((a+r)/(a-r))^d$, and the fringe function

G is of the form $\int_{R-r \le |x| \le R+r} g(x)\,dx$, where $|g(x)| \le A/(R-r)^d$, so that $|G| \le \Omega_d A((R+r)/(R-r))^d$. Thus $|F| \le 3^d \Omega_d A$ and $|G| \le 3^d \Omega_d A$.

Finally, Hörmander's condition implies that

$$\left| \frac{1}{2} \int_{a+r \le |x| \le R-r} e^{-i\langle x,\xi \rangle} (K(x) - K(x-z))\,dx \right| \le B/2.$$

$\square$

Suppose now that g is a smooth function of compact support. Then

$$(K_\epsilon \star g)(x) = \int_{|y|>1} g(x-y)K(y)\,dy + \int_{1 \ge |y| > \epsilon} (g(x-y) - g(x))K(y)\,dy.$$

The first integral defines a function in L^p, for all $1 < p \le \infty$, while

$$|(g(x-y) - g(x))K(y)| \le A \|g'\|_\infty /|y|^{d-1},$$

since $|g(x-y) - g(x)| \le \|g'\|_\infty |y|$, and so the second integral, which vanishes outside a compact set, converges uniformly as $\epsilon \to 0$. Thus for such g, $T_\epsilon(f)$ converges pointwise and in L^p norm as $\epsilon \to 0$.

Corollary 11.7.1 *If $f \in L^2$ then $K_\epsilon(f) = K_\epsilon \star f$ converges in L^2 norm, to $K(f)$ say, as $\epsilon \to 0$.*

For $\|K_\epsilon(f)\|_2 \le B \|f\|_2$, and so the result follows from Theorem 8.4.1.

11.8 Singular integral operators on $L^p(\mathbf{R}^d)$ for $1 \le p < \infty$

We now follow the proof of Theorem 11.3.1 to establish the following.

Theorem 11.8.1 *T_ϵ is of weak type $(1,1)$, with a constant independent of ϵ.*

Proof As before, a scaling argument shows that it is enough to show that K_1 is of weak type $(1,1)$.

Suppose that $f \in L^1(\mathbf{R}^d)$, that $f \ge 0$ and that $\alpha > 0$. As in Theorem 11.3.1, we consider the dyadic filtration of $\mathbf{R}^d$, define the stopping time τ, and define the good part g and the bad part b of f. Then $\|g\|_1 = \|f\|_1$, $\|b\|_1 \le 2 \|f\|_1$ and $\|g\|_\infty \le 2^d \alpha$. Then $\int g^2 \, d\lambda \le (4^d + 1)\alpha \|f\|_1$, so that $\|K_1(f)\|_2^2 \le (4^d + 1)B\alpha \|f\|_1$ and $\lambda(|K_1(g)| > \alpha/2) \le 4(4^d + 1)B \|f\|_1 /\alpha$.

What about b? Here we take F_k to be the cube with the same centre x_k as E_k, but with side $2^{d/2}$ as big. This ensures that if $x \notin F_k$ and $y \in E_k$

then $|x - x_k| \geq 2|y - x_k|$. As in Theorem 11.3.1, it is enough to show that $\int_{C(F_k)} |K_1(b_k)| \, d\lambda \leq B \, \|b_k\|_1$. We use Hörmander's condition:

$$\int_{C(F_k)} |K_1(b_k)| \, d\lambda = \int_{C(F_k)} \left| \int_{E_k} K_1(x - y) b_k(y) \, dy \right| \, dx$$

$$= \int_{C(F_k)} \left| \int_{E_k} (K_1(x - y) - K_1(x - x_k)) b_k(y) \, dy \right| \, dx$$

$$\leq \int_{C(F_k)} \int_{E_k} |K_1(x - y) - K_1(x - x_k)| \, |b_k(y)| \, dy \, dx$$

$$= \int_{E_k} \left(\int_{C(F_k)} |K_1(x - y) - K_1(x - x_k)| \, dx \right) |b_k(y)| \, dy$$

$$\leq B \, \|b_k\|_1 .$$

Compare this calculation with the calculation that occurs at the end of the proof of Theorem 11.3.1. □

Using the Marcinkiewicz interpolation theorem and duality, we have the following corollary.

Corollary 11.8.1 *For $1 < p < \infty$ there exists a constant C_p such that if $f \in L^p(\mathbf{R}^d)$ then $\|K_\epsilon(f)\|_p \leq C_p \|f\|_p$, and $K_\epsilon(f)$ converges in L^p norm to $K(f)$, as $\epsilon \to 0$.*

What about convergence almost everywhere? Here we need a d-dimensional version of Cotlar's inequality.

Proposition 11.8.1 *Suppose that T is a regular Calderón–Zygmund kernel. There exists a constant C such that if $f \in L^p(\mathbf{R}^d)$, where $1 < p < \infty$, then $K^*(f) = \sup_{\epsilon > 0} |K_\epsilon(f)| \leq m(K(f)) + Cm(f)$.*

This can be proved in the following way. Let ϕ be a *bump function*: a smooth bell-shaped function on $\mathbf{R}^d$ with $\|\phi\|_1 = 1$ which vanishes outside the unit ball of $\mathbf{R}^d$. Let $\phi_\epsilon(x) = \epsilon^{-d} \phi(x/\epsilon)$, for $\epsilon > 0$. Then $\phi_\epsilon \star K(f) = K(\phi_\epsilon) \star f$, so that, by Theorem 8.11.2, $\sup_{\epsilon > 0} |K(\phi_\epsilon) \star f| \leq m(T(f))$. Straightforward calculations now show that there exists D such that

$$|K_1(x) - K(\phi)(x)| \leq D \min(1, |x|^{-(d+1)}) = L_1(x), \text{ say.}$$

Then, by scaling,

$$\sup_{\epsilon > 0} |T_\epsilon(f) - T(\phi_\epsilon) \star f| \leq \sup_{\epsilon > 0} |L_\epsilon \star f| \leq \|L\|_1 \, m(f),$$

and Cotlar's inequality follows from this.

The proof of convergence almost everywhere now follows as in the one-dimensional case.

11.9 Notes and remarks

The results of this chapter are only the beginning of a very large subject, the study of harmonic analysis on Euclidean space, and on other Riemannian manifolds. An excellent introduction is given by Duoandikoetxea [Duo 01]. After several decades, the books by Stein [Stei 70] and Stein and Weiss [StW 71] are still a valuable source of information and inspiration. If you still want to know more, then turn to the encyclopedic work [Stei 93].

Exercises

11.1 Use contour integration and Jordan's lemma to show that

$$\hat{P}_t(\xi) = e^{-2\pi t|\xi|} \quad \text{and} \quad \hat{Q}_t(\xi) = -i\,\text{sgn}\,(\xi)e^{-2\pi t|\xi|}.$$

11.2 Prove parts (i) and (ii) of Theorem 11.7.1.

11.3 Let $R_j(x) = c_d x_j/|x|^{d+1}$, where c_d is a normalizing constant, be the jth Riesz kernel.

(i) Verify that R_j is a regular Calderón–Zygmund kernel.

(ii) Observe that the vector-valued kernel $R = (R_1, \ldots, R_d)$ is rotational invariant. Deduce that the Fourier transform $\hat{R}$ is rotational-invariant. Show that $\hat{R}_j(\xi) = -ib_d\xi_j/|\xi|$. In fact, c_d is chosen so that $b_d = 1$.

Let T_j be the singular integral operator defined by R_j.

(iii) Show that $\sum_{j=1}^{d} T_j^2 = -I$.

(iv) Suppose that $f_0 \in L^2(\mathbf{R}^d)$, and that $f_j = T_j(f_0)$. Let $u_j(x,t) = P_t(f_j)$, for $0 \le j \le d$. For convenience of notation, let $x_0 = t$. Show that the functions u_j satisfy the generalized Cauchy–Riemann equations

$$\sum_{j=0}^{d} \frac{\partial u_j}{\partial x_j} = 0, \quad \frac{\partial u_j}{\partial x_k} = \frac{\partial u_k}{\partial x_j},$$

for $0 \le j, k \le d$. These equations are related to Clifford algebras, and the Dirac operator. For more on this, see [GiM 91].

(v) Suppose that $1 < p < \infty$. Show that there exists a constant A_p such that if f is a smooth function of compact support on $\mathbf{R}^d$ then

$$\left\| \frac{\partial^2 f}{\partial x_j \partial x_k} \right\|_p \leq A_p \|\Delta f\|_p,$$

where Δ is the Laplacian.

$$\left[\text{Show that } \frac{\partial^2 f}{\partial x_j \partial x_{k_p}} = -T_j T_k \Delta f. \right]$$

For more on this, see [Stei 70] and [GiM 91].

12

Khintchine's inequality

12.1 The contraction principle

We now turn to a topic which will recur for the rest of this book. Let $(F, \|.\|_F)$ be a Banach space (which may well be the field of scalars). Let $\omega(F)$ denote the space of all infinite sequences in F, and let $\omega_d(E)$ denote the space of all sequences of length d in F. Then $D_2^{\mathbf{N}}$ acts on $\omega(F)$; if $\omega \in D_2^{\mathbf{N}}$ and $x = (x_n) \in \omega(F)$ we define $x(\omega)$ by setting $x(\omega)_n = (\epsilon_n(\omega)x_n)$. Similarly D_2^d acts on $\omega_d(F)$. In general, we shall consider the infinite case (although the arguments usually concern only finitely many terms of the sequence), and leave the reader to make any necessary adjustments in the finite case.

First we consider the case where F is a space of random variables. Suppose that $X = (X_n)$ is a sequence of random variables, defined on a probability space $(\Omega, \Sigma, \mathbf{P})$ (disjoint from $D_2^{\mathbf{N}}$), and taking values in a Banach space $(E, \|.\|_E)$. In this case we can consider $\epsilon_n X_n$ as a random variable defined on $\Omega \times D_2^{\mathbf{N}}$. We say that X is a *symmetric sequence* if the distribution of $X(\omega)$ is the same as that of X for each $\omega \in D_2^{\mathbf{N}}$. This says that each X_n is symmetric, and more. We shall however be largely concerned with independent sequences of random variables. If the (X_n) is an independent sequence, it is symmetric if and only if each X_n is symmetric.

If (X_n) is a symmetric sequence and if (η_n) is a Bernoulli sequence of random variables, independent of the X_n, then (X_n) and $(\eta_n X_n)$ have the same distribution, and in the real case, this is the same as the distribution of $(\epsilon_n |X_n|)$.

Symmetric sequences of random variables have many interesting properties which we now investigate. We begin with the contraction principle. This name applies to many inequalities, but certainly includes those in the next proposition.

Proposition 12.1.1 (The contraction principle) *(i) Suppose that (X_n) is a symmetric sequence of random variables, taking values in a Banach* space E. If $\lambda = (\lambda_n)$ is a bounded sequence of real numbers then

$$\left\| \sum_{n=1}^{N} \lambda_n X_n \right\|_p \leq \|\lambda\|_\infty \left\| \sum_{n=1}^{N} X_n \right\|_p$$

for $1 \leq p < \infty$.

(ii) Suppose that (X_n) and (Y_n) are symmetric sequences of real random variables defined on the same probability space $(\Omega_1, \Sigma_1, \mathbf{P}_1)$, that $|X_n| \leq |Y_n|$ for each n, and that (u_n) is a sequence in a Banach space $(E, \|.\|_E)$. Then

$$\left\| \sum_{n=1}^{N} X_n u_n \right\|_p \leq \left\| \sum_{n=1}^{N} Y_n u_n \right\|_p$$

for $1 \leq p < \infty$.

(iii) Suppose that (X_n) is a symmetric sequence of real random variables and that $\|X_n\|_1 \geq 1/C$ for all n. Suppose that (ϵ_n) is a Bernoulli sequence of random variables and that (u_n) is a sequence in a Banach space $(E, \|.\|_E)$. Then

$$\left\| \sum_{n=1}^{N} \epsilon_n u_n \right\|_p \leq C \left\| \sum_{n=1}^{N} X_n u_n \right\|_p$$

for $1 \leq p < \infty$.

Proof (i) We can suppose that $\|\lambda\|_\infty = 1$. Consider the mapping $T : \lambda \to \sum_{n=1}^{N} \lambda_n X_n$ from l_∞^N into $L^p(\Omega)$. Then $T(\lambda)$ is a convex combination of $\{T(\epsilon) : \epsilon_n = \pm 1\}$, and so

$$\left\| \sum_{n=1}^{N} \lambda_n X_n \right\|_p = \|T(\lambda)\|_p$$

$$\leq \max\{\|T(\epsilon)\|_p : \epsilon_n = \pm 1\} = \left\| \sum_{n=1}^{N} X_n \right\|_p .$$

(ii) Suppose that (ϵ_n) is a sequence of Bernoulli random variables on a separate space $\Omega_2 = D_2^N$. Then

$$\left\| \sum_{n=1}^{N} X_n u_n \right\|_p^p = \mathbf{E}_1 \left(\left\| \sum_{n=1}^{N} X_n u_n \right\|_E^p \right)$$

$$= \mathbf{E}_1 \mathbf{E}_2 \left(\left\| \sum_{n=1}^{N} \epsilon_n |X_n| u_n \right\|_E^p \right)$$

$$\leq \mathbf{E}_1 \mathbf{E}_2 \left(\left\| \sum_{n=1}^{N} \epsilon_n |Y_n| u_n \right\|_E^p \right) \quad \text{(by (i))}$$

$$= \mathbf{E}_1 \left(\left\| \sum_{n=1}^{N} Y_n u_n \right\|_E^p \right) = \left\| \sum_{n=1}^{N} Y_n u_n \right\|_p^p .$$

(iii) Again suppose that (X_n) are random variables on $(\Omega_1, \Sigma_1, \mathbf{P}_1)$ and that (ϵ_n) is a sequence of Bernoulli random variables on a separate space $\Omega_2 = D_2^{\mathbf{N}}$. Then

$$\left\| \sum_{n=1}^{N} \epsilon_n u_n \right\|_p^p = \mathbf{E}_2 \left(\left\| \sum_{n=1}^{N} \epsilon_n u_n \right\|_E^p \right)$$

$$\leq \mathbf{E}_2 \left(\left\| \sum_{n=1}^{N} C \epsilon_n \mathbf{E}_1(|X_n|) u_n \right\|_E^p \right) \quad \text{(by (i))}$$

$$\leq \mathbf{E}_2 \left(\mathbf{E}_1 \left(\left\| \sum_{n=1}^{N} C \epsilon_n |X_n| u_n \right\|_E \right)^p \right)$$

(by the mean-value inequality)

$$\leq \mathbf{E}_2 \mathbf{E}_1 \left(\left\| \sum_{n=1}^{N} C \epsilon_n |X_n| u_n \right\|_E^p \right) \quad \text{(by Proposition 5.5.1)}$$

$$= C^p \left\| \sum_{n=1}^{N} X_n u_n \right\|_p^p .$$

$\square$

12.2 The reflection principle, and Lévy's inequalities

The next result was originally due to Paul Lévy, in the scalar-valued case.

Theorem 12.2.1 (The reflection principle; Lévy's inequalities) *Suppose that (X_n) is a symmetric sequence of random variables taking values in a Banach space $(E \|.\|_E)$. Let $S_m = X_1 + \cdots + X_m$, and let $S^* = \sup_m \|S_m\|_E$.*

(i) If S_m converges to S almost everywhere then $\mathbf{P}(S^* > t) \leq 2\mathbf{P}(\|S\|_E > t)$, for $t > 0$.

(ii) If Λ is an infinite set of natural numbers, and $S_\Lambda^* = \sup_{\lambda \in \Lambda} \|S_\lambda\|_E$, then $\mathbf{P}(S^* > t) \leq 2\mathbf{P}(S_\Lambda^* > t)$, for $t > 0$.

Proof We use a stopping time argument. Let $\tau = \inf\{j: \ \|S_j\|_E > t\}$ (we set $\tau = \infty$ if $S^* \leq t$). Let A_m be the event $(\tau = m)$. The events A_m are disjoint, and $(S^* > t) = \cup_{m=1}^\infty A_m$, so that $\mathbf{P}(S^* > t) = \sum_{m=1}^\infty \mathbf{P}(A_m)$.

(i) Let $B = (\|S\|_E > t)$. Note that $B = \underline{\lim}(\|S_j\|_E > t)$. We shall use the fact that

$$S_n = \tfrac{1}{2}(S + (2S_n - S)) = \tfrac{1}{2}([S_n + (S - S_n)] + [S_n - (S - S_n)]),$$

so that

$$\|S_n\|_E \leq \max \left(\|S_n + (S - S_n)\|_E, \|S_n - (S - S_n)\|_E\right)$$
$$= \max \left(\|S\|_E, \|S_n - (S - S_n)\|_E\right).$$

Let $C_n = (\|S_n - (S - S_n)\|_E > t)$. Then

$$A_n = (A_n \cap B) \cup (A_n \cap C_n),$$

so that $\mathbf{P}(A_n) \leq \mathbf{P}(A_n \cap B) + \mathbf{P}(A_n \cap C_n)$. We shall show that these two summands are equal.

If $j > n$, then

$$\mathbf{P}(A_n \cap (\|S_j\|_E > t)) = \mathbf{P}(A_n \cap (\|S_n + (S_j - S_n)\|_E > t))$$
$$= \mathbf{P}(A_n \cap (\|S_n - (S_j - S_n)\|_E > t)),$$

by symmetry. Since

$$A_n \cap B = \underline{\lim}_{j\to\infty}(A_n \cap (\|S_j\|_E > t))$$

and

$$A_n \cap C_n = \underline{\lim}_{j\to\infty}(A_n \cap (\|S_n - (S_j - S_n)\|_E > t)),$$

$\mathbf{P}(A_n \cap B) = \mathbf{P}(A_n \cap C_n)$; thus $\mathbf{P}(A_n) \leq 2\mathbf{P}(A_n \cap B)$. Adding,

$$\mathbf{P}(S^* > t) \leq 2\mathbf{P}(B) = 2\mathbf{P}(\|S\|_E > t).$$

(ii) Let $E = (S_\Lambda^* > t)$, and let

$$E_n = (\sup\{\|S_\lambda\|_E: \ \lambda \in \Lambda, \lambda \geq n\} > t)$$
$$F_n = (\sup\{\|2S_n - S_\lambda\|_E: \ \lambda \in \Lambda, \lambda \geq n\} > t).$$

Then, arguing as before, $A_n = (A_n \cap E_n) \cup (A_n \cap F_n)$ and $\mathbf{P}(A_n \cap E_n) = \mathbf{P}(A_n \cap F_n)$, so that

$$\mathbf{P}(A_n) \leq 2\mathbf{P}(A_n \cap E_n) \leq 2\mathbf{P}(A_n \cap E).$$

Adding, $\mathbf{P}(S^* > t) \leq 2\mathbf{P}(E) = 2\mathbf{P}(S_\Lambda^* > t)$. $\qquad\square$

The reflection principle has many important consequences.

Corollary 12.2.1 *If (X_n) is a symmetric sequence of random variables then $\sum_{n=1}^{\infty} X_n$ converges almost everywhere if and only if it converges in probability.*

Proof Since a sequence which converges almost everywhere converges in probability, we need only prove the converse. Suppose then that (S_n) converges in probability to S. First we show that, given $\epsilon > 0$, there exists N such that $\mathbf{P}(\sup_{n \geq N} \|S_n - S_N\|_E > \epsilon) < \epsilon$. There is a subsequence (S_{n_k}) which converges almost everywhere to S. Let $A_K = (\sup_{k \geq K} \|S_{n_k} - S\|_E \leq \epsilon)$. Then (A_K) is an increasing sequence, whose union contains the set on which S_{n_k} converges to S, and so there exists K such that $\mathbf{P}(\sup_{k \geq K} \|S_{n_k} - S\|_E > \epsilon) < \epsilon/4$. Let $N = n_K$. We discard the first N terms: let $Y_j = X_{N+j}$, let $m_k = n_{K+k} - N$, let $\Lambda = \{m_k \colon k \in \mathbf{N}\}$ and let $Z_k = Y_{m_{k-1}+1} + \cdots + Y_{m_k}$. The sequences (Y_j) and (Z_k) are symmetric. Let $T_j = \sum_{i=1}^{j} Y_i$ and let $U_k = T_{m_k} = \sum_{l=1}^{k} Z_l$. Then $T_j \to S - S_N$ in probability, and $U_k \to S - S_N$ almost everywhere. Then, applying the reflection principle twice,

$$\mathbf{P}(\sup_{n \geq N} \|S_n - S_N\|_E > \epsilon) = \mathbf{P}(T^* > \epsilon) \leq 2\mathbf{P}(T_\Lambda^* > \epsilon)$$

$$= 2\mathbf{P}(U^* > \epsilon) \leq 4\mathbf{P}(\|S - S_N\|_E > \epsilon) < \epsilon.$$

We now use the first Borel–Cantelli lemma. Let (ϵ_r) be a sequence of positive numbers for which $\sum_{r=1}^{\infty} \epsilon_r < \infty$. We can find an increasing sequence (N_r) such that, setting $B_r = (\sup_{n > N_r} \|S_n - S_{N_r}\|_E > \epsilon_r)$, $\mathbf{P}(B_r) < \epsilon_r$. Thus the probability that B_r happens infinitely often is zero: S_n converges almost everywhere. $\square$

Corollary 12.2.2 *If (X_n) is a symmetric sequence of random variables for which $\sum_{n=1}^{\infty} X_n$ converges almost everywhere to S, and if $S \in L^p(E)$, where $0 < p < \infty$, then $S^* \in L^p$ and $\sum_{n=1}^{\infty} X_n$ converges to S in L^p norm.*

Proof

$$\mathbf{E}(S^*)^p = p \int_0^{\infty} t^{p-1} \mathbf{P}(S^* > t) \, dt$$

$$\leq 2p \int_0^{\infty} t^{p-1} \mathbf{P}(\|S\|_E > t) \, dt = 2\mathbf{E}(\|S\|_E)^p.$$

Since $\|S_n - S\|_E^p \leq (2S^*)^p$ and $\|S_n - S\|_E^p \to 0$ almost everywhere, $\mathbf{E}(\|S_n - S\|_E^p) \to 0$ as $n \to \infty$, by the dominated convergence theorem. $\square$

Corollary 12.2.3 *Suppose that (X_n) is a symmetric sequence of random variables for which $\sum_{n=1}^{\infty} X_n$ converges almost everywhere to S. Then, for each subsequence (X_{n_k}), $\sum_{k=1}^{\infty} X_{n_k}$ converges almost everywhere. Further, if $S \in L^p(E)$, where $0 < p < \infty$, then $\sum_{k=1}^{\infty} X_{n_k}$ converges in L^p norm.*

Proof Let $X_n' = X_n$, if $n = n_k$ for some k, and let $X_n' = -X_n$ otherwise. Then (X_n') has the same distribution as (X_n), and so it has the same convergence properties. Let $Y_n = \frac{1}{2}(X_n + X_n')$. Then $\sum_{n=1}^{\infty} Y_n = \sum_{k=1}^{\infty} X_{n_k}$, from which the result follows. $\square$

12.3 Khintchine's inequality

Let us now consider possibly the simplest example of a symmetric sequence. Let $X_n = \epsilon_n a_n$, where (a_n) is a sequence of real numbers and (ϵ_n) is a sequence of Bernoulli random variables. If $(a_n) \in l_1$, so that $\sum_n a_n$ converges absolutely, then $\sum_n \epsilon_n(\omega)a_n$ converges for all ω, and the partial sums s_n converge in norm in $L^{\infty}(D_2^{\mathbf{N}})$. On the other hand, if $(a_n) \in c_0$ and $(a_n) \notin l_1$ then $\sum_n \epsilon_n(\omega)a_n$ converges for some, but not all, ω. What more can we say?

First, let us consider the case where $p = 2$. Since

$$\mathbf{E}(\epsilon_m \epsilon_n) = \mathbf{E}(1) = 1 \text{ if } m = n, \quad \mathbf{E}(\epsilon_m \epsilon_n) = \mathbf{E}(\epsilon_m)\mathbf{E}(\epsilon_n) = 0 \text{ otherwise,}$$

(ϵ_n) is an orthonormal sequence in $L^2(\Omega)$. Thus $\sum_{n=1}^{\infty} \epsilon_n a_n$ converges in L^2 norm if and only if $(a_n) \in l_2$. If this is so then $\|\sum_{n=1}^{\infty} \epsilon_n a_n\|_2 = \|(a_n)\|_2$; further, the series converges almost everywhere, by Corollary 12.2.1 (or by the martingale convergence theorem). Thus things behave extremely well.

We now come to Khintchine's inequality, which we prove for finite sums. This does two things. First, it determines what happens for other values of p. Second, and perhaps more important, it gives information about the Orlicz norms $\|.\|_{\exp}$ and $\|.\|_{\exp^2}$, and the distribution of the sum.

Theorem 12.3.1 (Khintchine's inequality) *There exist positive constants A_p and B_p, for $0 < p < \infty$, such that if $a_1, \ldots, a_N$ are real numbers and $\epsilon_1, \ldots, \epsilon_N$ are Bernoulli random variables, then*

$$A_p \|s_N\|_p \leq \sigma \leq B_p \|s_N\|_p,$$

where $s_N = \sum_{n=1}^N \epsilon_n a_n$ and $\sigma^2 = \|s_N\|_2^2 = \sum_{n=1}^N a_n^2$.

If $0 < p \le 2$, we can take $A_p = 1$ and $B_p \le 3^{1/p-1/2}$. If $2 \le p < \infty$ we can take $A_p \sim (e/p)^{1/2}$ as $p \to \infty$, and $B_p = 1$.

If t is real then $\mathbf{E}(e^{ts_N}) \le e^{t^2\sigma^2/2}$. Further, $\mathbf{E}(e^{s_N^2/4\sigma^2}) \le 2$ and $\mathbf{P}(|S_N| > \beta) \le 2e^{-\beta^2/2\sigma^2}$, for $\beta > 0$.

Proof This proof was given by Khintchine and independently, in a slightly different form, by Littlewood. The inclusion mapping $L_q \to L_p$ is norm decreasing for $0 < p < q < \infty$, and so $\|s_N\|_p \le \sigma$ for $0 < p < 2$ and $\sigma \le \|s_N\|_p$ for $2 < p < \infty$. Thus we can take $A_p = 1$ for $0 < p \le 2$ and $B_p = 1$ for $2 < p < \infty$. The interest lies in the other inequalities. First we consider the case where $2 < p < \infty$. If $2k - 2 < p < 2k$, where $2k$ is an even integer, then $\|s_N\|_{2k-2} \le \|s_N\|_p \le \|s_N\|_{2k}$. Thus it is sufficient to establish the existence and asymptotic properties of A_{2k}, where $2k$ is an even integer. In this case,

$$\left\| \sum_{n=1}^N \epsilon_n a_n \right\|_{2k}^{2k} = \mathbf{E}\left(\sum_{n=1}^N \epsilon_n a_n \right)^{2k}$$

$$= \sum_{j_1+\cdots+j_N=2k} \frac{(2k)!}{j_1!\cdots j_N!} a_1^{j_1} \cdots a_N^{j_N} \mathbf{E}(\epsilon_1^{j_1} \cdots \epsilon_N^{j_N})$$

$$= \sum_{j_1+\cdots+j_N=2k} \frac{(2k)!}{j_1!\cdots j_N!} a_1^{j_1} \cdots a_N^{j_N} \mathbf{E}(\epsilon_1^{j_1}) \cdots \mathbf{E}(\epsilon_N^{j_N}),$$

by independence. Now $\mathbf{E}(\epsilon_n^{j_n}) = \mathbf{E}(1) = 1$ if j_n is even, and $\mathbf{E}(\epsilon_n^{j_n}) = \mathbf{E}(\epsilon_n) = 0$ if j_n is odd. Thus many of the terms in the sum are 0, and

$$\left\| \sum_{n=1}^N \epsilon_n a_n \right\|_{2k}^{2k} = \sum_{k_1+\cdots+k_N=k} \frac{(2k)!}{(2k_1)!\cdots(2k_N)!} a_1^{2k_1} \cdots a_N^{2k_N}.$$

But $(2k_1)!\cdots(2k_n)! \ge 2^{k_1}k_1!\cdots 2^{k_N}k_N! = 2^k k_1!\cdots k_N!$, and so

$$\left\| \sum_{n=1}^N \epsilon_n a_n \right\|_{2k}^{2k} \le \frac{(2k)!}{2^k k!} \sum_{k_1+\cdots+k_N=k} \frac{k!}{(k_1)!\cdots(k_N)!} a_1^{2k_1} \cdots a_N^{2k_N}$$

$$= \frac{(2k)!}{2^k k!} \sigma^{2k}.$$

Thus we can take $A_{2k} = ((2k)!/2^k k!)^{-1/2k}$. Note that $A_{2k} \ge 1/\sqrt{2k}$, and that $A_{2k} \sim (e/2k)^{1/2}$ as $k \to \infty$, by Stirling's formula.

Then, since $\mathbf{E}(S_N^n) = 0$ if n is odd,

$$\mathbf{E}(e^{ts_N}) = \sum_{n=0}^{\infty} \frac{t^n \mathbf{E}(s_N^n)}{n!} = \sum_{k=0}^{\infty} \frac{t^{2k} \mathbf{E}(s_N^{2k})}{(2k)!}$$

$$\leq \sum_{k=0}^{\infty} \frac{t^{2k}}{(2k)!} \frac{(2k)! \sigma^{2k}}{k! 2^k} = e^{t^2 \sigma^2/2}.$$

Similarly,

$$\mathbf{E}(e^{s_N^2/4\sigma^2}) = \sum_{k=0}^{\infty} \frac{\mathbf{E}(s_N^{2k})}{2^{2k} \sigma^{2k} k!} \leq \sum_{k=0}^{\infty} \frac{(2k)!}{2^{3k}(k!)^2} \leq 2,$$

since $(2k)! \leq 2^{2k}(k!)^2$.

Further, by Markov's inequality,

$$\mathbf{P}(|s_N| > \beta) = 2\mathbf{P}(s_N > \beta) = 2e^{-t\beta}\mathbf{E}(e^{ts_N}) \leq 2e^{-t\beta}e^{t^2\sigma^2/2}.$$

Setting $t = \beta/\sigma^2$, we obtain the final inequality.

We now consider the case where $0 < p \leq 2$. Here we use Littlewood's inequality. Note that the argument above shows that we can take $A_4 = 3^{1/4}$. Suppose that $0 < p < 2$. Let $\theta = (4 - 2p)/(4 - p)$, so that $1/2 = (1-\theta)/p + \theta/4$. Then, by Littlewood's inequality,

$$\sigma = \|s_N\|_2 \leq \|s_N\|_p^{(1-\theta)} \|s_N\|_4^{\theta} \leq 3^{\theta/4}\sigma^{\theta} \|s_N\|_p^{(1-\theta)},$$

so that $\sigma \leq 3^{1/p-1/2}\|s_N\|_p$, and we can take $B_p = 3^{1/p-1/2}$. In particular we can take $B_1 = \sqrt{3}$.

This part of the argument is due to Littlewood; unfortunately, he made a mistake in his calculations, and obtained $B_1 = \sqrt{2}$. This is in fact the best possible constant (take $N = 2$, $a_1 = a_2 = 1$), but this is much harder to prove. We shall do so later (Theorem 13.3.1). □

12.4 The law of the iterated logarithm

Why did Khintchine prove his inequality? In order to answer this, let us describe another setting in which a Bernoulli sequence of random variables occurs. Take $\Omega = [0, 1)$, with Lebesgue measure. If $x \in [0, 1)$, let $x = 0 \cdot x_1 x_2 \ldots$ be the binary expansion of x (disallowing recurrent 1s). Let $r_j(x) = 2x_j - 1$, so that $r_j(x) = 1$ if $x_j = 1$ and $r_j(x) = -1$ if $x_j = 0$. the functions r_j are the *Rademacher functions*; considered as random variables on Ω, they form a Bernoulli sequence of random variables. They are closely connected to the dyadic filtration of $[0, 1)$; the Rademacher function r_j is measurable with respect to the finite σ-field Σ_j generated by the intervals $[k/2^j, (k+1)/2^j)$, for $0 \leq k < 2^j - 1$. Suppose now that $x = 0.x_1 x_2 \ldots$ is a number in $[0, 1)$, in its binary expansion (disallowing recurrent 1s). Let $t_n(x)$ be the number of times that 1 occurs in $x_1, \ldots, x_n$, and let $a_n(x) = t_n(x)/n$. We say that x is 2-*normal* if $a_n(x) \to \frac{1}{2}$ as $n \to \infty$. In 1909, Borel proved his normal numbers theorem, the first of all the *strong laws of large numbers*. In its simplest form, this says that almost every number in $[0, 1)$ is 2-normal. We can express this in terms of the Rademacher functions, as follows. Let $s_n(x) = \sum_{j=1}^{n} r_j(x)$; then $s_n(x)/n \to 0$ for almost all x. Once Borel's theorem had been proved, the question was raised: how does the sequence $(t_n(x) - \frac{1}{2})$ behave as $n \to \infty$? Equivalently, how does the sequence $(s_n(x))$ behave? Hardy and Littlewood gave partial answers, but in 1923, Khintchine [Khi 23] proved the following.

Theorem 12.4.1 (Khintchine's law of the iterated logarithm) *For* $n \geq 3$, *let* $L_n = (2n \log \log n)^{1/2}$. *If* (r_n) *are the Rademacher functions and* $s_n = \sum_{j=1}^{n} r_j$ *then*

$$\limsup_{n \to \infty} |s_n(x)/L_n| \leq 1 \quad \textit{for almost all } x \in [0, 1).$$

Proof The proof that follows is essentially the one given by Khinchine, although he had to be rather more ingenious, since we use the reflection principle, which had not been proved in 1923. Suppose that $\lambda > 1$. We need to show that for almost all x, $|s_n(x)| > \lambda L_n$ for only finitely many n, and we shall use the first Borel–Cantelli lemma to do so.

Let $\alpha = \lambda^{1/2}$, so that $1 < \alpha < \lambda$. Let n_k be the least integer greater than α^k. The sequence n_k is eventually strictly increasing – there exists k_0 such that $n_k > n_{k-1} > 3$ for $k > k_0$. Let

$$B_k = \left(\sup_{n_{k-1} < n \leq n_k} |s_n| > \lambda L_n \right), \quad \text{for } k \geq k_0.$$

Now $L_{n_k}/L_{n_{k-1}} \to \sqrt{\alpha}$ as $k \to \infty$, and so there exists $k_1 \geq k_0$ so that $L_{n_k} \leq \alpha L_{n_{k-1}}$ for $k \geq k_1$. Thus if $k > k_1$ and $n_{k-1} < n \leq n_k$ then $\lambda L_n \geq \lambda L_{n_{k-1}} \geq \alpha L_{n_k}$, and so

$$B_k \subseteq \left(\sup_{n_{k-1} < n \leq n_k} |s_n| > \alpha L_{n_k} \right) \subseteq \left(s_{n_k}^* > \alpha L_{n_k} \right),$$

so that, since $\mathbf{E}(s_{n_k}^2) = n_k$,

$$
\begin{aligned}
\mathbf{P}(B_k) &\leq \mathbf{P}(s_{n_k}^* > \alpha L_{n_k}) \\
&\leq 2\mathbf{P}(|s_{n_k}| > \alpha L_{n_k}) \quad \text{(by the reflection principle)} \\
&\leq 4e^{-\lambda \log\log n_k} \quad \text{(by Khintchine's inequality)} \\
&\leq 4e^{-\lambda \log(k \log \alpha)} \quad \text{(by the choice of } n_k) \\
&= 4\left(\frac{1}{k \log \alpha}\right)^{\lambda},
\end{aligned}
$$

and so $\sum_{k=k_1}^{\infty} \mathbf{P}(B_k) < \infty$. Thus for almost all x, $|s_n(x)| \leq \lambda L_n$ for all but finitely many n. $\qquad\square$

Later Khintchine and Kolmogoroff showed that this is just the right answer:

$$
\limsup_{n \to \infty} |s_n(x)/L_n| = 1 \quad \text{for almost all } x \in [0, 1).
$$

We shall however not prove this; a proof, in the spirit of the above argument, using a more detailed version of the De Moivre central limit theorem that we shall prove in the next chapter, is given in [Fel 70], Theorem VIII.5.

12.5 Strongly embedded subspaces

We have proved Khintchine's inequality for finite sums. From this, it is a straightforward matter to prove the following result for infinite sums.

Theorem 12.5.1 *Let S be the closed linear span of the orthonormal sequence $(\epsilon_n)_{n=1}^{\infty}$ in $L^2(D_2^{\mathbf{N}})$, and suppose that $f \in S$. If $0 < p < 2$, then $\|f\|_p \leq \|f\|_2 \leq B_p \|f\|_p$, if $2 < p < \infty$ then $A_p \|f\|_p \leq \|f\|_2 \leq \|f\|_p$, and $\|f\|_{\exp^2} \leq 2\|f\|_2 \leq 2\|f\|_{\exp^2}$. Further, $P(|f| > \beta) \leq 2e^{-\beta^2/2\|f\|_2^2}$.*

Proof The details are left to the reader. $\qquad\square$

The fact that all these norms are equivalent on S is remarkable, important, and leads to the following definition. A closed linear subspace S of a Banach function space $X(E)$ is said to be *strongly embedded* in $X(E)$ if whenever $f_n \in S$ and $f_n \to 0$ in measure (or in probability) then $\|f_n\|_{X(E)} \to 0$.

Proposition 12.5.1 *If S is strongly embedded in $X(E)$ and $X(E) \subseteq Y(E)$ then the norms $\|.\|_{X(E)}$ and $\|.\|_{Y(E)}$ are equivalent on S, and S is strongly embedded in $Y(E)$.*

Proof A simple application of the closed graph theorem shows that the inclusion mapping $X(E) \to Y(E)$ is continuous. If $f_n \in S$ and $\|f_n\|_{Y(E)} \to 0$ then $f_n \to 0$ in measure, and so $\|f_n\|_{X(E)} \to 0$. Thus the inverse mapping is continuous on S, and the norms are equivalent on S. It now follows immediately that S is strongly embedded in Y. $\qquad\square$

Proposition 12.5.2 *Suppose that $\mu(\Omega) = 1$ and that $1 \leq p < q < \infty$. If S is a closed linear subspace of $L^q(E)$ on which the $L^p(E)$ and $L^q(E)$ norms are equivalent, then S is strongly embedded in $L^q(E)$.*

Proof We denote the norms on $L^p(E)$ and $L^q(E)$ by $\|.\|_p$ and $\|.\|_q$. There exists C_p such that $\|f\|_q \leq C \|f\|_p$ for $f \in S$. We shall show that there exists $\epsilon_0 > 0$ such that if $f \in S$ then

$$\mu(|f| \geq \epsilon_0 \|f\|_q) \geq \epsilon_0.$$

Suppose that $f \in S$, that $\epsilon > 0$ and that $\mu(|f| \geq \epsilon \|f\|_q) < \epsilon$ for some $\epsilon > 0$. We shall show that ϵ must be quite big. Let $L = (|f| \geq \epsilon \|f\|_q)$. Then

$$\|f\|_p^p = \int_L |f|^p \, d\mu + \int_{\Omega \setminus L} |f|^p \, d\mu \leq \int_L |f|^p \, d\mu + \epsilon^p \|f\|_q^p .$$

We apply Hölder's inequality to the first term. Define t by $p/q + 1/t = 1$. Then

$$\int_L |f|^p \, d\mu \leq \left(\int_L |f|^q \, d\mu \right)^{p/q} (\mu(L))^{1/t} \leq \epsilon^{1/t} \|f\|_q^p .$$

Consequently

$$\|f\|_p \leq \left(\epsilon^p + \epsilon^{1/t} \right)^{1/p} \|f\|_q \leq C_p \left(\epsilon^p + \epsilon^{1/t} \right)^{1/p} \|f\|_p .$$

Thus $\epsilon > \epsilon_0$, for some ϵ_0 which depends only on C_p, p and q. Thus if $f \in S$, $\mu(|f| \geq \epsilon_0 \|f\|_q) \geq \epsilon_0$.

Suppose now that $f_n \to 0$ in probability. Let $\eta > 0$. Then there exists n_0 such that $\mu(|f_n| \geq \epsilon_0 \eta) < \epsilon_0 / 2$ for $n \geq n_0$, and so $\epsilon_0 \|f_n\|_q \leq \epsilon_0 \eta$ for $n \geq n_0$. Consequently $\|f_n\|_q < \eta$ for $n \geq n_0$. $\qquad\square$

Corollary 12.5.1 *The space S of Theorem 12.5.1 is strongly embedded in $L_{\exp 2}$, and in each of the L^p spaces.*

Proof S is certainly strongly embedded in L^p, for $1 \leq p < \infty$; since the norms $\|.\|_p$ and $\|.\|_{\exp 2}$ are equivalent on S, it is strongly embedded in $L_{\exp 2}$. $\qquad\square$

Combining this with Corollary 12.2.1, we have the following.

Corollary 12.5.2 *Suppose that (a_n) is a real sequence. The following are equivalent:*

(i) $\sum_{n=1}^{\infty} a_n^2 < \infty$;

(ii) $\sum_{n=1}^{\infty} a_n \epsilon_n$ *converges in probability;*

(iii) $\sum_{n=1}^{\infty} a_n \epsilon_n$ *converges almost everywhere;*

(iv) $\sum_{n=1}^{\infty} a_n \epsilon_n$ *converges in L^p norm for some $0 < p < \infty$;*

(v) $\sum_{n=1}^{\infty} a_n \epsilon_n$ *converges in L^p norm for all $0 < p < \infty$;*

(vi) $\sum_{n=1}^{\infty} a_n \epsilon_n$ *converges in $L_{\exp 2}$ norm.*

12.6 Stable random variables

Are there other natural examples of strongly embedded subspaces? A real-valued random variable X is a *standard real Gaussian random variable* if it has density function $(1/2\pi)^{-1/2} e^{-t^2/2}$, and a complex-valued random variable X is a *standard complex Gaussian random variable* if it has density function $(1/2\pi)e^{-|z|^2}$. Each has mean 0 and variance $\mathbf{E}(|X|^2) = 1$. If (X_n) is a sequence of independent standard Gaussian random variables and $(a_1, \ldots, a_N)$ are real numbers then $S_N = \sum_{n=1}^{N} a_n X_n$ is a normal random variable with mean 0 and variance

$$\sigma^2 = \mathbf{E}\left(\left(\sum_{n=1}^{N} a_n X_n\right)^2\right) = \sum_{n=1}^{N} |a_n|^2;$$

that is, S_N/σ is a standard Gaussian random variable. Thus if $0 < q < \infty$ then

$$\mathbf{E}(|S_N|^q) = \sigma^q \sqrt{\frac{2}{\pi}} \int_0^{\infty} t^q e^{-t^2/2} \, dt$$

$$= \sigma^q \sqrt{\frac{2^q}{\pi}} \int_0^{\infty} u^{(q-1)/2} e^{-u} \, du$$

$$= \sqrt{\frac{2^q}{\pi}} \Gamma((q+1)/2) \sigma^q.$$

Thus if S is the closed linear span of (X_n) in L^2 then all the L^p norms on S are multiples of the L^2 norm, and the mapping $(a_n) \to \sum_{n=1}^{\infty} a_n X_n$ is a scalar multiple of an isometry of l_2 into $L^p(\Omega)$. Similarly, if $\|S_N\|_2 = \sqrt{3/8}$ then $\mathbf{E}(e^{S_n^2}) = 2$, so that in general $\|S_N\|_{\exp 2} = \sqrt{8/3}\,\|S_N\|_2$, the mapping $(a_n) \to \sum_{n=1}^{\infty} a_n X_n$ is a scalar multiple of an isometry of l_2 into $L_{\exp 2}$, and the image of l_2 is strongly embedded in $L_{\exp 2}$.

Here is another example. A real random variable X is said to have the Cauchy distribution with parameter a if it has probability density function $a/\pi(t^2 + a^2)$. If so, then it has characteristic function $\mathbf{E}(e^{iXt}) = e^{-|at|}$. X is not integrable, but is in $L^q(\Omega)$, for $0 < q < 1$. Now let (X_n) be an independent sequence of random variables, each with the Cauchy distribution, with parameter 1. If $(a_1, \ldots, a_N)$ are real numbers then $S_N = \sum_{n=1}^{N} a_n X_n$ is a Cauchy random variable with parameter $\|(a_n)\|_1$, so that $S_N / \|(a_n)\|_1$ is a Cauchy random variable with parameter 1. Thus the mapping $(a_n) \rightarrow \sum_{n=1}^{\infty} a_n X_n$ is a scalar multiple of an isometry of l_1 into $L^q(\Omega)$, for $0 < q < 1$, and the image of l_1 is strongly embedded in $L^q(\Omega)$, for $0 < q < 1$.

These examples are special cases of a more general phenomenon. If X is a standard real Gaussian random variable then its characteristic function $\mathbf{E}(e^{itX})$ is $e^{-t^2/2}$, while if X has Cauchy distribution with density $1/\pi(x^2+1)$ then its characteristic function is $e^{-|t|}$. In fact, for each $0 < p < 2$ there exists a random variable X with characteristic function $e^{-|t|^p/p}$; such a random variable is called a *symmetric p-stable* random variable. X is not in $L^p(\Omega)$, but $X \in L^q(\Omega)$ for $0 < q < p$. If (X_n) is an independent sequence of random variables, each with the same distribution as X, and if $a_1, \ldots, a_N$ are real, then $S_N / \|(a_n)\|_p = (\sum_{n=1}^{N} a_n X_n) / \|(a_n)\|_p$ has the same distribution as X; thus if $0 < q < p$, the mapping $(a_n) \rightarrow \sum_{n=1}^{\infty} a_n X_n$ is a scalar multiple of an isometry of l_p into $L^q(\Omega)$, and the image of l_p is strongly embedded in $L^q(\Omega)$, for $0 < q < p$.

12.7 Sub-Gaussian random variables

Recall that Khintchine's inequality shows that if $S_N = \sum_{n=1}^{N} a_n \epsilon_n$ then its moment generating function $\mathbf{E}(e^{tX})$ satisfies $\mathbf{E}(e^{tX}) \leq e^{\sigma^2 t^2/2}$. On the other hand, if X is a random variable with a Gaussian distribution with mean 0 and variance $\mathbf{E}(X^2) = \sigma^2$, its moment generating function $E(e^{tX})$ is $e^{\sigma^2 t^2/2}$. This led Kahane [Kah 85] to make the following definition. A random variable X is *sub-Gaussian, with exponent b*, if $\mathbf{E}(e^{tX}) \leq e^{b^2 t^2/2}$ for $-\infty < t < \infty$.

The next result gives basic information about sub-Gaussian random variables.

Theorem 12.7.1 *Suppose that X is a sub-Gaussian random variable with exponent b. Then*

(i) $P(X > R) \leq e^{-R^2/2b^2}$ and $P(X < -R) \leq e^{-R^2/2b^2}$ for each $R > 0$;

(ii) $X \in L_{\exp^2}$ and $\|X\|_{\exp^2} \leq 2b$;

(iii) X is integrable, $\mathbf{E}(X) = 0$, and $\mathbf{E}(X^{2k}) \leq 2^{k+1}k!b^{2k}$ for each positive integer k.

Conversely if X is a real random variable which satisfies (iii) then X is sub-Gaussian with exponent $2\sqrt{2}b$.

Proof (i) By Markov's inequality, if $t > 0$ then

$$e^{tR}\mathbf{P}(X > R) \leq \mathbf{E}(e^{tX}) \leq e^{b^2t^2/2}.$$

Setting $t = R/b^2$, we see that $\mathbf{P}(X > R) \leq e^{-R^2/2b^2}$. Since $-X$ is also sub-Gaussian with exponent b, $\mathbf{P}(X < -R) \leq e^{-R^2/2b^2}$ as well.

(ii)

$$\mathbf{E}(e^{X^2/4b^2}) = \frac{1}{2b^2} \int_0^\infty t e^{t^2/4b^2} \mathbf{P}(|X| > t)\, dt$$
$$\leq \frac{1}{b^2} \int_0^\infty t e^{-t^2/4b^2}\, dt = 2.$$

(iii) Since $X \in L_{\exp^2}$, X is integrable. Since $tx \leq e^{tx} - 1$, $t\mathbf{E}(X) \leq e^{b^2t^2/2} - 1$, from which it follows that $\mathbf{E}(X) \leq 0$. Since $-X$ is sub-Gaussian, $\mathbf{E}(X) \geq 0$ as well. Thus $\mathbf{E}(X) = 0$.

Further,

$$E(X^{2k}) = 2k \int_0^\infty t^{2k-1}\mathbf{P}(|X| > t)\, dt$$
$$\leq 2.2k \int_0^\infty t^{2k-1}e^{-t^2/2b^2}\, dt$$
$$= (2b^2)^k 2k \int_0^\infty s^{k-1}e^{-s}\, ds = 2^{k+1}k!b^{2k}.$$

Finally, suppose that X is a real random variable which satisfies (iii). If $y > 0$ and $k \geq 1$ then

$$\frac{y^{2k+1}}{(2k+1)!} \leq \frac{y^{2k}}{(2k)!} + \frac{y^{2k+2}}{(2k+2)!},$$

so that

$$\mathbf{E}(e^{tX}) \leq 1 + \sum_{n=2}^{\infty} \mathbf{E}\left(\frac{|tX|^n}{n!}\right)$$

$$\leq 1 + 2\sum_{k=1}^{\infty} \mathbf{E}\left(\frac{|tX|^{2k}}{(2k)!}\right)$$

$$\leq 1 + 4\sum_{k=1}^{\infty} \frac{k!(2b^2t^2)^k}{(2k)!}$$

$$\leq 1 + \sum_{k=1}^{\infty} \frac{(4b^2t^2)^k}{k!} = e^{4b^2t^2},$$

since $2(k!)^2 \leq (2k)!$ $\square$

Note that this theorem shows that if X is a bounded random variable with zero expectation then X is sub-Gaussian.

If $X_1, \ldots, X_N$ are independent sub-Gaussian random variables with exponents $b_1, \ldots, b_N$ respectively, and $a_1, \ldots, a_N$ are real numbers, then

$$\mathbf{E}(e^{t(a_1X_1+\cdots+a_NX_N)}) = \prod_{n=1}^{N} \mathbf{E}(e^{ta_nX_n}) \leq \prod_{n=1}^{N} e^{a_n^2 b_n^2/2},$$

so that $a_1X_1 + \cdots + a_NX_N$ is sub-Gaussian, with exponent $(a_1^2b_1^2 + \cdots + a_N^2b_N^2)^{1/2}$. We therefore obtain the following generalization of Khinchine's inequality.

Proposition 12.7.1 *Suppose that (X_n) is a sequence of independent identically distributed sub-Gaussian random variables with exponent b, and let S be their closed linear span in L^2. Then S is strongly embedded in $L_{\exp^2}$.*

12.8 Kahane's theorem and Kahane's inequality

We now turn to the vector-valued case. We restrict our attention to an independent sequence of symmetric random variables, taking values in the unit ball of a Banach space E.

Theorem 12.8.1 *Suppose that (X_n) is an independent sequence of symmetric random variables, and suppose that $\sum_{n=1}^{\infty} X_n$ converges almost everywhere to S. Let $S^* = \sup_n \|S_n\|_E$. Then, if $t > 0$,*

$$\mathbf{P}(S^* > 2t + 1) \leq 4(\mathbf{P}(S^* > t))^2.$$

Proof Once again, we use a stopping time argument. Let $T = \inf\{j\colon \|S_j\| > t\}$ and let $A_m = (T = m)$. Fix an index k, and consider the event $B_k = (\|S_k\|_E > 2t + 1)$. Clearly $B_k \subseteq (T \leq k)$, and so

$$\mathbf{P}(B_k) = \sum_{j=1}^{k} \mathbf{P}(A_j \cap B_k).$$

But if $\omega \in A_j$ then $\|S_{j-1}(\omega)\|_E \leq t$, so that $\|S_j(\omega)\|_E \leq t + 1$. Thus if $\omega \in A_j \cap B_k$, $\|S_k - S_j(\omega)\|_E > t$. Using the fact that A_j and $S_k - S_j$ are independent, we therefore have

$$\mathbf{P}(A_j \cap B_k) \leq \mathbf{P}(A_j \cap (\|S_k - S_j\|_E > t)) = \mathbf{P}(A_j)\mathbf{P}(\|S_k - S_j\|_E > t).$$

Applying the reflection principle to the sequence $(S_k - S_j, S_j, 0, 0, \ldots)$, we see that

$$\mathbf{P}(\|S_k - S_j\|_E > t) \leq 2\mathbf{P}(\|S_k\|_E > t) \leq 2\mathbf{P}(S^* > t).$$

Substituting and adding,

$$\mathbf{P}(B_k) = \sum_{j=1}^{k} \mathbf{P}(A_j \cap B_k) \leq 2\Big(\sum_{j=1}^{k} \mathbf{P}(A_k)\Big)\mathbf{P}(S^* > t) \leq 2(\mathbf{P}(S^* > t))^2.$$

Using the reflection principle again,

$$\mathbf{P}(\sup_{1 \leq n \leq k} \|S_n\|_E > 2t + 1) \leq 2\mathbf{P}(B_k) \leq 4(\mathbf{P}(S^* > t))^2.$$

Letting $k \to \infty$, we obtain the result. $\qquad\square$

Theorem 12.8.2 (Kahane's Theorem) *Suppose that (X_n) is an independent sequence of symmetric random variables, taking values in the unit ball of a Banach space E. If $\sum_{n=1}^{\infty} X_n$ converges almost everywhere to S then $S^* \in L_{\exp}$, $\mathbf{E}(e^{\alpha S^*}) < \infty$, for each $\alpha > 0$, and $\sum_{n=1}^{\infty} X_n$ converges to S in $L_{\exp}$ norm.*

Proof Suppose that $\alpha > 0$. Choose $0 < \theta < 1$ so that $e^{\alpha\theta} < 3/2$ and $e^{4\alpha\theta} < 1/2$. Since $S_n \to S$ almost everywhere, there exists N such that $\mathbf{P}(\|S - S_N\|_E > \theta) < \theta/8$. Let $Z_n = X_{N+n}$, let $R_k = \sum_{j=1}^{k} Z_j$, let $R = \sum_{j=1}^{\infty} Z_j$, and let $R^* = \sup_k \|R_k\|_E$. We shall show that $\mathbf{E}(e^{\alpha R^*}) \leq 2$, so that $R^* \in L_{\exp}$ and $\|R^*\|_{\exp} \leq 1/\alpha$. Since $S^* \leq N + R^*$, it follows that $S^* \in L_{\exp}$, that $\|S^*\|_{\exp} \leq \|N\|_{\exp} + \|R^*\|_{\exp} \leq (N/\log 2) + 1/\alpha$ and that $\mathbf{E}(e^{\alpha S^*}) \leq e^{\alpha N}\mathbf{E}(e^{\alpha R^*}) \leq 2e^{\alpha N}$. Further, since $\|S_n - S\|_E \leq 2R^*$ for

$n \geq N$, $\|S_n - S\|_{\exp} \leq 2/\alpha$ for $n \geq N$. Since this holds for any $\alpha > 0$, $S_n \to S$ in $L_{\exp}$ norm.

It remains to show that $\mathbf{E}(e^{\alpha R^*}) \leq 2$. Since $R = S - S_N$, $\mathbf{P}(\|R\|_E > \theta) < \theta/8$, and so by the reflection principal $\mathbf{P}(R^* > \theta) < \theta/4$. Let $\phi = \theta + 1$, let $t_0 = \theta = \phi - 1$, and let $t_r = 2^r\phi - 1$, for $r \in \mathbf{N}$. Then $t_{r+1} = 2t_r + 1$; applying Theorem 12.8.1 inductively, we find that

$$\mathbf{P}(R^* > t_r) \leq \frac{\theta 2^r}{4}.$$

Then, since $e^{2\alpha\phi}\theta < \frac{1}{2}$,

$$\mathbf{E}(e^{\alpha R^*}) \leq e^{\alpha t_0}\mathbf{P}(R^* \leq t_0) + \sum_{r=0}^{\infty} e^{\alpha t_{r+1}}\mathbf{P}(t_r < R^* \leq t_{r+1})$$

$$\leq e^{\alpha\theta} + \sum_{r=0}^{\infty} e^{\alpha t_{r+1}}\mathbf{P}(R^* > t_r)$$

$$\leq \frac{3}{2} + \frac{1}{4}\sum_{r=1}^{\infty} e^{\alpha(2^{r+1}\phi - 1)}\theta 2^r$$

$$= \frac{3}{2} + \frac{e^{-\alpha}}{4}\sum_{r=1}^{\infty} \left(e^{2\alpha\phi}\theta\right)^{2^r}$$

$$< \frac{3}{2} + \frac{1}{4}\sum_{r=1}^{\infty} 2^{-2^r} < 2.$$

$\square$

Corollary 12.8.1 $S \in L^p(\Omega)$, *for* $0 < p < \infty$, *and* $S_n \to S$ *in* L^p *norm.*

Corollary 12.8.2 *Suppose that* (ϵ_n) *is a Bernoulli sequence of random variables, and that E is a Banach space. Let*

$$S = \left\{\sum_{n=1}^{\infty} \epsilon_n x_n \colon x_n \in E, \sum_{n=1}^{\infty} \epsilon_n x_n \text{ converges almost everywhere}\right\}.$$

Then S is strongly embedded in $L_{\exp}(E)$.

Proof Take $a = 1$ and $\theta = e^{-5}$, so that $e^{\theta} < 3/2$ and $e^4\theta < 1/2$. If $s = \sum_{n=1}^{\infty} \epsilon_n x_n \in S$, then $\|s\|_1 < \infty$. Suppose that $\|s\|_1 \leq \theta^2/8$. Then $\|x_n\| \leq 1$ for each n, and so we can apply the theorem. Also $\mathbf{P}(\|s\|_E > \theta) \leq \theta/8$, by Markov's inequality, and the calculations of the theorem then show that $\|s\|_{\exp} \leq 1$. This shows that S is strongly embedded in $L_{\exp}$, and the final inequality follows from this. $\square$

Corollary 12.8.3 (Kahane's inequality) *If $1 < p < q$ then there exists a constant K_{pq} such that if $u_1, \ldots, u_n \in E$ then*

$$\left\| \sum_{n=1}^{N} \epsilon_n u_n \right\|_q \leq K_{pq} \left\| \sum_{n=1}^{N} \epsilon_n u_n \right\|_p .$$

We shall prove a more general form of Kahane's inequality in the next chapter.

12.9 Notes and remarks

Spellings of Khintchine's name vary. I have followed the spelling used in his seminal paper [Khi 23]. A similar remark applies to the spelling of Kolmogoroff.

For more details about p-stable random variables, see [Bre 68] or [ArG 80].

We have discussed Khintchine's use of his inequality. But why did Littlewood prove it? We shall discuss this in Chapter 18.

Exercises

12.1 Suppose that $L_\Phi(\Omega, \Sigma, \mu)$ is an Orlicz space and that $f \in L_\Phi$. Suppose that g is a measurable function for which $\mu(|g| > t) \leq 2\mu(|f| > t)$ for all $t > 0$. Show that $g \in L_\Phi$ and $\|g\|_\Phi \leq 2\|f\|_\Phi$.

Hint: Consider the functions g_1 and g_{-1} defined on $\Omega \times D_2$ as

$$g_1(\omega, 1) = g(\omega), \quad g_1(\omega, -1) = 0,$$
$$g_{-1}(\omega, 1) = 0, \quad g_{-1}(\omega, -1) = g(\omega).$$

12.2 Let

$$A_n = \left(\frac{1}{2^n}, \frac{1}{2^n} + \frac{1}{2^{n+1}} \right), \quad B_n = \left(\frac{1}{2^n} + \frac{1}{2^{n+1}}, \frac{1}{2^{n-1}} \right),$$

and let $X_n = n(I_{A_n} - I_{B_n})$. Show that (X_n) is a symmetric sequence of random variables defined on $(0, 1]$, equipped with Lebesgue measure. Let $S_n = \sum_{j=1}^{n} X_j$ and $S = \sum_{j=1}^{\infty} X_j$. Show that $S^* = |S|$, and that $S^* \in L_{\exp}$. Show that $S_n \to S$ pointwise, but that $\|S - S_n\|_{\exp} = 1/\log 2$, so that $S_n \nrightarrow S$ in norm. Compare this with Corollary 12.2.2.

12.3 Suppose that $a_1, \ldots, a_N$ are real numbers with $\sum_{n=1}^{N} a_n^2 = 1$. Let $f = \sum_{n=1}^{N} \epsilon_n a_n$ and let $g = \prod_{n=1}^{N} (1 + i\epsilon_n a_n)$.

(a) Use the arithmetic-mean geometric mean inequality to show that $\|g\|_\infty \leq \sqrt{e}$.

(b) Show that $\mathbf{E}(fg) = i$.

(c) Show that we can take $B_1 = \sqrt{e}$ in Khintchine's inequality.

12.4   Suppose that X is a random variable with Cauchy distribution with parameter a. Show that $\mathbf{E}(e^{iXt}) = e^{-|at|}$. [This is a standard exercise in the use of the calculus of residues and Jordan's lemma.]

12.5   Suppose that F is a strongly embedded subspace of $L^p(\Omega)$, where $2 < p < \infty$. Show that F is isomorphic to a Hilbert space, and that F is complemented in $L^q(\Omega)$ (that is, there is a continuous linear projection of $L^q(\Omega)$ onto F) for $p' \leq q \leq p$.

13

Hypercontractive and logarithmic Sobolev inequalities

13.1 Bonami's inequality

In the previous chapter, we proved Kahane's inequality, but did not estimate the constants involved. In order to do this, we take a different approach. We start with an inequality that seems banal, and has an uninformative proof, but which turns out to have far-reaching consequences. Throughout this chapter, we set $r_p = 1/\sqrt{p-1}$, for $1 < p < \infty$.

Proposition 13.1.1 (Bonami's inequality) *Let*

$$F_p(x, y) = (\tfrac{1}{2}(|x + r_p y|^p + |x - r_p y|^p))^{1/p},$$

where $x, y \in \mathbf{R}$. Then $F_p(x, y)$ is a decreasing function of p on $(1, \infty)$.

Proof By homogeneity, we can suppose that $x = 1$. We consider three cases.

First, suppose that $1 < p < q \le 2$ and that $0 \le |r_p y| \le 1$. Using the binomial theorem and the inequality $(1 + x)^\alpha \le 1 + \alpha x$ for $0 < \alpha \le 1$, and putting $\alpha = p/q$, we find that

$$F_q(1, y) = \left(1 + \sum_{k=1}^{\infty} \binom{q}{2k} \left(\frac{y^2}{q-1}\right)^k\right)^{1/q}$$

$$\le \left(1 + \frac{p}{q} \sum_{k=1}^{\infty} \binom{q}{2k} \left(\frac{y^2}{q-1}\right)^k\right)^{1/p}.$$

206

Now

$$\frac{p}{q}\binom{q}{2k}\left(\frac{1}{q-1}\right)^k = \frac{p}{q}\frac{q(q-1)\cdots(q-2k+1)}{(2k)!(q-1)^k}$$
$$= \frac{p(2-q)\cdots(2k-1-q)}{(2k)!(q-1)^{k-1}}$$
$$\leq \frac{p(2-p)\cdots(2k-1-p)}{(2k)!(p-1)^{k-1}} = \binom{p}{2k}\left(\frac{1}{p-1}\right)^k.$$

Thus

$$F_q(1,y) \leq \left(1 + \sum_{k=1}^{\infty}\binom{p}{2k}\left(\frac{y^2}{p-1}\right)^k\right)^{1/p} = F_p(1,y).$$

Second, suppose that $1 < p < q \leq 2$ and that $|r_p y| \geq 1$. We use the fact that if $0 < s, t < 1$ then $1 - st > s - t$ and $1 + st > s + t$. Set $\lambda = r_q/r_p$ and $\mu = 1/|r_p y|$. Then, using the first case,

$$F_q(1,y) = (\tfrac{1}{2}(|1 + \lambda r_p y|^q + |1 - \lambda r_p y|^q))^{1/q}$$
$$= \frac{1}{\mu}(\tfrac{1}{2}(|\lambda + \mu|^q + |\lambda - \mu|^q))^{1/q}$$
$$\leq \frac{1}{\mu}(\tfrac{1}{2}(|1 + \lambda\mu|^q + |1 - \lambda\mu|^q))^{1/q}$$
$$\leq \frac{1}{\mu}(\tfrac{1}{2}(|1 + \mu|^p + |1 - \mu|^p))^{1/p} = F_p(1,y).$$

Again, let $\lambda = r_q/r_p = \sqrt{(p-1)/(q-1)}$. Note that we have shown that the linear mapping $K \in L(L^p(D_2), L^q(D_2))$ defined by

$$K(f)(x) = \int_{D_2} k(x,y)f(y)\,d\mu(y),$$

where $k(1,1) = k(-1,-1) = 1 + \lambda$ and $k(1,-1) = k(-1,1) = 1 - \lambda$, is norm-decreasing.

Third, suppose that $2 \leq p < q < \infty$. Then $1 < q' < p' \leq 2$ and $\lambda^2 = (p-1)/(q-1) = (q'-1)/(p'-1)$, so that K is norm-decreasing from $L^{q'}$ to $L^{p'}$. But k is symmetric, and so $K' = K$ is norm-decreasing from $L^{p'}$ to L^q. $\qquad\square$

Next we extend this result to vector-valued functions.

Corollary 13.1.1 *Let*

$$F_p(x,y) = (\tfrac{1}{2}(\|x + r_p y\|^p + \|x - r_p y\|^p))^{1/p},$$

where x and y are vectors in a normed space $(E, \|.\|_E)$. Then $F_p(x, y)$ is a decreasing function of p on $(1, \infty)$.

Proof We need the following lemma.

Lemma 13.1.1 *If x and z are vectors in a normed space and $-1 \le \lambda < 1$ then*

$$\|x + \lambda z\| \le \frac{1}{2}(\|x + z\| + \|x - z\|) + \frac{\lambda}{2}(\|x + z\| - \|x - z\|).$$

Proof Since

$$x + \lambda z = \left(\frac{1 + \lambda}{2}\right)(x + z) + \left(\frac{1 - \lambda}{2}\right)(x - z),$$

we have

$$\|x + \lambda z\| \le \left(\frac{1 + \lambda}{2}\right)\|x + z\| + \left(\frac{1 - \lambda}{2}\right)\|x - z\|$$

$$= \frac{1}{2}(\|x + z\| + \|x - z\|) + \frac{\lambda}{2}(\|x + z\| - \|x - z\|).$$

$\square$

We now prove the corollary. Let us set $s = x + r_p y$, $t = x - r_p y$ and $\lambda = r_q/r_p$, so that $0 < \lambda < 1$.

$$\left(\tfrac{1}{2}\left(\|x + r_q y\|^q + \|x - r_q y\|^q\right)\right)^{1/q}$$

$$= \left(\tfrac{1}{2}\left(\|x + \lambda r_p y\|^q + \|x - \lambda r_p y\|^q\right)\right)^{1/q}$$

$$\le \left(\tfrac{1}{2}\left([\tfrac{1}{2}(\|s\| + \|t\|) + (\lambda/2)(\|s\| - \|t\|)]^q\right.\right.$$

$$\left.\left. + [\tfrac{1}{2}(\|s\| + \|t\|) - (\lambda/2)(\|s\| - \|t\|)]^q\right)\right)^{1/q}$$

$$\le \left(\tfrac{1}{2}\left([\tfrac{1}{2}(\|s\| + \|t\|) + \tfrac{1}{2}(\|s\| - \|t\|)]^p\right.\right.$$

$$\left.\left. + [\tfrac{1}{2}(\|s\| + \|t\|) - \tfrac{1}{2}(\|s\| - \|t\|)]^p\right)\right)^{1/p}$$

$$= \left(\tfrac{1}{2}(\|s\|^p + \|t\|^p)\right)^{1/p}$$

$$= \left(\tfrac{1}{2}\left(\|x + r_p y\|^p + \|x - r_p y\|^p\right)\right)^{1/p}.$$

$\square$

We now extend Bonami's inequality.

Theorem 13.1.1 (Bonami's Theorem) *Suppose that $1 < p < q < \infty$, and that $\{x_A : A \subseteq \{1, \ldots, N\}\}$ is a family of vectors in a normed space*

$(E, \|.\|_E)$. *Then*

$$\left\| \sum_A r_q^{|A|} w_A x_A \right\|_{L^q(E)} \leq \left\| \sum_A r_p^{|A|} w_A x_A \right\|_{L^p(E)},$$

where the w_A are Walsh functions.

Proof We prove the result by induction on N. The result is true for $N = 1$, by Corollary 13.1.1. Suppose that the result is true for $N - 1$. We can write $D_2^N = D_2^{N-1} \times D_2^{\{N\}}$, and $\mathbf{P}_N = \mathbf{P}_{N-1} \times \mathbf{P}_{\{N\}}$. Let $P(N-1)$ denote the set of subsets of $\{1, \ldots, N-1\}$ and let $P(N)$ denote the set of subsets of $\{1, \ldots, N\}$. If $B \in P(N-1)$, let $B^+ = B \cup \{N\}$, so that $P(N) = P(N-1) \cup \{B^+ \colon B \in P(N-1)\}$. Let

$$u_p = \sum_{B \in P(N-1)} r_p^{|B|} w_B x_B \quad \text{and} \quad v_p = \sum_{B \in P(N-1)} r_p^{|B|} w_B x_{B^+},$$

so that $\sum_{A \in P(N)} r_p^{|A|} w_A x_A = u_p + \epsilon_N r_p v_p$; let u_q and v_q be defined similarly. Then we need to show that

$$\|u_q + \epsilon_N r_q v_q\|_{L^q(E)} \leq \|u_p + \epsilon_N r_p v_p\|_{L^p(E)}.$$

Now, by the inductive hypothesis, for each $\omega \in D_2^N$,

$$\mathbf{E}_{N-1} \left(\|u_q + \epsilon_N(\omega) r_q v_q\|_E^q \right)^{1/q}$$

$$= \left(\mathbf{E}_{N-1} \left(\left\| \sum_{B \in P(N-1)} r_q^{|B|} (x_B + \epsilon_N(\omega) r_q y_B) \right\|_E^q \right) \right)^{1/q}$$

$$\leq \left(\mathbf{E}_{N-1} \left(\left\| \sum_{B \in P(N-1)} r_p^{|B|} (x_B + \epsilon_N(\omega) r_q y_B) \right\|_E^p \right) \right)^{1/p}$$

$$= \mathbf{E}_{N-1} (\|u_p + \epsilon_N(\omega) r_q v_p\|_E^p)^{1/p}.$$

Thus, using Corollary 5.4.2 and the result for $n = 1$,

$$\|u_q + \epsilon_N r_q v_q\|_{L^q(E)} = (\mathbf{E}_{\{N\}} (\mathbf{E}_{N-1} (\|u_q + \epsilon_N r_q v_q\|_E^q)))^{1/q}$$

$$\leq (\mathbf{E}_{\{N\}} (\mathbf{E}_{N-1} (\|u_p + \epsilon_N r_q v_p\|_E^p)))^{1/p}$$

$$\leq (\mathbf{E}_{N-1} (\mathbf{E}_{\{N\}} (\|u_p + \epsilon_N r_q v_p\|_E^q)^{p/q}))^{1/p}$$

$$\leq (\mathbf{E}_{N-1} (\mathbf{E}_{\{N\}} (\|u_p + \epsilon_N r_p v_p\|_E^p)))^{1/p}$$

$$= \|u_p + \epsilon_N r_p v_p\|_{L^p(E)}.$$

$\square$

13.2 Kahane's inequality revisited

We have the following generalization of Kahane's inequality (which corresponds to the case $n = 1$). Let W_n denote the set of Walsh functions w_A with $|A| = n$ and let $H_n(E)$ be the closed linear span of random vectors of the form $w_A u_A$, with $|A| = n$.

Theorem 13.2.1 *Suppose that (u_k) is a sequence in a Banach space E and that (w_{A_k}) is a sequence of distinct elements of W_n. Then if $1 < p < q$*

$$\left\| \sum_{k=1}^{K} w_{A_k} u_k \right\|_{L^q(E)} \leq \left(\frac{q-1}{p-1} \right)^{n/2} \left\| \sum_{k=1}^{K} w_{A_k} u_k \right\|_{L^p(E)}.$$

Thus H_n is strongly embedded in L_p for all $1 < p < \infty$. Further $H_1(E)$ is strongly embedded in $L_{\exp^2}(E)$ and $H_2(E)$ is strongly embedded in $L_{\exp}(E)$.

Proof If $S_K = \sum_{k=1}^{K} \epsilon_k u_k$ and $\|S_K\|_2 \leq 1/(2\sqrt{e})$ then

$$\mathbf{E}(e^{\|S_K\|^2}) = \sum_{j=0}^{\infty} \frac{\mathbf{E}(\|S_K\|^{2j})}{j!} \leq \sum_{j=0}^{\infty} \frac{(2j)^j}{2^{2j} e^j j!} \leq \sum_{j=0}^{\infty} \frac{1}{2^j} = 2,$$

since $j^j \leq e^j j!$ (Exercise 3.5).

Similarly, if $T_K = \sum_{k=1}^{K} w_{A_k} u_k$ with $|A_k| = 2$ for all k and $\|T_K\|_2 \leq 1/e$ then

$$\mathbf{E}(e^{\|T_K\|}) = \sum_{j=0}^{\infty} \frac{\mathbf{E}(\|T_K\|^{j})}{j!} \leq \sum_{j=0}^{\infty} \frac{j^j}{e^j j!} \leq \sum_{j=0}^{\infty} \frac{1}{2^j} = 2.$$

□

We also have the following result in the scalar case.

Corollary 13.2.1 *span $\{H_k \colon k \leq n\}$ is strongly embedded in L^p for all $1 < p < \infty$.*

Proof Since the spaces H_k are orthogonal, if $f = f_0 + \cdots + f_n$ and $q > 2$ then

$$\|f\|_q \leq \sum_{j=1}^{n} \|f_j\|_q$$

$$\leq \sum_{j=1}^{n} (q-1)^{j/2} \|f_j\|_2$$

$$\leq \left(\sum_{j=1}^{n} (q-1)^j \right)^{1/2} \left(\sum_{j=1}^{n} \|f_j\|_2^2 \right)^{1/2} \leq \sqrt{\frac{(q-1)^{n+1}}{q-2}} \cdot \|f\|_2 .$$

$\square$

13.3 The theorem of Latała and Oleszkiewicz

Theorem 13.2.1 gives good information about what happens for large values of p (which is the more important case), but does not deal with the case where $p = 1$. We do however have the following remarkable theorem relating the $L^1(E)$ and $L^2(E)$ norms of Bernoulli sums, which not only shows that $\sqrt{2}$ is the best constant in Khintchine's inequality but also shows that the same constant works in the vector-valued case.

Theorem 13.3.1 (Latała–Oleszkiewicz [La O 94]) *Let* $S_d = \sum_{i=1}^{d} \epsilon_i a_i$, *where* $\epsilon_1, \ldots, \epsilon_d$ *are Bernoulli random variables and* $a_1, \ldots, a_d$ *are vectors in a normed space* E. *Then* $\|S_d\|_{L^2(E)} \leq \sqrt{2} \|S_d\|_{L^1(E)}$.

Proof The Walsh functions form an orthonormal basis for $L^2(D_2^d)$, so that if $f \in C_{\mathbf{R}}(D_2^d)$ then

$$f = \sum_A \hat{f}_A w_A = \mathbf{E}(f) + \sum_{i=1}^{d} \hat{f}_i \epsilon_i + \sum_{|A|>1} \hat{f}_A w_A,$$

and $\|f\|_2^2 = \langle f, f \rangle = \sum_A \hat{f}_A^2$.

We consider a graph with vertices the elements of D_2^d and edges the set of pairs

$$\{(\omega, \eta): \ \omega_i \neq \eta_i \text{ for exactly one } i\}.$$

If (ω, η) is an edge, we write $\omega \sim \eta$. We use this to define the *graph Laplacian* of f as

$$L(f)(\omega) = \tfrac{1}{2} \sum_{\{\eta:\eta\sim\omega\}} (f(\eta) - f(\omega)),$$

and the *energy* $\mathcal{E}(f)$ of f as $\mathcal{E}(f) = - \langle f, L(f) \rangle$. Let us calculate the Laplacian for the Walsh functions. If $\omega \sim \eta$ and $\omega_i \neq \eta_i$, then

$$w_A(\omega) = w_A(\eta) \quad \text{if } i \notin A,$$
$$w_A(\omega) = -w_A(\eta) \quad \text{if } i \in A,$$

so that $L(w_A) = -|A|w_A$. Thus the Walsh functions are the eigenvectors of L, and L corresponds to differentiation. Further,

$$-L(f) = \sum_{i=1}^{d} \hat{f}_i \epsilon_i + \sum_{|A|>1} |A|\hat{f}_A w_A,$$

so that

$$\mathcal{E}(f) = \sum_{i=1}^{d} \hat{f}_i^2 + \sum_{|A|>1} |A|\hat{f}_A^2.$$

Thus

$$2\,\|f\|_2^2 = \langle f, f \rangle \leq \mathcal{E}(f) + 2(\mathbf{E}(f))^2 + \sum_{i=1}^{d} \hat{f}_i^2.$$

We now embed D_2^d as the vertices of the unit cube of l_∞^d. Let $f(x) = \|x_1 a_1 + \cdots + x_d a_d\|$, so that $f(\omega) = \|S_d(\omega)\|$, $\langle f, f \rangle = \|S_d\|_{L^2(E)}^2$, and $\mathbf{E}(f) = \|S_d\|_{L^1(E)}$. Since f is an even function, $\hat{f}_i = 0$ for $1 \leq i \leq d$, and since f is convex and positive homogeneous,

$$\frac{1}{d} \sum_{\{\eta:\eta\sim\omega\}} f(\eta) \geq f\left(\frac{1}{d} \sum_{\{\eta:\eta\sim\omega\}} \eta\right) = f\left(\frac{d-2}{d}\omega\right) = \frac{d-2}{d}f(\omega),$$

by Jensen's inequality. Consequently,

$$-Lf(\omega) \leq \tfrac{1}{2}(df(\omega) - (d-2)f(\omega)) = f(\omega),$$

so that $\mathcal{E}(f) \leq \|f\|_2^2$ and $2\,\|f\|_2^2 \leq \|f\|_2^2 + 2(\mathbf{E}(f))^2$. Thus $\|S_d\|_{L^2(E)} \leq \sqrt{2}\,\|S_d\|_{L^1(E)}$. $\square$

13.4 The logarithmic Sobolev inequality on D_2^d

The introduction of the Laplacian in the proof of Theorem 13.3.1 indicates that the results that we have proved are related to semigroup theory. Let $P_t = e^{tL}$; then $(P_t)_{t \geq 0}$ is a semigroup of operators on $C_{\mathbf{R}}(D_2^d)$ with infinitesimal generator L. Then $P_t(w_A) = e^{-t|A|}w_A$, and so Bonami's theorem shows that if $1 < p < \infty$ and $q(t) = 1 + (p-1)e^{2t}$ then

$$\|P_t(f)\|_{q(t)} \leq \|f\|_p .$$

This inequality is known as the *hypercontractive inequality*.

The hypercontractive inequality is closely related to the *logarithmic Sobolev inequality*, which is obtained by differentiation. Suppose that f is a nonnegative function on D_2^d. We define its entropy, $\mathrm{Ent}(f)$, as

$$\mathrm{Ent}(f) = \mathbf{E}(f \log f) - \|f\|_1 \log \|f\|_1 .$$

[We set $0 \log 0 = 0$, since $x \log x \to 0$ as $x \searrow 0$.] Since the function $x \log x$ is strictly convex, it follows from Jensen's inequality that $\mathrm{Ent}(f) \geq 0$, with equality if and only if f is constant. If $\|f\|_1 = 1$ then $\mathrm{Ent}(f) = \mathbf{E}(f \log f)$, and generally $\mathrm{Ent}(\alpha f) = \alpha \mathrm{Ent}(f)$ for $\alpha > 0$. This entropy is a relative entropy, related to the entropy of information theory in the following way. Recall that the *information entropy* $\mathrm{ent}(\nu)$ of a probability measure ν on D_2^d is defined as $-\sum_{\omega \in D_2^d} \nu(\omega) \log_2 \nu(\omega)$. Thus $\mathrm{ent}(\mathbf{P}_d) = d$ (where $\mathbf{P}_d$ is Haar measure), and, as we shall see, $\mathrm{ent}(\nu) \leq \mathrm{ent}(\mathbf{P}_d)$ for any other probability measure ν on D_2^d. Now if $f \geq 0$ and $\|f\|_1 = 1$ then f defines a probability measure $f \, d\mathbf{P}_d$ on D_2^d which gives the point ω probability $f(\omega)/2^d$. Thus

$$\mathrm{ent}(f \, d\mathbf{P}_d) = -\sum_{\omega \in D_2^d} \frac{f(\omega)}{2^d} \log_2 \left(\frac{f(\omega)}{2^d}\right) = d - \frac{\mathrm{Ent}(f)}{\log 2} .$$

Thus $\mathrm{Ent}(f)$ measures how far the information entropy of $f \, d\mathbf{P}_d$ falls below the maximum entropy d.

Theorem 13.4.1 (The logarithmic Sobolev inequality) *If $f \in C_{\mathbf{R}}(D_2^d)$ then $\mathrm{Ent}(f^2) \leq 2\mathcal{E}(f)$.*

Proof Take $p = 2$ and set $q(t) = 1 + e^{2t}$. Since $P_t(w_A) = e^{-t|A|}w_A$, $dP_t(w_A)/dt = -|A|e^{-t|A|}w_A = LP_t(w_A)$, and so by linearity $dP_t(f)/dt = LP_t(f)$. Suppose that $\|f\|_2 = 1$. Then $\|P_t(f)\|_{q(t)} \leq 1$, so that $(d/dt)\mathbf{E}(P_t(f)^{q(t)}) \leq 0$

at $t = 0$. Now

$$\frac{d}{dt}(P_t(f)^{q(t)}) = P_t(f)^{q(t)}\frac{d}{dt}\log(P_t(f)^{q(t)}) = P_t(f)^{q(t)}\frac{d}{dt}(q(t)\log(P_t(f)))$$

$$= P_t(f)^{q(t)}\frac{dq(t)}{dt}\log(P_t(f)) + P_t(f)^{q(t)-1}q(t)LP_t(f)$$

$$= 2e^{2t}P_t(f)^{q(t)}\log(P_t(f)) + (1 + e^{2t})P_t(f)^{q(t)-1}LP_t(f).$$

Taking expectations, and setting $t = 0$, we see that

$$0 \geq \mathbf{E}(f^2\log(f^2)) + 2\mathbf{E}(fL(f)) = \mathrm{Ent}(f^2) - 2\mathcal{E}(f).$$

$\square$

We can use the logarithmic Sobolev inequality to show that certain functions are sub-Gaussian. Let $\eta_i \in D_2^d$ be defined by $(\eta_i)_i = -1$, $(\eta_i)_j = 1$, otherwise. If $f \in C_{\mathbf{R}}(D_2^d)$ and $\omega \in D_2^d$, define the *gradient* $\nabla f(\omega) \in \mathbf{R}^d$ by setting $\nabla f(\omega)_i = f(\eta_i\omega) - f(\omega)$. Then

$$|\nabla f(\omega)|^2 = \sum_{i=1}^{d}(f(\eta_i\omega) - f(\omega))^2 = \sum_{\{\eta:\eta\sim\omega\}}(f(\eta) - f(\omega))^2.$$

Note that

$$\mathcal{E}(f) = \frac{1}{2^d}\sum_{\omega}\sum_{\{\eta:\eta\sim\omega\}}(f(\omega) - f(\eta))f(\omega)$$

$$= \frac{1}{2^{d+1}}\left(\sum_{\omega}\sum_{\{\eta:\eta\sim\omega\}}(f(\omega) - f(\eta))f(\omega) + \sum_{\eta}\sum_{\{\omega:\omega\sim\eta\}}(f(\eta) - f(\omega))f(\eta)\right)$$

$$= \tfrac{1}{2}\mathbf{E}(|\nabla f|^2).$$

Theorem 13.4.2 *Suppose that* $\mathbf{E}(f) = 0$ *and that* $|\nabla(f)(\omega)| \leq 1$ *for all* $\omega \in D_2^d$. *Then* f *is sub-Gaussian with exponent* $1/\sqrt{2}$: *that is,* $\mathbf{E}(e^{\lambda f}) \leq e^{-\lambda^2/4}$, *for all real* λ.

Proof It is clearly sufficient to consider the case where $\lambda > 0$. Let $H(\lambda) = \mathbf{E}(e^{\lambda f})$. First we show that $\mathbf{E}(|\nabla(e^{\lambda f/2})|^2) \leq \lambda^2 H(\lambda)/4$. Using the mean value theorem to establish the first inequality,

$$\mathbf{E}(|\nabla(e^{\lambda f/2})|^2) = \frac{1}{2^d} \sum_\omega \left(\sum_{\{\eta:\eta\sim\omega\}} (e^{\lambda f(\eta)/2} - e^{\lambda f(\omega)/2})^2 \right)$$

$$= \frac{2}{2^d} \sum_\omega \left(\sum \{(e^{\lambda f(\eta)/2} - e^{\lambda f(\omega)/2})^2 : \eta\sim\omega, f(\eta) < f(\omega)\} \right)$$

$$\leq \frac{\lambda^2}{2.2^d} \sum_\omega \left(\sum \{(f(\eta) - f(\omega))^2 e^{\lambda f(\omega)} : \eta\sim\omega, f(\eta) < f(\omega)\} \right)$$

$$\leq \frac{\lambda^2}{4.2^d} \sum_\omega \left(\sum_{\{\eta:\eta\sim\omega\}} (f(\eta) - f(\omega))^2 \right) e^{\lambda f(\omega)}$$

$$= \frac{\lambda^2}{4} \mathbf{E}(\|\nabla(f)\|_2^2 e^{\lambda f}) \leq \frac{\lambda^2}{4} \mathbf{E}(e^{\lambda f}) = \frac{\lambda^2 H(\lambda)}{4}.$$

Thus, applying the logarithmic Sobolev inequality,

$$\mathrm{Ent}(e^{\lambda f}) \leq 2\mathcal{E}(e^{\lambda f/2}) = \mathbf{E}(|\nabla(e^{\lambda f/2})|^2) \leq \frac{\lambda^2 H(\lambda)}{4}.$$

But

$$\mathrm{Ent}(e^{\lambda f}) = \mathbf{E}(\lambda f e^{\lambda f}) - H(\lambda) \log H(\lambda) = \lambda H'(\lambda) - H(\lambda) \log H(\lambda),$$

so that

$$\lambda H'(\lambda) - H(\lambda) \log H(\lambda) \leq \frac{\lambda^2 H(\lambda)}{4}.$$

Let $K(\lambda) = (\log H(\lambda))/\lambda$, so that $e^{\lambda K(\lambda)} = \mathbf{E}(e^{\lambda f})$. Then

$$K'(\lambda) = \frac{H'(\lambda)}{\lambda H(\lambda)} - \frac{\log H(\lambda)}{\lambda^2} \leq \frac{1}{4}.$$

Now as $\lambda \to 0$, $H(\lambda) = 1 + \lambda \mathbf{E}(f) + O(\lambda^2) = 1 + O(\lambda^2)$, so that $\log H(\lambda) = O(\lambda^2)$, and $K(\lambda) \to 0$ as $\lambda \to 0$. Thus $K(\lambda) = \int_0^\lambda K'(s)\, ds \leq \lambda/4$, and $H(\lambda) = \mathbf{E}(e^{\lambda f}) \leq e^{\lambda^2/4}$. □

Corollary 13.4.1 *If $r > 0$ then $\mathbf{P}(f \geq r) \leq e^{-r^2}$.*

This leads to a 'concentration of measure' result. Let h be the Hamming metric on D_2^d, so that $h(\omega, \eta) = \frac{1}{2} \sum_{i=1}^d |\omega_i - \eta_i|$, and $\omega \sim \eta$ if and only if $h(\omega, \eta) = 1$. If A is a non-empty subset of D_2^d, let $h_A(\omega) = \inf\{h(\omega, \eta) : \eta \in A\}$.

Corollary 13.4.2 *Suppose that* $\mathbf{P}(A) > 1/e$. *Then* $\mathbf{E}(h_A) \leq \sqrt{d}$. *Let* $A_s = \{\omega : h(\omega, A) \leq s\}$. *If* $t > 1$ *then* $\mathbf{P}(A_{t\sqrt{d}}) \geq 1 - e^{-(t-1)^2}$.

Proof Let $g(\omega) = h_A(\omega)/\sqrt{d}$. Then $|g(\omega) - g(\eta)| \leq d(\omega, \eta)/\sqrt{d}$, so that $|\nabla(g)(\omega)| \leq 1$ for each $\omega \in D_2^d$. Applying Corollary 13.4.1 to $\mathbf{E}(g) - g$ with $r = 1$, we see that $\mathbf{P}(g \leq \mathbf{E}(g) - 1) \leq 1/e$. But $\mathbf{P}(g \leq 0) > 1/e$, so that $\mathbf{E}(g) \leq 1$. Now apply Corollary 13.4.1 to $g - \mathbf{E}(g)$, with $r = t - 1$:

$$1 - \mathbf{P}(A_{t\sqrt{d}}) = \mathbf{P}(g > t) \leq \mathbf{P}(g - \mathbf{E}(g) > t - 1) \leq e^{-(t-1)^2}.$$

$\square$

13.5 Gaussian measure and the Hermite polynomials

Although, as we have seen, analysis on the discrete space D_2^d leads to interesting conclusions, it is natural to want to obtain similar results on Euclidean space. Here it turns out that the natural underlying measure is not Haar measure (that is, Lebesgue measure) but is Gaussian measure. In this setting, we can obtain logarithmic Sobolev inequalities, which correspond to the Sobolev inequalities for Lebesgue measure, but have the great advantage that they are not dependent on the dimension of the space, and so can be extended to the infinite-dimensional case.

First, let us describe the setting in which we work. Let γ_1 be the probability measure on the line $\mathbf{R}$ given by

$$d\gamma_1(x) = \frac{1}{\sqrt{2\pi}} e^{-x^2/2} \, dx,$$

and let ξ_1 be the random variable $\xi_1(x) = x$, so that ξ_1 is a standard Gaussian or normal random variable, with mean 0 and variance $\mathbf{E}(\xi_1^2) = 1$. Similarly, let γ_d be the probability measure on $\mathbf{R}^d$ given by

$$d\gamma_d(x) = \frac{1}{(2\pi)^{d/2}} e^{-|x|^2/2} \, dx,$$

and let $\xi_i(x) = x_i$, for $1 \leq i \leq d$. Then $(\xi_1, \ldots, \xi_d)$ is a sequence of independent standard Gaussian random variables. More generally, a closed linear subspace H of $L^2(\Omega)$ is a *Gaussian Hilbert space* if each $f \in H$ has a centred Gaussian distribution (with variance $\|f\|_2^2$). As we have seen, H is then strongly embedded in $L_{\exp^2}$. If, as we shall generally suppose, H is separable and (f_i) is an orthonormal basis for H, then (f_i) is a sequence of independent standard Gaussian random variables.

We shall discuss in some detail what happens in the one-dimensional case, and then describe how the results extend to higher dimensions. The sequence of functions $(1, x, x^2, \ldots)$ is linearly independent, but not orthogonal, in $L^2(\gamma_1)$; we apply Gram–Schmidt orthonormalization to obtain an orthonormal sequence $(\tilde{h}_n)$ of polynomials. We shall see that these form an orthonormal basis of $L^2(\gamma_1)$. Each $\tilde{h}_n$ is a polynomial of degree n, and we can choose it so that its leading coefficient is positive. Let us then write $\tilde{h}_n = c_n h_n$, where $c_n > 0$ and h_n is a monic polynomial of degree n (that is, the coefficient of x^n is 1). The next proposition enables us to recognize h_n as the nth Hermite polynomial.

Proposition 13.5.1 *Define the nth Hermite polynomial as*

$$h_n(x) = (-1)^n e^{x^2/2} (\frac{d}{dx})^n e^{-x^2/2}.$$

Then

$$h_n(x) = (x - \frac{d}{dx})h_{n-1}(x) = (x - \frac{d}{dx})^n 1.$$

Each h_n is a monic polynomial of degree n, (h_n) is an orthogonal sequence in $L^2(\gamma_1)$, and $\|h_n\|_2 = (n!)^{1/2}$.

Proof Differentiating the defining relation for h_{n-1}, we see that $dh_{n-1}(x)/dx = xh_{n-1}(x) - h_n(x)$, which gives the first assertion, and it follows from this that h_n is a monic polynomial of degree n. If $m \leq n$, then, integrating by parts m times,

$$\int x^m h_n(x) \, d\gamma_1(x) = \frac{(-1)^n}{\sqrt{2\pi}} \int_{-\infty}^{\infty} x^m (\frac{d}{dx})^n e^{-x^2/2} \, dx$$

$$= \frac{(-1)^{n-m}m!}{\sqrt{2\pi}} \int_{-\infty}^{\infty} (\frac{d}{dx})^{n-m} e^{-x^2/2} \, dx$$

$$= \begin{cases} 0 & \text{if } m < n, \\ n! & \text{if } m = n. \end{cases}$$

Thus (h_n) is orthogonal to all polynomials of lower degree; consequently (h_n) is an orthogonal sequence in $L^2(\gamma_1)$. Finally,

$$\|h_n\|_2^2 = \langle h_n, x^n \rangle + \langle h_n, h_n - x^n \rangle = n!$$

$\square$

Corollary 13.5.1 *We have the following relations:*

$$(i) \quad h_n(x) = i^n e^{x^2/2} \int_{\mathbf{R}} u^n e^{-iux} \, d\gamma_1(u) = \frac{1}{\sqrt{2\pi}} \int_{-\infty}^{\infty} (x+iy)^n e^{-y^2/2} \, dy.$$

$$(ii) \quad \frac{dh_n}{dx}(x) = n h_{n-1}(x).$$

$$(iii) \quad \int \left(\frac{dh_n}{dx}\right)^2 d\gamma_1 = n(n!), \quad \int \frac{dh_n}{dx}\frac{dh_m}{dx} \, d\gamma_1 = 0 \text{ for } m \neq n.$$

Proof The first equation of (i) follows by repeatedly applying the operator $x - d/dx$ to the equation $1 = e^{x^2/2} \int_{\mathbf{R}} e^{-iux} \, d\gamma_1(u)$. Making the change of variables $y = u + ix$ (justified by Cauchy's theorem), we obtain the second equation. Differentiating under the integral sign (which is easily seen to be valid), we obtain (ii), and (iii) follows from this, and the proposition. $\qquad\square$

Proposition 13.5.2 *The polynomial functions are dense in $L^p(\gamma_1)$, for $0 < p < \infty$.*

Proof We begin by showing that the exponential functions are approximated by their power series expansions. Let $e_n(\lambda x) = \sum_{j=0}^{n} (\lambda x)^n/n!$ Then

$$|e^{\lambda x} - e_n(\lambda x)|^p = |\sum_{j=n+1}^{\infty} (\lambda x)^n/n!|^p \leq e^{p|\lambda x|},$$

and $\int e^{p|\lambda x|} \, d\gamma_1(x) < \infty$, so that by the theorem of dominated convergence $\int |e^{\lambda x} - e_n(\lambda x)|^p \, d\gamma_1 \to 0$ as $n \to \infty$, and so $e_n(\lambda x) \to e^{\lambda x}$ in $L^p(\gamma_1)$.

Now suppose that $1 \leq p < \infty$ and that $f \in L^p(\gamma_1)$ is not in the closure of the polynomial functions in $L^p(\gamma_1)$. Then by the separation theorem there exists $g \in L^{p'}(\gamma_1)$ such that $\int fg \, d\gamma_1 = 1$ and $\int qg \, d\gamma_1 = 0$ for every polynomial function q. But then $\int e^{\lambda x} g(x) \, d\gamma_1(x) = 0$ for all λ, so that

$$\frac{1}{\sqrt{2\pi}} \int_{-\infty}^{\infty} e^{-isx} g(x) e^{-x^2/2} \, dx = \int e^{-isx} g(x) \, d\gamma_1(x) = 0,$$

so that the Fourier transform of $g(x)e^{-x^2/2}$ is zero, and so $g = 0$, giving a contradiction.

Thus the polynomial functions are dense in $L^p(\gamma_1)$, for $1 \leq p < \infty$. Since $L^1(\gamma_1)$ is dense in $L^p(\gamma_1)$ for $0 < p < 1$, the polynomial functions are dense in these spaces too. $\qquad\square$

Corollary 13.5.2 *The functions* $(\tilde{h}_n)$ *form an orthonormal basis for* $L^2(\gamma_1)$.

It is worth noting that this is a fairly sophisticated proof, since it uses the theorem of dominated convergence, and Fourier transforms. It is possible to give a more elementary proof, using the Stone–Weierstrass theorem, but this is surprisingly complicated.

13.6 The central limit theorem

We wish to establish hypercontractive and logarithmic Sobolev inequalities in this Gaussian setting. We have seen that in D_2^d these inequalities are related to a semigroup of operators. The same is true in the Gaussian case, where the semigroup is the *Ornstein–Uhlenbeck semigroup* $(P_t)_{t\geq0}$ acting on $L^2(\gamma_1)$:

$$\text{if } f = \sum_{n=0}^{\infty} f_n \tilde{h}_n(\xi), \text{ then } P_t(f) = \sum_{n=0}^{\infty} e^{-nt} f_n \tilde{h}_n(\xi).$$

There are now two ways to proceed. The first is to give a careful direct analysis of the Ornstein–Uhlenbeck semigroup; but this would take us too far into semigroup theory. The second, which we shall follow, is to use the central limit theorem to carry results across from the D_2^d case. For this we only need the simplest form of the central limit theorem, which goes back to the work of De Moivre, in the eighteenth century.

A function g defined on $\mathbf{R}$ is *of polynomial growth* if there exist $C > 0$ and $N \in \mathbf{N}$ such that $|f(x)| \leq C(1 + |x|^N)$, for all $x \in \mathbf{R}$.

Theorem 13.6.1 (De Moivre's central limit theorem) *Let* (ϵ_n) *be a sequence of Bernoulli random variables and let* $C_n = (\epsilon_1 + \cdots + \epsilon_n)/\sqrt{n}$. *Let* ξ *be a Gaussian random variable with mean 0 and variance 1. Then* $\mathbf{P}(C_n \leq t) \to \mathbf{P}(\xi \leq t)$ *for each* $t \in \mathbf{R}$, *and if* g *is a continuous function of polynomial growth then* $\mathbf{E}(g(C_n)) \to \mathbf{E}(g(\xi))$ *as* $n \to \infty$.

Proof We shall prove this for even values of n: the proof for odd values is completely similar. Fix m, and let $t_j = j/\sqrt{2m}$. The random variable C_{2m} takes values t_{2k}, for $-m \leq k \leq m$, and

$$\mathbf{P}(C_{2m} = t_{2k}) = \frac{1}{2^{2m}} \binom{2m}{m+k}.$$

First we show that we can replace the random variables (C_{2m}) by random variables (D_{2m}) which have density functions, and whose density functions are step functions. Let $I_{2k} = (t_{2k-1}, t_{2k+1}]$ and let D_{2m} be the random variable which has density

$$p_{2m}(t) = \sqrt{\frac{m}{2}\frac{1}{2^{2m}}}\binom{2m}{m+k} \text{ if } t \in I_{2k} \text{ for some } -m \leq k \leq m,$$

$$= 0 \text{ otherwise.}$$

Thus $\mathbf{P}(C_{2m} \in I_{2k}) = \mathbf{P}(D_{2m} \in I_{2k})$. The random variables C_{2m} are all sub-Gaussian, with exponent 1, and so $\mathbf{P}(|C_{2m}| > R) \leq 2e^{-R^2}$, and if $m \geq 2$ then $\mathbf{P}(|D_{2m}| > R+1) \leq 2e^{-R^2}$. Thus if g is a continuous function of polynomial growth and $\epsilon > 0$ there exists $R > 0$ such that

$$\int_{|C_{2m}|>R} |g(C_{2m})|\, d\mathbf{P} \leq \frac{\epsilon}{3} \quad \text{and} \quad \int_{|D_{2m}|>R} |g(D_{2m})|\, d\mathbf{P} \leq \frac{\epsilon}{3}$$

for all m. On the other hand, it follows from the uniform continuity of g on $[-R, R]$ that there exists m_0 such that

$$\left| \int_{|C_{2m}|>R} g(C_{2m})\, d\mathbf{P} - \int_{|D_{2m}|>R} g(D_{2m})\, d\mathbf{P} \right| \leq \frac{\epsilon}{3}$$

for $m \geq m_0$. Thus $\mathbf{E}(g(C_{2m})) - \mathbf{E}(g(D_{2m})) \to 0$ as $m \to \infty$. Similarly, $\mathbf{P}(C_{2m} \leq t) - \mathbf{P}(D_{2m} \leq t) \to 0$ as $m \to \infty$. It is therefore sufficient to prove the result with the random variables (D_{2m}) in place of (C_{2m}).

First we show that $p_{2m}(t) \to e^{-t^2/2}/C$ (where C is the constant in Stirling's formula) as $m \to \infty$. Applying Stirling's formula (Exercise 13.1),

$$p_{2m}(0) = \sqrt{\frac{m}{2}\frac{(2m)!}{2^{2m}(m!)^2}} \to 1/C.$$

If $t > 0$ and $m \geq 2t^2$ then $t \in I_{2k_t}$ for some k_t with $|k_t| \leq m/2$. Then

$$p_{2m}(t) = p_{2m}(0)\frac{(m-1)\dots(m-k_t)}{(m+1)\dots(m+k_t)} = p_{2m}(0)\frac{(1-1/m)\dots(1-k_t/m)}{(1+1/m)\dots(1+k_t/m)}.$$

Let

$$r_{2m}(t) = \log\left(\frac{(1-1/m)\dots(1-k_t/m)}{(1+1/m)\dots(1+k_t/m)}\right)$$

$$= \sum_{j=1}^{k_t}\log(1-j/m) - \sum_{j=1}^{k_t}\log(1+j/m).$$

Since $|\log(1+x) - x| \leq x^2$ for $|x| < 1/2$,

$$|r_m(t) + k_t(k_t+1)/m| \leq k_t(k_t+1)(2k_t+1)/3m^2,$$

for large enough m. But $k_t^2/m \to t^2/2$ as $m \to \infty$, and so $r_{2m}(t) \to -t^2/2$

as $m \to \infty$. Thus $p_{2m}(t) \to e^{-t^2/2}/C$ as $m \to \infty$. By symmetry, the result also holds for $t < 0$.

Finally, p_{2m} is a decreasing function on $[0, \infty)$, so that the functions p_{2m} are uniformly bounded; further, if $t \geq 3$ and $m \geq 2$ then

$$p_{2m}(t) \leq (|t|/2)\mathbf{P}(D_{2m} > |t|/2) \leq |t|e^{-(|t|/2-1)^2}.$$

We apply the theorem of dominated convergence: if g is a continuous function of polynomial growth then

$$\mathbf{E}(g(D_{2m})) = \int_{-\infty}^{\infty} g(t)p_{2m}(t)\, dt \to \frac{1}{C}\int_{-\infty}^{\infty} g(t)e^{-t^2/2}\, dt = \mathbf{E}(g(\xi)).$$

In particular, taking $g = 1$, $1 = (1/C)\int_{-\infty}^{\infty} e^{-t^2/2}\, dt$, so that the constant C in Stirling's formula is $\sqrt{2\pi}$. Similarly,

$$\mathbf{P}(D_{2m} \leq t) = \int_{-\infty}^{t} p_{2m}(s)\, ds \to \frac{1}{\sqrt{2\pi}}\int_{-\infty}^{t} e^{-s^2/2}\, dt = \mathbf{P}(\xi \leq t).$$

$\square$

13.7 The Gaussian hypercontractive inequality

If f is a function on D_2^d and $\sigma \in \Sigma_d$, the group of permutations of $\{1, \ldots, d\}$, we set $f_\sigma(\omega) = f(\omega_{\sigma(1)}, \ldots, \omega_{\sigma(d)})$. Let

$$SL^2(D_2^d) = \{f \in L^2(D_2^d): \ f = f_\sigma \text{ for each } \sigma \in \Sigma_d\}.$$

Then $SL^2(D_2^d)$ is a $d+1$-dimensional subspace of $L^2(D_2^d)$, with orthonormal basis $(S_0^{(d)}, \ldots, S_d^{(d)})$, where

$$S_j^{(d)} = \left(\sum_{\{A:|A|=j\}} w_A\right) \Big/ \binom{d}{j}^{1/2}.$$

But span $(S_0^{(d)}, \ldots, S_j^{(d)})$ = span $(1, C_d, \ldots, C_d^j)$, where $C_d = S_1^{(d)} = (\sum_{i=1}^{d} \epsilon_i)/\sqrt{d}$. Thus $(1, C_d, \ldots, C_d^d)$ is also a basis for $SL^2(D_2^d)$, and there exists a non-singular upper-triangular matrix $H^{(d)} = (h_{k,j}^{(d)})$ such that

$$S_i^{(d)} = \sum_{j=0}^{k} h_{k,j}^{(d)} C_d^j = h_i^{(d)}(C_d),$$

where $h_i^{(d)}(x) = \sum_{j=0}^{i} h_{k,j}^{(d)}(x)$. With this notation, we have the following corollary of Bonami's theorem.

Corollary 13.7.1 *Suppose that $1 < p < q < \infty$, and that $(x_0, \ldots, x_N)$ is a sequence of vectors in a normed space $(E, \|.\|_E)$. If $d \geq N$ then*

$$\left\| \sum_{k=0}^{N} r_q^k h_k^{(d)}(C_d) \right\|_{L^q(E)} \leq \left\| \sum_{k=0}^{N} r_p^k h_k^{(d)}(C_d) \right\|_{L^p(E)}.$$

We now show that the polynomials $h_k^{(d)}$ converge to the normalized Hermite polynomial $\tilde{h}_k$ as $d \to \infty$.

Proposition 13.7.1 $h_{k,j}^{(d)} \to \tilde{h}_{k,j}$ *(the coefficient of x^j in the normalized Hermite polynomial $\tilde{h}_k$) as $d \to \infty$.*

Proof We prove this by induction on k. The result is certainly true when $k = 0$. Suppose that it is true for all $l < k$. Note that, since $\|C_d\|_2 = 1$, it follows from Khintchine's inequality that there exists a constant M_k such that $\mathbf{E}(|C_d|^k(1 + |C_d|^k)) \leq M_k$, for all d. It follows from the inductive hypothesis that given $\epsilon > 0$ there exists d_k such that $|h_l^{(d)}(x) - \tilde{h}_l(x)| < \epsilon(1 + |x|^k)/M_k$ for $l < k$ and $d \geq d_l$. Now it follows from orthogonality that

$$h_k^{(d)}(x) = x^k - \sum_{l=0}^{k-1} \left(\mathbf{E}(C_d^k h_l^{(d)}(C_d)) \right) h_l^{(d)}(x).$$

If $d \geq d_l$ then

$$|\mathbf{E}(C_d^k(h_l^{(d)}(C_d) - \tilde{h}_l(C_d)))| \leq \epsilon \mathbf{E}(|C_d^k(1 + |C_d|^k)|)/M_k \leq \epsilon,$$

and $\mathbf{E}(C_d^k \tilde{h}_l(C_d)) \to \mathbf{E}(\xi^k \tilde{h}_l(\xi))$, by De Moivre's central limit theorem, and so $\mathbf{E}(C_d^k h_l^{(d)}(C_d)) \to \mathbf{E}(\xi^k \tilde{h}_l(\xi))$ as $d \to \infty$. Consequently

$$h_k^{(d)}(x) \to x^k - \sum_{l=0}^{k-1} \mathbf{E}(\xi^k \tilde{h}_l(\xi))\tilde{h}_l(x) = h_k(x),$$

for each $x \in \mathbf{R}$, from which the result follows.   □

We now have the following consequence.

Theorem 13.7.1 *Suppose that $1 < p < q < \infty$ and that $\beta_0, \ldots, \beta_n$ are real numbers. Then*

$$\left\| \sum_{n=0}^{N} r_q^n \beta_n \tilde{h}_n \right\|_{L^q(\gamma_1)} \leq \left\| \sum_{n=0}^{N} r_p^n \beta_n \tilde{h}_n \right\|_{L^p(\gamma_1)},$$

where as before $r_p = 1/\sqrt{p-1}$ and $r_q = 1/\sqrt{q-1}$.

Proof Suppose that $\epsilon > 0$. As in Proposition 13.7.1, there exists d_0 such that

$$\left| |\sum_{n=0}^{N} r_p^n \beta_n h_n^{(d)}(x)|^p - |\sum_{n=0}^{N} r_p^n \beta_n \tilde{h}_n(x)|^p \right| \leq \epsilon(1 + |x|^{Np}),$$

for $d \geq d_0$, from which it follows that

$$\left\| \sum_{n=0}^{N} r_p^n \beta_n h_n^{(d)}(C_d) \right\|_p - \left\| \sum_{n=0}^{N} r_p^n \beta_n \tilde{h}_n(C_d) \right\|_p \to 0.$$

But

$$\left\| \sum_{n=0}^{N} r_p^n \beta_n \tilde{h}_n(C_d) \right\|_p \to \left\| \sum_{n=0}^{N} r_p^n \beta_n \tilde{h}_n(\xi) \right\|_p,$$

as $d \to \infty$, by De Moivre's central limit theorem. Thus

$$\left\| \sum_{n=0}^{N} r_p^n \beta_n h_n^{(d)}(C_d) \right\|_p \to \left\| \sum_{n=0}^{N} r_p^n \beta_n \tilde{h}_n(\xi) \right\|_p,$$

as $d \to \infty$. Similarly,

$$\left\| \sum_{n=0}^{N} r_q^n \beta_n \tilde{h}_n^{(d)}(C_d) \right\|_q \to \left\| \sum_{n=0}^{N} r_q^n \beta_n \tilde{h}_n(\xi) \right\|_q,$$

as $d \to \infty$, and so the result follows from Corollary 13.7.1. $\square$

We can interpret this inequality as a hypercontractive inequality. If $(P_t)_{t \geq 0}$ is the Ornstein–Uhlenbeck semigroup, if $1 < p < \infty$, if $q(t) = 1 + (p-1)e^{2t}$ and if $f \in L^p(\gamma_1)$, then $P_t(f) \in L^{q(t)}(\gamma_1)$, and $\|P_t(f)\|_{q(t)} \leq \|f\|_p$.

13.8 Correlated Gaussian random variables

Suppose now that ξ and η are standard Gaussian random variables with a joint normal distribution, whose correlation $\rho = \mathbf{E}(\xi\eta)$ satisfies $-1 < \rho < 1$. Then if we set $\xi_1 = \xi$ and $\xi_2 = (\eta - \rho\xi)/\tau$, where $\tau = \sqrt{1 - \rho^2}$, then ξ_1 and ξ_2 are independent standard Gaussian random variables, and $\eta = \rho\xi_1 + \tau\xi_2$. Let γ_2 be the joint distribution of (ξ_1, ξ_2). We can consider $L^2(\xi)$ and $L^2(\eta)$ as subspaces of $L^2(\gamma_2)$. Let π_η be the orthogonal projection of $L^2(\eta)$ onto $L^2(\xi)$; it is the conditional expectation operator $\mathbf{E}(\cdot|\xi)$.

Proposition 13.8.1 *Suppose that ξ and η are standard Gaussian random variables with a joint normal distribution, whose correlation $\rho = \mathbf{E}(\xi\eta)$ satisfies $-1 < \rho < 1$. Then $\pi_\eta(h_n(\eta)) = \rho^n h_n(\xi)$.*

Proof Since $P_\eta(f) = \sum_{m=0}^\infty \langle f, \tilde{h}_m(\xi) \rangle \tilde{h}_m(\xi)$, we must show that

$$\langle \tilde{h}_n(\eta), \tilde{h}_m(\xi) \rangle = \rho^n \quad \text{if } m = n,$$
$$= 0 \quad \text{otherwise.}$$

First observe that if $m < n$ then

$$\tilde{h}_m(\eta) = \tilde{h}_m(\rho\xi_1 + \tau\xi_2) = \sum_{j=0}^m p_j(\xi_2)\xi_1^j,$$

where each p_j is a polynomial of degree $m - j$, so that

$$\langle \tilde{h}_n(\eta), \tilde{h}_m(\xi) \rangle = \sum_{j=0}^m \left(\mathbf{E}_{\xi_1} \tilde{h}_n(\xi_1)\xi_1^j \right) (\mathbf{E}_{\xi_2} p_j(\xi_2)) = 0,$$

by the orthogonality of $\tilde{h}_n(\xi_1)$ and ξ_1^j. A similar result holds if $m > n$, by symmetry. Finally, if $m = n$ then $p_n(\xi_2) = \tilde{h}_m(\rho\xi_1)(0) = \rho^n/(n!)^{1/2}$, and so

$$\langle \tilde{h}_n(\eta), \tilde{h}_n(\xi) \rangle = \mathbf{E}_{\xi_1} (\rho^n/(n!)^{1/2})\tilde{h}_n(\xi_1)\xi_1^n = \rho^n.$$

$\square$

Corollary 13.8.1 *Let ξ_1 and ξ_2 be independent standard Gaussian random variables, and for $t \geq 0$ let $\eta_t = e^{-t}\xi_1 + (1 - e^{-2t})^{1/2}\xi_2$. If $f \in L^2(\gamma_1)$ then $P_t(f) = \mathbf{E}(f(\eta_t)|\xi_1)$ (where $(P_t)_{t\geq 0}$is the Ornstein–Uhlenbeck semigroup).*

This proposition enables us to prove the following fundamental result.

Theorem 13.8.1 *Suppose that ξ and η are standard Gaussian random variables with a joint normal distribution, whose correlation $\rho = \mathbf{E}(\xi\eta)$ satisfies $-1 < \rho < 1$, and suppose that $(p - 1)(q - 1) \geq \rho^2$. If $f \in L^2(\xi) \cap L^p(\xi)$ and $g \in L^2(\eta) \cap L^q(\eta)$ then*

$$|\mathbf{E}(fg)| \leq \|f\|_p \|g\|_q.$$

Proof By approximation, it is enough to prove the result for $f = \sum_{j=0}^m \alpha_j h_j(\xi)$ and $g = \sum_{k=0}^n \beta_j h_j(\eta)$ Let $e^{2t} = \rho^2$, and let $r = 1 + \rho^2(p' - 1)$. Note that $1 < r < q$ and that $p' = 1 + e^{2t}(r-1)$. Then

$$
\begin{aligned}
|\mathbf{E}(fg)| &= |\mathbf{E}(f\mathbf{E}(g|\xi)| \\
&\leq \|f\|_p \|\mathbf{E}(g|\xi)\|_{p'} \quad \text{(by Hölder's inequality)} \\
&= \|f\|_p \|P_t(g)\|_{p'} \\
&\leq \|f\|_p \|g\|_r \quad \text{(by hypercontractivity)} \\
&\leq \|f\|_p \|g\|_q .
\end{aligned}
$$

$\square$

The statement of this theorem does not involve Hermite polynomials. Is there a more direct proof? There is a very elegant proof by Neveu [Nev 76], using stochastic integration and the Itô calculus. This is of interest, since it is easy to deduce Theorem 13.7.1 from Theorem 13.8.1. Suppose that $1 < p < q < \infty$. Let $r = \sqrt{(p-1)/(q-1)}$, and let ξ and η be standard Gaussian random variables with a joint normal distribution, with correlation r. If $f(\xi) = \sum_{n=0}^{N} r_p^n \beta_n \tilde{h}_n(\xi)$ then $P_\eta(f(\xi)) = \sum_{n=0}^{N} r_q^n \beta_n \tilde{h}_n(\eta)$. There exists $g \in L_\eta^{q'}$ with $\|g\|_{q'} = 1$ such that $|\mathbf{E}(P_\eta(f)(\eta)g(\eta))| = \|P_\eta(f)\|_q$. Then

$$
\|P_\eta(f)\|_q = |\mathbf{E}(P_\eta(f)(\eta)g(\eta))| = |\mathbf{E}(f(\xi)g(\eta))| \leq \|f\|_p \|g\|_{q'} = \|f\|_p .
$$

13.9 The Gaussian logarithmic Sobolev inequality

We now turn to the logarithmic Sobolev inequality. First we consider the infinitesimal generator L of the Ornstein–Uhlenbeck semigroup. What is its domain $D(L)$? Since $(P_t(\tilde{h}_n) - \tilde{h}_n)/t \to -n\tilde{h}_n$, $\tilde{h}_n \in D(L)$ and $L(\tilde{h}_n) = -n\tilde{h}_n$. Let

$$
D = \left\{ f = \sum_{n=0}^{\infty} f_n \tilde{h}_n \in L^2(\gamma_1) : \sum_{n=0}^{\infty} n^2 f_n^2 < \infty \right\}.
$$

If $f \in D$ then, applying the mean value theorem term by term, $\|(P_t(f) - f)/t\|_2 \leq \sum_{n=0}^{\infty} n^2 f_n^2$, and so $f \in D(L)$, and $L(f) = -\sum_{n=0}^{\infty} n f_n \tilde{h}_n$. Conversely, if $f \in D(L)$ then

$$
\left\langle (P_t(f) - f)/t, \tilde{h}_n \right\rangle = ((e^{-nt} - 1)/t)f_n \to \left\langle L(f), \tilde{h}_n \right\rangle,
$$

so that $L(f) = -\sum_{n=0}^{\infty} n f_n \tilde{h}_n$, and $f \in D$. Thus $D = D(L)$. Further, if $f \in D(L)$ then

$$
\mathcal{E}(f) = -\langle f, L(f) \rangle = \sum_{n=0}^{\infty} n f_n^2 = \int_{-\infty}^{\infty} \left(\frac{df}{dx} \right)^2 d\gamma_1,
$$

where $df/dx = \sum_{n=1}^{\infty} \sqrt{n} f_n \tilde{h}_n \in L^2$ is the formal derivative of f.

We want to use De Moivre's central limit theorem. To this end, let us denote the infinitesimal generator of the semigroup acting on $SL^2(D_2^d)$ by L_d, and denote the entropy and the energy of $f(C_d)$ by Ent_d and $\mathcal{E}_d(f)$.

Proposition 13.9.1 *If f is a continuous function of polynomial growth which is in $D(L)$, then $\text{Ent}_d(f^2) \to \text{Ent}(f^2)$.*

Proof Since f^2 and $f^2 \log f^2$ are of polynomial growth,

$$\mathbf{E}((f(C_d))^2) \to \int f^2 \, d\gamma_1 \quad \text{and} \quad \mathbf{E}((f(C_d))^2 \log(f(C_d))^2) \to \int f^2 \log f^2 \, d\gamma_1$$

as $d \to \infty$; the result follows from this. $\qquad\square$

Theorem 13.9.1 *Suppose that $f \in L^2(\gamma_1)$ is differentiable, with a uniformly continuous derivative f'. Then $\mathcal{E}_d(f) \to \mathcal{E}(f)$ as $d \to \infty$.*

Proof The conditions ensure that $f' \in L^2(\gamma)$ and that $\mathcal{E}(f) = \int_{-\infty}^{\infty} (f')^2 \, d\gamma_1$. We shall prove the result for even values of d: the proof for odd values is completely similar. We use the notation introduced in the proof of De Moivre's central limit theorem.

Fix $d = 2m$. If $C_d(\omega) = t_{2k}$ then

$$L_d(f(C_d))(\omega) = \tfrac{1}{2} \left((m+k)f(t_{2k-2}) + (m-k)f(t_{2k+2}) - 2mf(t_{2k}) \right),$$

so that $\mathbf{E}(\langle f, L_d(f) \rangle) = \tfrac{1}{2}(J_1 + J_2)$, where

$$J_1 = \sum_k ((m-k)f(t_{2k})(f(t_{2k+2}) - f(t_{2k}))\mathbf{P}(C_d = t_{2k}))$$

and

$$J_2 = \sum_k ((m+k)f(t_{2k})(f(t_{2k-2}) - f(t_{2k}))\mathbf{P}(C_d = t_{2k}))$$

$$= -\sum_k ((m+k+1)f(t_{2k+2})(f(t_{2k+2}) - f(t_{2k}))\mathbf{P}(C_d = t_{2k+2})),$$

by a change of variables. Now

$$(m+k+1)\mathbf{P}(C_d = t_{2k+2}) = (m+k+1)\frac{(2m)!}{2^{2m}(m+k+1)!(m-k-1)!}$$

$$= (m-k)\frac{(2m)!}{2^{2m}(m+k)!(m-k)!}$$

$$= (m-k)\mathbf{P}(C_d = t_{2k}),$$

so that

$$\mathbf{E}(\langle f, L_d(f)\rangle) = -\tfrac{1}{2}\sum_k (m-k)(f(t_{2k+2}) - f(t_{2k}))^2 \mathbf{P}(C_d = t_{2k})$$

$$= -\sum_k \left(\frac{m-k}{m}\right)\left(\frac{f(t_{2k+2}) - f(t_{2k})}{t_{2k+2} - t_{2k}}\right)^2 \mathbf{P}(C_d = t_{2k}).$$

Given $\epsilon > 0$ there exists $\delta > 0$ such that $|(f(x+h) - f(x))/h - f'(x)| < \epsilon$ for $0 < |h| < \delta$, so that

$$|(f(x+h) - f(x))^2/h^2 - (f'(x))^2| < \epsilon(2|f'(x)| + \epsilon).$$

Also, $k/m = t_{2k}/\sqrt{d}$. Thus it follows that

$$|\mathbf{E}(\langle f, L_d(f)\rangle) + \mathbf{E}((f'(C_d))^2)| \le \epsilon(\mathbf{E}(2|f'(C_d)|) + \epsilon) + K_d,$$

where

$$K_d = \sum_k \frac{|k|}{m}|f'(t_{2k})|^2 \mathbf{P}(C_d = t_{2k}) = \frac{1}{\sqrt{d}}\mathbf{E}(|C_d|(f'(C_d))^2).$$

By De Moivre's central limit theorem, $\mathbf{E}(|C_d|(f'(C_d))^2) \to \mathbf{E}(|\xi|(f'(\xi))^2)$ as $d \to \infty$, so that $K_d \to 0$ as $d \to \infty$; further, $\mathbf{E}(f'(C_d))^2 \to \mathbf{E}((f')^2)$ as $d \to \infty$ and so $\mathcal{E}_d(f) \to \mathbf{E}((f')^2) = \mathcal{E}(f)$ as $d \to \infty$. $\qquad\square$

Corollary 13.9.1 (The Gaussian logarithmic Sobolev inequality)
Suppose that $f \in L^2(\gamma_1)$ is differentiable, with a uniformly continuous derivative f'. Then $\mathrm{Ent}(f^2) \le 2\mathcal{E}(f)$.

13.10 The logarithmic Sobolev inequality in higher dimensions

What happens in higher dimensions? We describe briefly what happens in $\mathbf{R}^d$; the ideas extend easily to the infinite-dimensional case. The measure γ_d is the d-fold product $\gamma_1 \times \cdots \times \gamma_1$. From this it follows that the polynomials in $x_1, \ldots, x_d$ are dense in $L^2(\mathbf{R}^d)$. Let P_n be the finite-dimensional subspace spanned by the polynomials of degree at most n, let p_n be the orthogonal projection onto P_n, let $\pi_n = p_n - p_{n-1}$ and let $H^{:n:} = \pi_n(L^2(\gamma_d))$. Then $L^2(\gamma) = \oplus_{n=0}^{\infty} H^{:n:}$. This orthogonal direct sum decomposition is the *Wiener chaos* decomposition; $H^{:n:}$ is the n-th *Wiener chaos*. If $x^{\alpha} = x_1^{\alpha_1}\ldots x_d^{\alpha_d}$, with $|\alpha| = \alpha_1 + \cdots + \alpha_d = n$, then $\pi_n(x^{\alpha}) = \prod_{i=1}^d h_{\alpha_i}(x_i)$. This is the *Wick product*: we write it as $:x^{\alpha}:$.

A more complicated, but essentially identical argument, using independent copies $C_{m,1}, \ldots, C_{m,d}$ of C_m, establishes the Gaussian version of Bonami's theorem.

Theorem 13.10.1 *Suppose that $1 < p < q < \infty$, and that $\{y_\alpha\}_{\alpha \in A}$ is a family of elements of a Banach space $(E, \|.\|_E)$, where A is a finite set of multi-indices $\alpha = (\alpha_1, \ldots, \alpha_d)$. Then*

$$\left\| \sum_{\alpha \in A} r_q^{|\alpha|} :x^\alpha: y_\alpha \right\|_{L^q(E)} \leq \left\| \sum_{\alpha \in A} r_p^{|\alpha|} :x^\alpha: y_\alpha \right\|_{L^p(E)}.$$

Proof The details are left to the reader. □

This result then extends by continuity to infinite sums, and to infinitely many independent Gaussian random variables.

The logarithmic Sobolev inequality also extends to higher dimensions. The Ornstein-Uhlenbeck semigroup acts on multinomials as follows: if $f = \sum_{\alpha \in A} f_\alpha :x^\alpha:$ then

$$P_t(f) = \sum_{\alpha \in A} e^{-|\alpha|t} f_\alpha :x^\alpha: \quad \text{and} \quad L(f) = -\sum_{\alpha \in A} |\alpha| f_\alpha :x^\alpha:$$

Then we have the following theorem.

Theorem 13.10.2 *Suppose that $f \in L^2(\gamma_d)$ has a uniformly continuous derivative ∇f, and that $\|f\|_{L^2(\gamma_d)} = 1$. Then*

$$0 \leq \int |f|^2 \log |f|^2 \, d\gamma_d \leq \int |\nabla f|^2 \, d\gamma_d.$$

This theorem and its corollary have the important property that the inequalities do not involve the dimension d; contrast this with the Sobolev inequality obtained in Chapter 5 (Theorem 5.8.1).

We also have the following consequence: the proof is the same as the proof of Theorem 13.4.2.

Theorem 13.10.3 *Suppose that $f \in L^2(\gamma_d)$ has a uniformly continuous derivative ∇f, that $\int_{\mathbf{R}^d} f \, d\gamma_d = 0$, and that $|\nabla(f)(x)| \leq 1$ for all $x \in \mathbf{R}^d$. Then f is sub-Gaussian with index $1/\sqrt{2}$: that is, $\int_{\mathbf{R}^d} (e^{\lambda f}) \, d\gamma_d \leq e^{-\lambda^2/4}$, for all real λ.*

Corollary 13.10.1 *If $r > 0$ then $\gamma_d(f \geq r) \leq e^{-r^2}$.*

If A is a closed subset of $\mathbf{R}^d$, and $s > 0$ we set $A_s = \{x : d(x, A) \leq s\}$.

Corollary 13.10.2 *Suppose that $\gamma_d(A) > 1/e$. Let If $s > 1$ then $\gamma_d(A_s) \geq 1 - e^{-(t-1)^2}$.*

Proof Let $g(x) = d(x, A)$. Then $|\nabla g(x)| \leq 1$ for each $x \notin A$, but g is not differentiable at every point of A. But we can approximate g uniformly by smooth functions g_n with $|\nabla g_n(x)| \leq 1$ for all x, and apply the argument of Corollary 13.4.2, to obtain the result. The details are again left to the reader. □

13.11 Beckner's inequality

Bonami's inequality, and the hypercontractive inequality, are essentially real inequalities. As Beckner [Bec 75] showed, there is an interesting complex version of the hypercontractive inequality.

Theorem 13.11.1 (Beckner's inequality) *Suppose that $1 < p < 2$, and let $s = \sqrt{p-1} = r_{p'}$, so that $0 < s < 1$. If a and b are complex numbers then*

$$\|a + \epsilon i s b\|_{p'} \leq \|a + \epsilon b\|_p .$$

Proof The result is trivially true if $a = 0$. Otherwise, by homogeneity, we can suppose that $a = 1$. Let $b = c + id$. Then $|1 + \epsilon i s b|^2 = |1 - \epsilon s d|^2 + s^2 c^2$, so that

$$
\begin{aligned}
\|1 + isb\|_{p'}^2 &= \left\|\, |1 + \epsilon i s b|^2 \,\right\|_{p'/2} \\
&= \left\|(1 - \epsilon s d)^2 + s^2 c^2\right\|_{p'/2} \\
&\leq \left\|(1 - \epsilon s d)^2\right\|_{p'/2} + s^2 c^2 \quad \text{(by Minkowski's inequality)} \\
&= \|1 - \epsilon s d\|_{p'} + s^2 c^2 \\
&\leq \|1 - \epsilon d\|_2 + s^2 c^2 \quad \text{(by the hypercontractive inequality)} \\
&= 1 + d^2 + s^2 c^2 \\
&= \|1 + \epsilon s c\|_2^2 + d^2 \\
&\leq \|1 + \epsilon c\|_p^2 + d^2 \quad \text{(by the hypercontractive inequality again)} \\
&= \left\|(1 + \epsilon c)^2\right\|_{p/2} + d^2 \\
&\leq \left\|(1 + \epsilon c)^2 + d^2\right\|_{p/2} \quad \text{(by the reverse Minkowski inequality)} \\
&= \|1 + \epsilon b\|_p^2 .
\end{aligned}
$$

□

Following through the second half of the proof of Bonami's inequality, and the proof of Theorem 13.7.1, we have the following corollary.

Corollary 13.11.1 (Beckner's theorem) *Suppose that $1 < p < 2$, and that $s = \sqrt{p-1}$.*

(i) If $\{z_A \colon A \subseteq \{1, \ldots, n\}\}$ is a family of complex numbers, then

$$\left\| \sum_A (is)^{|A|} w_A z_A \right\|_{L^{p'}} \leq \left\| \sum_A w_A z_A \right\|_{L^p},$$

where the w_A are Walsh functions.

(ii) If $f = \sum_{j=0}^n \beta_j h_j$ is a polynomial, let $M_{is}(f) = \sum_{j=0}^n (is)^j \beta_j h_j$. Then

$$\|M_{is}(f)\|_{L^{p'}(\gamma_1)} \leq \|f\|_{L^p(\gamma_1)}.$$

13.12 The Babenko–Beckner inequality

Beckner [Bec 75] used Corollary 13.11.1 to establish a stronger form of the Hausdorff–Young inequality. Recall that this says that the Fourier transform is a norm-decreasing linear map from $L^p(\mathbf{R})$ into $L^{p'}(\mathbf{R})$, for $1 < p \leq 2$, and that we proved it by complex interpolation. Can we do better? Babenko had shown that this was possible, and obtained the best possible result, when p' is an even integer. Beckner then obtained the best possible result for all $1 < p \leq 2$.

Theorem 13.12.1 (The Babenko–Beckner inequality) *Suppose that $1 < p \leq 2$. Let $n_p = p^{1/2p}$, $n_{p'} = (p')^{1/2p'}$ and let $A_p = n_p/n_{p'}$. If $f \in L^p(\mathbf{R})$ then its Fourier transform $\mathcal{F}(f)(u) = \int_{-\infty}^{\infty} e^{-2\pi i x u} f(x) \, dx$ satisfies $\|\mathcal{F}(f)\|_{p'} \leq A_p \|f\|_p$, and A_p is the best possible constant.*

Proof First let us show that we cannot do better than A_p. If $e(x) = e^{-\pi x^2}$, then $\mathcal{F}(e)(u) = e^{-\pi u^2}$. Since $\|e\|_p = 1/n_p$ and $\|e\|_{p'} = 1/n_{p'}$, $\|\mathcal{F}(e)\|_{p'} = A_p \|e\|_p$.

There is a natural isometry J_p of $L^p(\gamma_1)$ onto $L^p(\mathbf{R})$: if $f \in L^p(\gamma_1)$, let

$$J_p(f)(x) = n_p e^{-\pi x^2} f(\lambda_p x),$$

where $\lambda_p = \sqrt{2\pi p}$. Then

$$\|J_p(f)\|_p^p = \sqrt{p} \int_{-\infty}^{\infty} e^{-p\pi x^2} |f(\lambda_p x)|^p \, dx$$

$$= \frac{1}{\sqrt{2\pi}} \int_{-\infty}^{\infty} e^{-y^2/2} |f(y)|^p \, dy = \|f\|_{L^p(\gamma_1)}^p.$$

We therefore consider the operator $T_p = J_{p'}^{-1} \mathcal{F} J_p : L^p(\gamma_1) \to L^{p'}(\gamma_1)$. Let

$f_n = J_p(h_n)$. Then, since $(dh_n/dx)(x) = xh_n(x) - h_{n+1}(x)$,

$$\frac{df_n}{dx}(x) = -2\pi x f_n(x) + \lambda_p n_p e^{-\pi x^2} \frac{dh_n}{dx}(\lambda_p x)$$
$$= 2\pi(p-1)x f_n(x) - \lambda_p f_{n+1}(x);$$

thus we have the recurrence relation

$$\lambda_p f_{n+1}(x) = 2\pi(p-1)x f_n(x) - \frac{df_n}{dx}(x).$$

Now let k_n be the Fourier transform of f_n. Bearing in mind that if f is a smooth function of rapid decay and if $g(x) = xf(x)$ and $h(x) = (df/dx)(x)$ then

$$\mathcal{F}(g)(u) = \frac{i}{2\pi}\frac{d\mathcal{F}(f)}{du}(u) \text{ and } \mathcal{F}(h)(u) = 2\pi iu\mathcal{F}(f)(u),$$

we see that

$$\lambda_p k_{n+1}(u) = i(p-1)\frac{dk_n}{du}(u) - 2\pi iuk_n(u)$$
$$= -i\left(2\pi uk_n(u) - (p-1)\frac{dk_n}{du}(u)\right),$$

so that, since $\lambda_p s(p-1) = \lambda_{p'}$, we obtain the recurrence relation

$$\lambda_{p'} k_{n+1}(u) = -is\left(2\pi(p'-1)uk_n(u) - \frac{dk_n}{du}(u)\right)$$

where, as before, $s = \sqrt{p-1}$.

Now $f_0(x) = n_p e^{-\pi x^2}$, so that $k_0(u) = n_p e^{-\pi u^2} = A_p f_0(u)$. Comparing the recurrence relations for (f_n) and (k_n), we see that $k_n = A_p(-is)^n J_{p'}^{-1}(h_n)$, so that $T_p(h_n) = A_p(-is)^n h_n$. Thus $T_p = A_p M(is)$, and so, by Beckner's theorem, $\left\|T_p : L^p(\gamma_1) \to L^{p'}(\gamma_1)\right\| \leq A_p$. Since J_p and $J_{p'}$ are isometries, it follows that $\left\|\mathcal{F} : L^p(\mathbf{R}) \to L^{p'}(\mathbf{R})\right\| \leq A_p$. $\qquad\square$

An exactly similar argument establishes a d-dimensional version.

Theorem 13.12.2 (The Babenko–Beckner inequality) *Suppose that* $1 < p \leq 2$. *Let* $A_p = p^{1/2p}/p'^{1/2p'}$. *If* $f \in L^p(\mathbf{R}^d)$, *then its Fourier transform* $\hat{f}(u) = \int_{\mathbf{R}^d} e^{-2\pi i\langle x,u\rangle} f(x)\,dx$ *satisfies* $\left\|\hat{f}\right\|_{p'} \leq A_p^d \|f\|_p$, *and* A_p^d *is the best possible constant.*

13.13 Notes and remarks

Bonami's inequality was proved in [Bon 71]; it was used in her work on harmonic analysis on the group D_2^N. At about the same time, a similar inequality was proved by Nelson [Nel 73] in his work on quantum field theory, and the inequality is sometimes referred to as Nelson's inequality.

The relationship between the hypercontractive inequality and the logarithmic Sobolev inequality is an essential part of modern semigroup theory, and many aspects of the results that are proved in this chapter are clarified and extended in this setting. Accounts are given in [Bak 94] and [Gro 93]. An enjoyable panoramic view of the subject is given in [Ané 00].

A straightforward account of information and entropy is given in [App 96].

In his pioneering paper [Gro 75], Gross used the central limit theorem, as we have, to to establish Gaussian logarithmic Sobolev inequalities.

The book by Janson [Jan 97] gives an excellent account of Gaussian Hilbert spaces.

Exercises

13.1   Let

$$f_n(x) = (-1)^n e^{\pi x^2} \frac{d^n}{dx^n} (e^{-\pi x^2}).$$

Show that $(f_n)_{n=0}^\infty$ is an orthonormal sequence in $L^2(\mathbf{R})$, whose linear span is dense in $L^2(\mathbf{R})$. Find constants C_n such that $(\tilde{f}_n) = (C_n f_n)$ is an orthonormal basis for $L^2(\mathbf{R})$. Show that $\mathcal{F}(\tilde{f}_n) = i^n \tilde{f}_n$. Deduce the Plancherel theorem for $L^2(\mathbf{R})$: the Fourier transform is an isometry of $L^2(\mathbf{R})$ onto $L^2(\mathbf{R})$.

The idea of using the Hermite functions to prove the Plancherel theorem goes back to Norbert Wiener.

13.2   Calculate the constants given by the Babenko–Beckner inequality for various values of p, and compare them with those given by the Hausdorff–Young inequality.

14

Hadamard's inequality

14.1 Hadamard's inequality

So far, we have been concerned with inequalities that involve functions. In the next chapter, we shall turn to inequalities which concern linear operators. In the finite-dimensional case, this means considering matrices and determinants. Determinants, however, can also be considered as volume forms. In this chapter, we shall prove Hadamard's inequality [Had 93], which can usefully be thought of in this way. We shall also investigate when equality holds, in the real case: this provides a digression into number theory, and also has application to coding theory, which we shall also describe.

Theorem 14.1.1 (Hadamard's inequality) *Let $A = (a_{ij})$ be a real or complex $n \times n$ matrix. Then*

$$|\det A| \leq \prod_{j=1}^{n} \left(\sum_{i=1}^{n} |a_{ij}|^2 \right)^{1/2},$$

with equality if and only if either both sides are zero or $\sum_{i=1}^{n} a_{ij}\overline{a_{ik}} = 0$ for $j \neq k$.

Proof Let $a_j = (a_{ij})$ be the j-th column of A, considered as an element of the inner product space l_n^2. Then the theorem states that $|\det A| \leq \prod_{j=1}^{n} \|a_j\|$, with equality if and only if the columns are orthogonal, or one of them is zero.

The result is certainly true if $\det A = 0$. Let us suppose that $\det A$ is not zero. Then the columns of A are linearly independent, and we orthogonalize them. Let $E_j = \text{span}\,(a_1, \ldots, a_j)$, and let Q_j be the orthogonal projection of l_2^n onto $E_j^\perp$. Let $b_1 = a_1$ and let $b_j = Q_{j-1}(a_j)$, for $2 \leq j \leq n$. Then

$\|b_j\| \leq \|a_j\|$. On the other hand,

$$b_j = a_j - \sum_{i=1}^{j-1} \frac{\langle a_j, b_i \rangle}{\langle b_i, b_i \rangle} b_i$$

for $2 \leq j \leq n$, so that the matrix B with columns $b_1, \ldots, b_n$ is obtained from A by elementary column operations. Thus $\det B = \det A$. Since the columns of B are orthogonal, $B^*B = \mathrm{diag}(\|b_1\|^2, \ldots, \|b_n\|^2)$, so that

$$|\det A| = |\det B| = (\det(B^*B))^{1/2} = \prod_{j=1}^{n} \|b_j\| \leq \prod_{j=1}^{n} \|a_j\|.$$

We have equality if and only if $\|b_j\| = \|a_j\|$ for each j, which happens if and only if the columns of A are orthogonal. $\qquad\qquad\square$

The theorem states that the volume of a parallelopiped in l_2^n with given side lengths has maximal volume when the sides are orthogonal, and the proof is based on this.

14.2 Hadamard numbers

Hadamard's inequality has the following corollary.

Corollary 14.2.1 *Suppose that $A = (a_{ij})$ is a real or complex matrix and that $|a_{ij}| \leq 1$ for all i and j. Then $|\det A| \leq n^{n/2}$, and equality holds if and only if $|a_{ij}| = 1$ for all i and j and $\sum_{i=1}^{n} a_{ij}\overline{a_{ik}} = 0$ for $i \neq k$.*

It is easy to give examples where equality holds in the complex case, for any n; for example, set $a_{hj} = e^{2\pi i hj/n}$.

In the real case, it is a much more interesting problem to find examples where equality holds. An $n \times n$ matrix $A = (a_{ij})$ all of whose entries are 1 or -1, and which satisfies $\sum_{i=1}^{n} a_{ij}a_{ik} = 0$ for $i \neq k$ is called an *Hadamard matrix*, and if n is an integer for which an Hadamard matrix of order n exists, then n is called an *Hadamard number*. Note that the orthogonality conditions are equivalent to the condition that $AA' = nI_n$.

If $A = (a_{ij})$ and $B = (b_{i'j'})$ are Hadamard matrices of orders n and n' respectively, then it is easy to check that the Kronecker product, or tensor product,

$$K = A \otimes B = \left(k_{(i,i')(j,j')}\right) = (a_{ij}) \cdot (b_{i'j'})$$

is a Hadamard matrix of order nn'. Thus if n and n' are Hadamard numbers, then so is nn'. Now the 2×2 matrix $\begin{bmatrix} 1 & 1 \\ 1 & -1 \end{bmatrix}$ is an Hadamard matrix.

By repeatedly forming Kronecker products, we can construct Hadamard matrices of all orders 2^k.

Are there any other (essentially different) Hadamard matrices? Hadamard [Had 93] constructed Hadamard matrices of orders 12 and 20. Forty years later, Paley [Pal 33] gave a powerful way of constructing infinitely many new Hadamard matrices. Before we present Paley's result, let us observe that not every number can be an Hadamard number.

Proposition 14.2.1 *If $A = (a_{ij})$ is a Hadamard matrix of order n, where $n \geq 3$, then 4 divides n.*

Proof Let a, b, c be distinct columns. Then

$$\sum_{i=1}^{n}(a_i + b_i)(a_i + c_i) = \langle a + b, a + c \rangle = \langle a, a \rangle = n.$$

But each summand is 0 or 4, so that 4 divides n. $\qquad\square$

Theorem 14.2.1 (Paley [Pal 33]) *Suppose that $q = p^k$ is a prime power. If $q = 1(mod\ 4)$, then there is a symmetric Hadamard matrix of order $2(q + 1)$, while if $q = 3(mod\ 4)$ then there is a skew-symmetric matrix C of order $n = q + 1$ such that $I_n + C$ is an Hadamard matrix.*

In order to prove this theorem, we introduce a closely related class of matrices. An $n \times n$ matrix C is a *conference matrix* (the name comes from telephone network theory) if the diagonal entries c_{ii} are zero, all the other entries are 1 or -1 and the columns are orthogonal: $\sum_{i=1}^{n} c_{ij}c_{ik} = 0$ for $i \neq k$. Note that the orthogonality conditions are equivalent to the condition that $CC' = (n-1)I_n$.

Proposition 14.2.2 *If C is a symmetric conference matrix, then the matrix*

$$D = \begin{bmatrix} I_n + C & -I_n + C \\ -I_n + C & -I_n - C \end{bmatrix}$$

is a symmetric Hadamard matrix.

If C is a skew-symmetric conference matrix, then the matrix $I_n + C$ is an Hadamard matrix.

Proof If C is a symmetric conference matrix,

$$DD' = \begin{bmatrix} (I_n + C)^2 + (-I_n + C)^2 & ((I_n + C) + (-I_n - C))(-I_n + C) \\ ((I_n + C) + (-I_n - C))(-I_n + C) & (-I_n - C)^2 + (-I_n - C)^2 \end{bmatrix}$$

$$= \begin{bmatrix} 2I_n + 2C^2 & 0 \\ 0 & 2I_n + 2C^2 \end{bmatrix} = 2nI_{2n}.$$

If C is a skew-symmetric conference matrix, then

$$(I_n + C)(I_n + C)' = (I_n + C)(I_n - C) = I_n - C^2 = I_n + CC' = nI_n.$$

In order to prove Paley's theorem, we therefore need only construct conference matrices of order $q+1$ with the right symmetry properties. In order to do this, we use the fact that there is a finite field F_q with q elements. Let χ be the Legendre character on F_q:

$$\chi(0) = 0,$$
$$\chi(x) = 1 \text{ if } x \text{ is a non-zero square,}$$
$$\chi(x) = -1 \text{ if } x \text{ is not a square.}$$

We shall use the elementary facts that $\chi(x)\chi(y) = \chi(xy)$, that $\chi(-1) = 1$ if and only if $q = 1 \pmod 4$ and that $\sum_{x \in F_q} \chi(x) = 0$.

First we define a $q \times q$ matrix $A = (a_{xy})$ indexed by the elements of F_q: we set $a_{xy} = \chi(x-y)$. A is symmetric if $q = 1 \pmod 4$ and A is skew-symmetric if $q = 3 \pmod 4$.

We now augment A, by adding an extra row and column:

$$C = \begin{bmatrix} 0 & \chi(-1) & \cdots & \chi(-1) \\ 1 & & & \\ \vdots & & A & \\ 1 & & & \end{bmatrix}.$$

C has the required symmetry properties, and we shall show that it is a conference matrix. Since $\sum_{x \in F_q} \chi(x) = 0$, the first column is orthogonal to each of the others. If c_y and c_z are two other distinct columns, then

$$\langle c_y, c_z \rangle = 1 + \sum_{x \in F_q} \chi(x - y)\chi(x - z)$$

$$= 1 + \sum_{x \in F_q} \chi(x)\chi(x + y - z)$$

$$= 1 + \sum_{x \neq 0} (\chi(x))^2 \chi(1 + x^{-1}(y - z))$$

$$= \chi(1) + \sum_{x \neq 0} \chi(1 + x) = 0.$$

This completes the proof. $\qquad\qquad\qquad\qquad\qquad\qquad\qquad\qquad\square$

Paley's theorem implies that every multiple of four up to 88 is an Hadamard number. After another twenty-nine years, it was shown [BaGH 62] that 92 is an Hadamard number. Further results have been obtained, but it is still not known if every multiple of four is an Hadamard number.

14.3 Error-correcting codes

Hadamard matrices are useful for construction error-correcting codes. Suppose that Alice wants to send Bob a message, of some 10,000 characters, say. The characters of her message belong to the extended ASCII set of 256 characters, but she must send the message as a sequence of bits (0's and 1's). She could for example assign the numbers 0 to 255 to the ASCII characters in the usual way, and put each of the numbers in binary form, as a string of eight bits. Thus her message will be a sequence of 80,000 bits. Suppose however that the channel through which she send her message is a 'noisy' one, and that there is a probability $1/20$ that a bit is received incorrectly by Bob (a 0 being read as a 1, or a 1 being read as a 0), the errors occurring independently. Then for each character, there is probability about 0.34 that it will be misread by Bob, and this is clearly no good.

Suppose instead that Alice and Bob construct an Hadamard matrix H of order 128 (this is easily done, using the Kronecker product construction defined above, or the character table of F_{127}) and replace the -1's by 0's, to obtain a matrix K. They then use the columns of K and of $-K$ as codewords for the ASCII characters, so that each ASCII character has a codeword consisting of a string of 128 bits. Thus Alice sends a message of 1,280,000 bits. Different characters have different codewords, and indeed any two codewords differ in either 64 or 128 places. Bob decodes the message by replacing the strings of 128 bits by the ASCII character whose codeword it is (if no error has occurred in transmission), or by an ASCII character whose codeword differs in as few places as possible from the string of 128 bits. Thus Bob will only decode a character incorrectly if at least 32 errors have occurred in the transmission of a codeword. The probability of this happening is remarkably small. Let us estimate it approximately. The expected number of errors in transmitting a codeword is 6.4, and so the probability of the number of errors is distributed approximately as a Poisson distribution with parameter $\lambda = 6.4$. Thus the probability of 32 errors (or more) is about $e^{-\lambda}\lambda^{32}/32!$. Using Stirling's approximation for 32!, we see that this probability is about $e^{-\lambda}(e\lambda/32)^{32}/8\sqrt{\pi}$, which is a number of order 10^{-13}. Thus the probability that Bob will receive the message with any errors at all is about 10^{-9}, which is really negligible. Of course there is a price to pay: the message using the Hadamard matrix code is sixteen times as long as the message using the simple binary code.

14.4 Note and remark

An excellent account of Hadamard matrices and their uses is given in Chapter 18 of [vLW 92].

15

Hilbert space operator inequalities

15.1 Jordan normal form

We now turn to inequalities that involve linear operators. In this chapter, we consider operators between finite-dimensional complex vector spaces, which involve matrices and determinants, and operators between infinite dimensional complex Hilbert spaces. Let us spend some time setting the scene, and describing the sorts of problem that we shall consider.

First, suppose that E is a finite-dimensional complex vector space, and that T is an endomorphism of E: that is a linear mapping of E into itself. We describe without proof the results from linear algebra that we need; an excellent account is given in the book by Hirsch and Smale [HiS 74], although their terminology is slightly different from what follows. We consider the operator $\lambda I - T$; this is invertible if and only if $\chi_T(\lambda) = \det(\lambda I - T) \neq 0$. The polynomial χ_T is the *characteristic polynomial*; its roots $\lambda_1, \ldots, \lambda_d$ (repeated according to multiplicity, and arranged in decreasing absolute value) form the *spectrum* $\sigma(T)$. They are the singular points: if $\lambda \in \sigma(T)$ then $E_\lambda(T) = \{x : T(x) = \lambda x\}$ is a non-trivial linear subspace of E, so that λ is an eigenvalue, with eigenspace E_λ. Of equal interest are the subspaces

$$E_\lambda^{(k)}(T) = \{x : (T - \lambda I)^k(x) = 0\} \quad \text{and} \quad G_\lambda(T) = \bigcup_{k>1} E_\lambda^{(k)}(T).$$

$G_\lambda = G_\lambda(T)$ is a *generalized eigenspace*, and elements of G_λ are called *principal vectors*. If $\mu_1, \ldots, \mu_r$ are the distinct eigenvalues of T, then each G_{μ_s} is T-invariant, and E is the algebraic direct sum

$$E = G_{\mu_1} \oplus \cdots \oplus G_{\mu_r}.$$

Further, each generalized eigenspace G_λ can be written as a T-invariant direct sum

$$G_\lambda = H_1 \oplus \cdots \oplus H_l,$$

where each H_i has a basis $(h_1, \ldots, h_k)$, where $T(h_1) = \lambda h_1$ and $T(h_l) = \lambda h_l + h_{l-1}$ for $2 \leq l \leq k$. Combining all of these bases in order, we obtain a *Jordan basis* $(e_1, \ldots, e_d)$ for E; the corresponding matrix represents T in *Jordan normal form*. This basis has the important property that if $1 \leq k \leq d$ and $E_k = $ span $(e_1, \ldots, e_k)$ then E_k is T invariant, and $T_k = T_{|E_k}$ has eigenvectors $\lambda_1(T), \ldots, \lambda_k(T)$.

15.2 Riesz operators

Although we shall be concerned in this chapter with linear operators between Hilbert spaces, in later chapters we shall consider operators between Banach spaces. In this section, we consider endomorphisms of Banach spaces. Suppose then that T is a bounded endomorphism of a complex Banach space E. Then the spectrum $\sigma(T)$ of T, defined as

$$\{\lambda \in \mathbf{C} \colon \lambda I - T \text{ is not invertible}\},$$

is a non-empty closed subset of $\mathbf{C}$, contained in $\{\lambda \colon |\lambda| \leq \inf \|T^n\|^{1/n}\}$, and the *spectral radius* $r(T) = \sup\{|\lambda| \colon \lambda \in \sigma(T)\}$ satisfies the *spectral radius formula* $r(T) = \inf\{\|T^n\|^{1/n}\}$. The complement of the spectrum is called the *resolvent set* $\rho(T)$, and the operator $R_\lambda(T) = R_\lambda = (\lambda I - T)^{-1}$ defined on $\rho(T)$ is called the *resolvent* of T.

The behaviour of $\lambda I - T$ at a point of the spectrum can however be complicated; we restrict our attention to a smaller class of operators, the *Riesz* operators, whose properties are similar to those of operators on finite-dimensional spaces.

Suppose that $T \in L(E)$. T is a *Riesz operator* if

- $\sigma(T) \setminus \{0\}$ is either finite or consists of a sequence of points tending to 0.
- If $\mu \in \sigma(T) \setminus \{0\}$, then μ is an eigenvalue and the generalized eigenspace

$$G_\mu = \{x \colon (T - \mu I)^k(x) = 0 \text{ for some } k \in \mathbf{N}\}$$

is of finite dimension.
- If $\mu \in \sigma(T) \setminus \{0\}$, there is a T-invariant decomposition $E = G_\mu \oplus H_\mu$, where H_μ is a closed subspace of E and $T - \mu I$ is an isomorphism of H_μ onto itself.

We denote the corresponding projection of E onto G_μ with null-space H_μ by $P_\mu(T)$, and set $Q_\mu(T) = I - P_\mu(T)$.

If T is a Riesz operator and $\mu \in \sigma(T) \setminus \{0\}$, we call the dimension of G_μ the *algebraic multiplicity* $m_T(\mu)$ of μ. We shall use the following convention: we denote the distinct non-zero elements of $\sigma(T)$, in decreasing absolute value, by $\mu_1(T), \mu_2(T), \ldots$, and denote the non-zero elements of $\sigma(T)$, repeated according to algebraic multiplicity and in decreasing absolute value, by $\lambda_1(T), \lambda_2(T), \ldots$. (If $\sigma(T) \setminus \{0\} = \{\mu_1, \ldots, \mu_t\}$ is finite, then we set $\mu_s(T) = 0$ for $s > t$, and use a similar convention for $\lambda_j(T)$.)

Suppose that T is a Riesz operator and that $\mu \in \sigma(T) \setminus \{0\}$. Then μ is an isolated point of $\sigma(T)$. Suppose that $s > 0$ is sufficiently small that μ is the only point of $\sigma(T)$ in the closed disc $\{z : |z - \mu| \leq s\}$. Then it follows from the functional calculus that

$$P_\mu(T) = \frac{1}{2\pi i} \int_{|z-\mu|=s} R_z(T) \, dz.$$

This has the following consequence, that we shall need later.

Proposition 15.2.1 *Suppose that T is a Riesz operator on E and that* $|\mu_j(T)| > r > |\mu_{j+1}(T)|$. *Let*

$$J_r = G_{\mu_1} \oplus \cdots \oplus G_{\mu_j}, \quad K_r = H_{\mu_1} \cap \cdots \cap H_{\mu_j}.$$

Then $E = J_r \oplus K_r$. If Π_r is the projection of E onto K_r with null-space J_r then

$$\Pi_r(T) = \frac{1}{2\pi i} \int_{|z|=r} R_z(T) \, dz.$$

We denote the restriction of T to J_r by $T_{>r}$, and the restriction of T to K_r by $T_{<r}$. $T_{<r}$ is a Riesz operator with eigenvalues $\mu_{j+1}, \mu_{j+2}, \ldots$.

15.3 Related operators

Suppose that E and F are Banach spaces, and that $S \in L(E)$ and $T \in L(F)$. Following Pietsch [Pie 63], we say that S and T are *related* if there exist $A \in L(E, F)$, $B \in L(F, E)$ such that $S = BA$ and $T = AB$. This simple idea is extremely powerful, as the following proposition indicates.

Proposition 15.3.1 *Suppose that $S = BA$ and $T = AB$ are related.*

(i) $\sigma(S) \setminus \{0\} = \sigma(T) \setminus \{0\}$.

(ii) Suppose that $p(x) = xq(x) + \lambda$ is a polynomial with non-zero constant term λ. Let $N_S = \{y : p(S)y = 0\}$ and let $N_T = \{z : p(T)(z) = 0\}$. Then $A(N_S) \subseteq N_T$, and A is one-one on N_S.

Proof (i) Suppose that $\lambda \in \rho(S)$ and that $\lambda \neq 0$. Set $J_\lambda(T) = (AR_\lambda(S)B - I_F)/\lambda$. Then

$$(T - \lambda I_F)J_\lambda(T) = (A(BA - \lambda I_E)R_\lambda(S)B - AB + \lambda I_F)/\lambda = I_F,$$
$$J_\lambda(T)(T - \lambda I_F) = (AR_\lambda(S)(BA - \lambda I_E)B - AB + \lambda I_F)/\lambda = I_F,$$

so that $\lambda \in \rho(T)$ and $R_\lambda(T) = J_\lambda(T)$. Similarly if $\lambda \in \rho(T)$ and $\lambda \neq 0$ then $\lambda \in \rho(S)$.

(ii) Since $Ap(BA) = p(AB)A$, if $y \in N_S$ then $p(T)A(y) = Ap(S)(y) = 0$, and so $A(N_S) \subseteq N_T$. If $y \in N_S$ and $A(y) = 0$, then $p(S)(y) = \lambda y = 0$, so that $y = 0$. Thus A is one-one on N_S. $\qquad\square$

Since a similar result holds for $B(N_T)$, we have the following corollary.

Corollary 15.3.1 *If $S = BA$ and $T = AB$ are related Riesz operators and $\mu \in \sigma(S) \setminus \{0\}$ then $A(G_\mu(S)) = G_\mu(T)$ and $B(G_\mu(T)) = G_\mu(S)$. In particular, $m_S(\mu) = m_T(\mu)$.*

In fact, although we shall not need this, if $S \in L(E)$ and $T \in L(F)$ are related, and S is a Riesz operator, then T is a Riesz operator [Pie 63].

15.4 Compact operators

Are there enough examples of Riesz operators to make them important and interesting? To begin to answer this, we need to introduce the notion of a compact linear operator. A linear operator T from a Banach space $(E, \|.\|_E)$ to a Banach space $(F, \|.\|_F)$ is *compact* if the image $T(B_E)$ of the unit ball B_E of E is relatively compact in F: that is, the closure $\overline{T(B_E)}$ is a compact subset of F. Alternatively, T is compact if $T(B_E)$ is precompact: given $\epsilon > 0$ there exists a finite subset G in F such that $T(B_E) \subseteq \cup_{g \in G}(g + \epsilon B_F)$. It follows easily from the definition that a compact linear operator is bounded, and that its composition (on either side) with a bounded linear operator is again compact. Further the set $K(E, F)$ of compact linear operators from E to F is a closed linear subspace of the Banach space $L(E, F)$ of bounded linear operators from E to F, with the operator norm.

Theorem 15.4.1 *Suppose that $T \in L(E)$, where $(E, \|.\|_E)$ is an infinite-dimensional complex Banach space. If T^k is compact, for some k, then T is a Riesz operator.*

The proof of this result is unfortunately outside the scope of this book. A full account is given in [Dow 78], and details are also given, for example, in [DuS 88], Chapter VII.

Our task will be to establish inequalities which give information about the eigenvalues of a Riesz operator T in terms of other properties that it possesses. For example, $|\lambda_1| \leq r(T)$. The Jordan normal form gives exhaustive information about a linear operator on a finite-dimensional spaces, but the eigenvalues and generalized eigenspaces of a Riesz operator can give very limited information indeed. The simplest example of this phenomenon is given by the Fredholm integral operator

$$T(f)(x) = \int_0^x f(t)\,dt$$

on $L^2[0,1]$. T is a compact operator (Exercise 15.2). It follows from the Cauchy–Schwarz inequality that $|T(f)(x)| \leq x^{1/2}\|f\|_2 \leq \|f\|_2$, and arguing inductively,

$$|T^n(f)(x)| \leq (x^{n-1}/(n-1)!)\,\|f\|_2\,.$$

From this it follows easily that T has no non-zero eigenvalues, and indeed the spectral radius formula shows that $\sigma(T) = \{0\}$. We shall therefore also seek other parameters that give information about Riesz operators.

15.5 Positive compact operators

For the rest of this chapter, we shall consider linear operators between Hilbert spaces, which we denote as $H, H_0, H_1, \ldots$. We shall suppose that all these spaces are separable, so that they have countable orthonormal bases; this is a technical simplification, and no important features are lost.

We generalize the notion of a Hermitian matrix to the notion of a Hermitian operator on a Hilbert space. $T \in L(H)$ is *Hermitian* if $T = T^*$: that is, $\langle T(x), y \rangle = \langle x, T(y) \rangle$ for all $x, y \in H$. If T is Hermitian then $\langle T(x), x \rangle = \langle x, T(x) \rangle = \overline{\langle T(x), x \rangle}$, so that $\langle T(x), x \rangle$ is real. A Hermitian operator T is *positive*, and we write $T \geq 0$, if $\langle T(x), x \rangle \geq 0$ for all $x \in H$. If $S \in L(H)$ then $S + S^*$ and $i(S - S^*)$ are Hermitian, and S^*S is positive.

Proposition 15.5.1 *Suppose that $T \in L(H)$ is positive. Let $w = w(T) = \sup\{\langle T(x), x \rangle : \|x\| \leq 1\}$. Then $w = \|T\|$.*

Proof Certainly $w \leq \|T\|$. Let $v > w$. Then $vI - T \geq 0$, and so, if $x \in H$,

$$\langle (vI - T)T(x), T(x) \rangle \geq 0 \text{ and } \langle T(vI - T)(x), (vI - T)(x) \rangle \geq 0.$$

Adding, $\langle (vT - T^2)(x), vx \rangle \geq 0$, so that $v\,\langle T(x), x \rangle \geq \langle T^2(x), x \rangle = \|T(x)\|^2$. Thus $vw \geq \|T\|^2$, and $w \geq \|T\|$. $\qquad\square$

Proposition 15.5.2 *If $T \in L(H)$ is positive, then $w = \|T\| \in \sigma(T)$.*

Proof By the preceding proposition, there exists a sequence (x_n) of unit vectors in H such that $\langle T(x_n), x_n \rangle \to w$. Then

$$0 \leq \|T(x_n) - wx_n\|^2 = \|T(x_n)\|^2 - 2w \langle T(x_n), x_n \rangle + w^2$$
$$\leq 2w(w - \langle T(x_n), x_n \rangle) \to 0$$

as $n \to \infty$, so that $(T - wI)(x_n) \to 0$ as $n \to \infty$. $\qquad\square$

Just as a Hermitian matrix can be diagonalized, so can a compact Hermitian operator. We can deduce this from Theorem 15.4.1, but, since this theorem has been stated without proof, we prefer to give a direct proof, which corresponds to the proof of the finite-dimensional case.

Theorem 15.5.1 *Suppose that T is a positive compact operator on H. Then there exists an orthonormal sequence (x_n) in H and a decreasing finite or infinite sequence (s_n) of non-negative real numbers such that $T(x) = \sum_n s_n \langle x, x_n \rangle x_n$ for each $x \in H$. If the sequence is infinite, then $s_n \to 0$ as $n \to \infty$.*

Conversely, such a formula defines a positive element of $K(H)$.

Proof If $T = 0$ we can take any orthonormal sequence (x_n), and take $s_n = 0$. Otherwise, $\mu_1 = \|T\| > 0$, and, as in Proposition 15.5.2, there exists a sequence (x_n) of unit vectors in H such that $T(x_n) - \mu_1 x_n \to 0$. Since T is compact, there exists a subsequence (x_{n_k}) and an element y of H such that $T(x_{n_k}) \to y$. But then $\mu_1 x_{n_k} \to y$, so that $y \neq 0$, and $T(y) = \lim_{k \to \infty} T(\mu_1 x_{n_k}) = \mu_1 y$. Thus y is an eigenvector of T, with eigenvalue μ_1. Let E_{μ_1} be the corresponding eigenspace. Then E_{μ_1} is finite-dimensional; for, if not, there exists an infinite orthonormal sequence (e_n) in E_{μ_1}, and $(T(e_n)) = (\mu_1 e_n)$ has no convergent subsequence.

Now let $H_1 = E_{\mu_1}^{\perp}$. If $x \in H_1$ and $y \in E_{\mu_1}$ then

$$\langle T(x), y \rangle = \langle x, T(y) \rangle = \mu_1 \langle x, y \rangle = 0.$$

Since this holds for all $y \in E_{\mu_1}$, $T(x) \in H_1$. Let $T_1 = T_{|H_1}$. Then T_1 is a positive operator on H_1, and $\mu_2 = \|T_1\| < \mu_1$, since otherwise μ_1 would be an eigenvalue of T_1. We can therefore iterate the procedure, stopping if $T_k = 0$. In this latter case, we put together orthonormal bases of $E_{\mu_1}, \ldots, E_{\mu_{k-1}}$ to obtain a finite orthonormal sequence $(x_1, \ldots, x_N)$. If $x_n \in E_{\mu_j}$, set $s_n = \mu_j$. Then it is easy to verify that $T(x) = \sum_{n=1}^N s_n \langle x, x_n \rangle x_n$ for each $x \in H$.

If the procedure does not stop, we have an infinite sequence of orthogonal eigenspaces (E_{μ_k}), with $\mu_k > 0$. Again, we put together orthonormal bases of the E_{μ_k} to obtain an infinite orthonormal sequence (x_n), and if $x_n \in E_{\mu_k}$, set $s_n = \mu_k$. Then $T(x_n) = s_n x_n$, so that, since $(T(x_n))$ has a convergent subsequence, $s_n \to 0$.

If now $x \in H$ and $k \in \mathbf{N}$, we can write

$$x = \sum_{n=1}^{N_k} \langle x, x_n \rangle \, x_n + r_k,$$

where $N_k = \dim\,(E_{\mu_1} + \cdots + E_{\mu_k})$ and $r_k \in H_k$. Note that $\|r_k\| \leq \|x\|$. Then

$$T(x) = \sum_{n=1}^{N_k} \langle x, x_n \rangle \, T(x_n) + T(r_k) = \sum_{n=1}^{N_k} s_n \, \langle x, x_n \rangle \, x_n + T(r_k).$$

But $\|T(r_k)\| \leq \|T_k\| \, \|x\| = \mu_k \, \|x\| \to 0$ as $n \to \infty$, and so $T(x) = \sum_{n=1}^{\infty} s_n \langle x, x_n \rangle \, x_n$.

For the converse, let $T_{(k)}(x) = \sum_{n=1}^{k} s_n \langle x, x_n \rangle \, x_n$. Each $T_{(k)}$ is a finite rank operator, and $T_{(k)}(x) \to T(x)$ as $k \to \infty$. Suppose that $\epsilon > 0$. There exists N such that $s_N < \epsilon/2$. $T_{(N)}(B_H)$ is a bounded finite-dimensional set, and so is precompact: there exists a finite set F in H such that $T_{(N)}(B_H) \subseteq \cup_{f \in F}(f + (\epsilon/2)B_H)$. But if $x \in B_H$ then $\|T(x) - T_{(N)}(x)\| < \epsilon/2$, and so $T(B_H) \subseteq \cup_{f \in F}(f + \epsilon B_H)$: T is compact. $\qquad\square$

15.6 Compact operators between Hilbert spaces

We now use Theorem 15.5.1 to give a representation theorem for compact linear operators between Hilbert spaces.

Theorem 15.6.1 *Suppose that $T \in K(H_1, H_2)$. Then there exist orthonormal sequences (x_n) in H_1 and (y_n) in H_2, and a finite or infinite decreasing null-sequence (s_n) of positive real numbers such that $T(x) = \sum_n s_n \langle x, x_n \rangle \, y_n$ for each $x \in H_1$.*

Conversely, such a formula defines an element of $K(H_1, H_2)$.

Proof The operator T^*T is a positive compact operator on H_1, and so there exist an orthonormal sequence (x_n) in H_1, and a finite or infinite decreasing sequence (t_n) of positive real numbers such that $T^*T(x) = \sum_n t_n \langle x, x_n \rangle \, y_n$ for each $x \in H_1$. For each n, let $s_n = \sqrt{t_n}$ and let $y_n = T(x_n)/t_n$, so that $T(x_n) = y_n$. Then

$$\langle y_n, y_n \rangle = \langle T(x_n)/t_n, T(x_n)/t_n \rangle = \langle T^*T(x_n), x_n \rangle \, / s_n = 1,$$

and

$$\langle y_n, y_m \rangle = \langle T(x_n)/t_n, T(x_m)/t_m \rangle = \langle T^*T(x_n), x_m \rangle /t_n t_m = 0$$

for $m \neq n$, so that (y_n) is an orthonormal sequence. The rest of the proof is just as the proof of Theorem 15.5.1. $\qquad \square$

We write $T = \sum_{n=1}^{\infty} s_n \langle \cdot, x_n \rangle y_n$ or $T = \sum_{n=1}^{N} s_n \langle \dots, x_n \rangle y_n$.

We can interpret this representation of T in the following way. Suppose that $T = \sum_{n=1}^{\infty} s_n \langle \cdot, x_n \rangle y_n \in K(H_1, H_2)$. Then $T^* = \sum_{n=1}^{\infty} s_n \langle \cdot, y_n \rangle x_n \in K(H_2, H_1)$, and $T^*T = \sum_{n=1}^{\infty} s_n^2 \langle \dots, x_n \rangle x_n \in K(H_1)$. Then $|T| = \sum_{n=1}^{\infty} s_n \langle \cdot, x_n \rangle x_n \in K(H_1)$ is the positive square root of T^*T, and $T = U|T|$, where $U(x) = \sum_{n=1}^{\infty} \langle x, x_n \rangle y_n$ is a partial isometry of H_1 into H_2, mapping the closed linear span K of (x_n) isometrically onto the closed linear span L of (y_n), and mapping $K^\perp$ to 0.

We leave the reader to formulate and prove the corresponding finite-dimensional version of Theorem 15.6.1.

15.7 Singular numbers, and the Rayleigh–Ritz minimax formula

Suppose that $T = \sum_{n=1}^{\infty} s_n(T) \langle \cdot, x_n \rangle y_n \in K(H_1, H_2)$, where (x_n) and (y_n) are orthonormal sequences in H_1 and H_2 respectively, and $(s_n(T))$ is a decreasing sequence of non-negative real numbers. The numbers $s_n(T)$ are called the *singular numbers* of T, and can be characterized as follows.

Theorem 15.7.1 (The Rayleigh–Ritz minimax formula) *Suppose that* $T = \sum_{n=1}^{\infty} s_n(T) \langle \cdot, x_n \rangle y_n \in K(H_1, H_2)$, *where (x_n) and (y_n) are orthonormal sequences in H_1 and H_2 respectively, and $(s_n(T))$ is a decreasing sequence of non-negative real numbers. Then*

$$s_n(T) = \inf \left\{ \left\| T_{|J^\perp} \right\| : \dim J < n \right\}$$
$$= \inf \{ \sup \{ \|T(x)\| : \|x\| \leq 1, x \in J^\perp \} : \dim J < n \},$$

and the infimum is achieved.

Proof Let $r_n = \inf \{ \left\| T_{|J^\perp} \right\| : \dim J < n \}$. If $K_{n-1} = \text{span}\ (x_1, \dots, x_{n-1})$, then $s_n(T) = \left\| T_{|K_{n-1}} \right\|$, and so $s_n(T) \geq r_n$. On the other hand, suppose that J is a subspace with $\dim J = j < n$. If $x \in K_n = \text{span}\ (x_1, \dots, x_n)$, then $\|T(x)\| \geq s_n(T) \|x\|$. Let $D = K_n + J$, let $L = J^\perp \cap D$ and let $d = \dim D$.

Then $\dim L = d - j$ and $\dim (K_n + L) \leq d$, so that

$$\dim (K_n \cap L) = \dim K_n + \dim L - \dim (K_n + L)$$
$$\geq n + (d - j) - d = n - j > 0.$$

Thus there exists $x \in K_n \cap L$ with $\|x\| = 1$, and then $\left\|T_{|J^\perp}\right\| \geq \|T(x)\| \geq s_n(T)$, so that $r_n \geq s_n(T)$. Finally, the infimum is achieved on $K_{n-1}^\perp$. $\square$

Proposition 15.7.1 *(i) If $A \in L(H_0, H_1)$ and $B \in L(H_2, H_3)$ then $s_n(BTA) \leq \|A\| . \|B\| . s_n(T)$.*

(ii) If $S, T \in K(H_1, H_2)$ then $s_{n+m-1}(S + T) \leq s_m(S) + s_n(T)$.

(iii) Suppose that (T_k) is a sequence in $K(H_1, H_2)$ and that $T_k \to T$ in operator norm. Then $s_n(T_k) \to s_n(T)$ as $k \to \infty$, for each n.

Proof (i) follows immediately from the Rayleigh–Ritz minimax formula.

(ii) There exist subspaces J_S of dimension $m - 1$ and J_T of dimension $n - 1$ such that $\left\|S_{|J_S^\perp}\right\| = s_m(S)$ and $\left\|T_{|J_T^\perp}\right\| = s_m(T)$. Let $K = J_S + J_T$. Then $\dim K < m + n - 1$ and

$$s_{m+n-1}(S + T) \leq \left\|(S + T)_{|K^\perp}\right\| \leq \left\|S_{|K^\perp}\right\| + \left\|T_{|K^\perp}\right\| \leq s_m(S) + s_n(T).$$

(iii) Suppose that $\epsilon > 0$. Then there exists k_0 such that $\|T - T_k\| < \epsilon$, for $k \geq k_0$. If K is any subspace of H_1 of dimension less than n and $x \in K^\perp$,

$$\|T(x)\| \geq \|T_k(x)\| - \epsilon \|x\|,$$

so that $s_n(T) \geq s_n(T_k) - \epsilon$ for $k \geq k_0$. On the other hand, if $k \geq k_0$ there exists a subspace K_k with $\dim K_k = n - 1$ such that $\left\|(T_k)_{|K_k^\perp}\right\| = s_n(T_k)$, and so $\left\|T_{|K_k^\perp}\right\| \leq s_n(T_k) + \epsilon$ for $k \geq k_0$. Thus $s_n(T) \leq s_n(T_k) + \epsilon$ for $k \geq k_0$. $\square$

We again leave the reader to formulate and prove the corresponding finite-dimensional versions of Theorem 15.7.1 and Proposition 15.7.1.

15.8 Weyl's inequality and Horn's inequality

We have now set the scene. Suppose that $T \in K(H)$. On the one hand, T is a Riesz operator, and we can consider its eigenvalues $(\lambda_i(T))$, repeated according to their algebraic multiplicities. On the other hand we can write $T = \sum_{n=1}^\infty s_n(T) \langle \cdot, x_n \rangle y_n$, where $(s_n(T))$ are the singular numbers of T. How are they related?

Theorem 15.8.1 *(i) Suppose that $T \in L(l_2^n)$ is represented by the matrix A. There exist unitary matrices U and V such that $A = U \, diag(s_1(T), \ldots, s_n(T))V$.*

Thus

$$|\det A| = \left| \prod_{j=1}^{n} \lambda_j(T) \right| = \prod_{j=1}^{n} s_j(T).$$

(ii) **(Weyl's inequality I)** *Suppose that $T \in K(H)$. Then*

$$\left| \prod_{j=1}^{J} \lambda_j(T) \right| \leq \prod_{j=1}^{J} s_j(T).$$

(iii) **(Horn's inequality I)** *Suppose that $T_k \in K(H_{k-1}, H_k)$ for $1 \leq k \leq K$. Then*

$$\prod_{j=1}^{J} s_j(T_K \cdots T_1) \leq \prod_{k=1}^{K} \prod_{j=1}^{J} s_j(T_k).$$

Proof (i) follows immediately from the finite-dimensional version of Theorem 15.6.1 and the change-of-basis formula for matrices.

(ii) We can suppose that $\lambda_J \neq 0$. Then, by the remarks at the end of Section 1, there exists a J-dimensional T-invariant subspace H_J for which $\tilde{T} = T_{|H_J}$ has eigenvalues $\lambda_1(T), \ldots, \lambda_J(T)$. Let I_J be the inclusion: $H_J \to H$, and let P_J be the orthogonal projection $H \to H_J$. Then $s_j(\tilde{T}) = s_j(P_J T I_J) \leq s_j(T)$. Thus

$$\left| \prod_{j=1}^{J} \lambda_j(T) \right| = \prod_{j=1}^{J} s_j(\tilde{T}) \leq \prod_{j=1}^{J} s_j(T).$$

(iii) Again, we can suppose that $s_J(T_K \cdots T_1) \neq 0$. Let $T_K \ldots T_1 = \sum_{n=1}^{\infty} s_n(T_K \ldots T_1 \langle \cdot, x_n \rangle y_n$, and let $V_0 = $ span $(x_1, \ldots, x_J)$. Let $V_k = T_k \ldots T_1(V_0)$, so that $T_k(V_{k-1}) = V_k$. Let $\tilde{T}_k = T_{k|V_{k-1}}$. Since $s_J(T_K \ldots T_1) \neq 0$, $\dim(V_k) = J$, for $0 \leq k \leq K$; let W_k be an isometry from l_2^J onto V_k.

$$
\begin{array}{ccccccc}
H_0 & \xrightarrow{T_0} & H_1 & \xrightarrow{T_1} & \cdots & \xrightarrow{T_K} & H_K \\
\subseteq\uparrow & & \subseteq\uparrow & & & & \subseteq\uparrow \\
V_0 & \xrightarrow{\tilde{T}_0} & V_1 & \xrightarrow{\tilde{T}_1} & \cdots & \xrightarrow{\tilde{T}_K} & V_K \\
W_0\uparrow & & W_1\uparrow & & & & W_K\uparrow \\
l_2^J & & l_2^J & & & & l_2^J
\end{array}
$$

Let A_k be the matrix representing $W_k^{-1}\tilde{T}_k W_{k-1}$. Then $A_K \ldots A_1$ represents $(T_K \ldots T_1)_{|V_0}$, so that

$$\prod_{j=1}^{J} s_j(T_K \ldots T_1) = |\det(A_K \ldots A_1)| = \prod_{k=1}^{K} |\det A_k|$$

$$= \prod_{k=1}^{K}\prod_{j=1}^{J} s_j(\tilde{T}_k) \leq \prod_{k=1}^{K}\prod_{j=1}^{J} s_j(T_k)$$

$\square$

Weyl [Wey 49] proved his inequality by considering alternating tensor products, and also proved the first part of the following corollary. As Pólya [Pól 50] observed, the inequality above suggests that majorization should be used; let us follow Pólya, as Horn [Hor 50] did when he proved the second part of the corollary.

Corollary 15.8.1 *Suppose that ϕ is an increasing function on $[0, \infty)$ and that $\phi(e^t)$ is a convex function of t.*

*(i) (**Weyl's inequality II**) Suppose that $T \in K(H)$. Then*

$$\sum_{j=1}^{J} \phi(|\lambda_j(T)|) \leq \sum_{j=1}^{J} \phi(s_j(T)), \quad \text{for each } J.$$

In particular,

$$\sum_{j=1}^{J} |\lambda_j(T)|^p \leq \sum_{j=1}^{n} (s_j(T))^p, \quad \text{for } 0 < p < \infty, \text{ for each } J.$$

Suppose that $(X, \|.\|_X)$ is a symmetric Banach sequence space. If $(s_j(T)) \in X$ then $(\lambda_j(T)) \in X$ and $\|(\lambda_j(T))\|_X \leq \|(s_j(T))\|_X$.

*(ii) (**Horn's inequality II**) Suppose that $T_k \in K(H_{k-1}, H_k)$ for $1 \leq k \leq K$. Then*

$$\sum_{j=1}^{J} \phi(s_j(T_K \cdots T_1)) \leq \sum_{j=1}^{J} \phi\left(\prod_{k=1}^{K} s_j(T_k)\right), \quad \text{for each } J.$$

In particular,

$$\sum_{j=1}^{J} (s_j(T_K \cdots T_1))^p \leq \sum_{j=1}^{n} \left(\prod_{k=1}^{K} s_j(T_k)\right)^p, \quad \text{for } 0 < p < \infty, \text{ for each } j.$$

Suppose that $(X, \|.\|_X)$ is a symmetric Banach sequence space. If

$(\prod_{k=1}^{K} s_j(T_k)) \in X$ *then* $(s_j(T_K \cdots T_1)) \in X$ *and* $\|(s_j(T_K \cdots T_1))\|_X \leq$ $\left\|(\prod_{k=1}^{K} s_j(T_k))\right\|_X$.

Proof These results follow from Proposition 7.6.3. ☐

15.9 Ky Fan's inequality

The Muirhead maximal numbers $(\lambda_k^{\dagger}(T))$ and $(s_k^{\dagger}(T))$ play as important role in operator theory as they do for sequences. We now characterize $s_k^{\dagger}$ in terms of the trace of a matrix. Let us recall the definition. Suppose that E is a finite dimensional vector space, with basis $(e_1, \ldots, e_n)$, and dual basis $(\phi_1, \ldots, \phi_n)$. Then if $T \in L(E)$, we define the *trace* of T, $tr(T)$, to be $tr(T) = \sum_{j=1}^{n} \phi_j(T(e_j))$. Thus if T is represented by the matrix (t_{ij}), then $tr(T) = \sum_{j=1}^{n} t_{jj}$. The trace is independent of the choice of basis, and is equal to $\sum_{j=1}^{n} \lambda_j$, where the λ_j are the roots of the characteristic polynomial, counted according to multiplicity. The trace also has the following important commutation property: if F is another finite-dimensional vector space, not necessarily of the same dimension, and $S \in L(E, F)$, $T \in L(F, E)$ then $tr(ST) = tr(TS)$; for if S and T are represented by matrices (s_{ij}) and (t_{jk}), then $Tr(ST) = \sum_i \sum_j s_{ij} t_{ji} = tr(TS)$.

Theorem 15.9.1 (Ky Fan's theorem) *Suppose that* $T \in K(H_1, H_2)$. *Then*

$$s_k^{\dagger}(T) = (1/k) \sup\{|tr(ATB): \ A \in L(H_2, l_2^k), B \in L(l_2^k, H_1), \|A\| \leq 1, \|B\| \leq 1\}.$$

Proof Suppose that $T = \sum_{n=1}^{\infty} s_n(T) \langle \cdot, x_n \rangle y_n$. Define $A \in L(H_2, l_2^k)$ by setting $A(z) = (\langle z, y_j \rangle)_{j=1}^k$, and define $B \in L(l_2^k, H_1)$ by setting $B(v) = \sum_{j=1}^{k} v_j x_j$. Then $\|A\| \leq 1$ and $\|B\| = 1$. The operator $ATB \in L(l_2^k)$ is represented by the matrix $\mathrm{diag}(s_1(T), \ldots, s_k(T))$, so that $s_k^{\dagger}(T) = (1/k)tr(ATB)$.

On the other hand, suppose that $A \in L(H_2, l_2^k)$, that $B \in L(l_2^k, H_1)$, and that $\|A\| \leq 1$ and $\|B\| \leq 1$. Let $A(y_j) = (a_{lj})_{l=1}^k$ and let $\langle B(e_i), x_j \rangle = b_{ji}$. Then

$$ATB(e_i) = A\left(\sum_{j=1}^{k} s_j(T) b_{ji} y_j\right) = \left(\sum_{j=1}^{k} a_{lj} s_j(T) b_{ji}\right)_{l=1}^{k},$$

so that

$$\mathrm{tr}(ATB) = \sum_{i=1}^{k}\left(\sum_{j=1}^{k} a_{ij}s_j(T)b_{ji}\right) = \sum_{j=1}^{k}\left(\sum_{i=1}^{k} a_{ij}b_{ji}\right) s_j(T).$$

Now

$$\langle BA(y_j), x_j\rangle = \left\langle B\left(\sum_{i=1}^{k} a_{ij}e_i\right), x_j\right\rangle = \sum_{i=1}^{k} a_{ij}b_{ji},$$

and

$$|\langle BA(y_j), x_j\rangle| \le \|B\|\cdot\|A\|\cdot\|y_j\|\cdot\|x_j\| \le 1,$$

so that $|\sum_{i=1}^{k} a_{ij}b_{ji}| \le 1$, and $(1/k)|\mathrm{tr}(ATB)| \le s_k^{\dagger}(T)$. $\qquad\square$

Corollary 15.9.1 (Ky Fan's inequality) *If $S, T \in K(H_1, H_2)$ then*

$$s_k^{\dagger}(S + T) \le s_k^{\dagger}(S) + s_k^{\dagger}(T).$$

15.10 Operator ideals

We are now in a position to extend the results about symmetric Banach sequence spaces to ideals of operators. Suppose that $(X, \|.\|_X)$ is a symmetric Banach sequence space contained in c_0. We define the *Banach operator ideal* $S_X(H_1, H_2)$ to be

$$S_X(H_1, H_2) = \{T \in K(H_1, H_2)\colon (s_n(T)) \in X\},$$

and set $\|T\|_X = \|(s_n(T))\|_X$. If $X = l_p$, we write $S_p(H_1, H_2)$ for $S_X(H_1, H_2)$ and denote the norm by $\|.\|_p$.

Theorem 15.10.1 $S_X(H_1, H_2)$ *is a linear subspace of $K(H_1, H_2)$, and $\|.\|_X$ is a norm on it, under which it is complete. If $T \in S_X(H_1, H_2)$, $A \in L(H_2, H_3)$ and $B \in L(H_0, H_1)$ then $ATB \in S_X(H_0, H_3)$, and $\|ATB\|_X \le \|A\|\cdot\|T\|_X\cdot\|B\|$.*

Proof Ky Fan's inequality says that $(s_n(S + T)) \prec_w (s_n(S) + s_n(T))$. If $S, T \in S_X$ then $(s_n(S)+s_n(T)) \in X$, and so by Corollary 7.4.1 $(s_n(S+T)) \in X$, and $\|(s_n(S + T))\|_X \le \|(s_n(S))\|_X + \|(s_n(T))\|_X$. Thus $S+T \in S_X$ and $\|S + T\|_X \le \|S\|_X + \|T\|_X$.

Since $\|\alpha S\| = |\alpha|\,\|S\|_X$, it follows that $S_X(H_1, H_2)$ is a linear subspace of $K(H_1, H_2)$, and that $\|.\|_X$ is a norm on it.

Completeness is straightforward.   If (T_n) is a Cauchy sequence in $S_X(H_1, H_2)$ then (T_n) is a Cauchy sequence in operator norm, and so converges in this norm to some $T \in K(H_1, H_2)$. Then $s_k(T_n) \to s_k(T)$, for each k, by Corollary 15.7.1, and so $T \in S_X(H_1, H_2)$, and $\|T\|_X \leq$ $\sup \|T_n\|_X$, by Fatou's Lemma (Proposition 6.1.1). Similarly, $\|T - T_n\| \leq$ $\sup_{m \geq n} \|T_m - T_n\|_X \to 0$ as $n \to \infty$.

The final statement also follows from Corollary 15.7.1. $\square$

The final statement of Theorem 15.10.1 explains why $S_X(H_1, H_2)$ is called an ideal. The ideal property is very important; for example, we have the following result, which we shall need later.

Proposition 15.10.1 *Suppose that $S_X(H)$ is a Banach operator ideal, and that $r > 0$. The set*

$$O_X^{(r)}(H) = \{T \in S_X(H): \{z: |z| = r\} \cap \sigma(T) = \emptyset\}$$

is an open subset of $S_X(H)$, and the map $T \to T_{<r}$ is continuous on it.

Proof Suppose that $T \in O_X^{(r)}(H)$. Let $M_T = \sup_{|z|=r} \|R_z(T)\|$. If $\|S-T\|_X <$ $1/M_T$ then $\|S - T\| < 1/M_T$, so that if $|z| = r$ then $zI - S$ is invertible and $\|R_S(z)\| \leq 2M(T)$. Thus $S \in O_X^{(r)}(H)$, and $O_X^{(r)}(H)$ is open. Further, we have the resolvent equation

$$SR_z(S) - TR_z(T) = zR_z(S)(S - T)R_z(T),$$

so that, using Proposition 15.2.1,

$$\|S_{<r} - T_{<r}\|_X = \left\| \frac{1}{2\pi} \int_{|z|=r} SR_z(S) - TR_z(T) \, dz \right\|_X \leq 2r M_T^2 \|S - T\|_X .$$

$\square$

Ky Fan's theorem allows us to establish the following characterization of $S_X(H_1, H_2)$.

Proposition 15.10.2 *Suppose that X is a symmetric Banach sequence space and that $T = \sum_{n=1}^{\infty} s_n(T) \langle \cdot, x_n \rangle y_n \in K(H_1, H_2)$. Then $T \in S_X(H_1, H_2)$ if and only if $(\langle T(e_j), f_j \rangle) \in X$ for all orthonormal sequences (e_j) and (f_j) in H_1 and H_2, respectively. Then*

$$\|T\|_X = \sup\{\|(\langle T(e_j), f_j \rangle)\|_X : (e_j), (f_j) \text{ orthonormal in } H_1, H_2 \text{ respectively}\}.$$

Proof The condition is certainly sufficient, since $(s_n(T)) = (\langle T(x_n), y_n \rangle)$.

Suppose that $T \in S_X(H_1, H_2)$ and that (e_j) and (f_j) are orthonormal sequences in H_1 and H_2, respectively. Let us set $y_j = \langle T(e_j), f_j \rangle$. We arrange $y_1, \ldots, y_k$ in decreasing absolute value: there exists a one-one mapping $\tau : \{1, \ldots, k\} \to \mathbf{N}$ such that $|y_{\tau(j)}| = y_j^*$ for $1 \leq j \leq k$. Define $A \in L(H_2, l_2^k)$ by setting

$$A(z)_j = \overline{\operatorname{sgn} y_{\tau(j)}} \langle z, f_{\tau(j)} \rangle,$$

and define $B \in L(l_2^k, H_1)$ by setting $B(v) = \sum_{j=1}^k v_j e_{\tau(j)}$. Then $\|A\| \leq 1$ and $\|B\| = 1$, and $\operatorname{tr}(ATB) = \sum_{j=1}^k y_j^*$. But $|\operatorname{tr}(ATB)| \leq k s_k^\dagger(T)$, by Ky Fan's theorem, and so $(y_j) \prec_w (s_j(T))$. Thus $(\langle T(e_j), f_j \rangle) \in X$ and $\|(\langle T(e_j), f_j \rangle)\|_X \leq \|T\|_X$. $\qquad\square$

We can use Horn's inequality to transfer inequalities from symmetric sequence spaces to operator ideals. For example, we have the following, whose proof is immediate.

Proposition 15.10.3 *(i) (Generalized Hölder's inequality) Suppose that $0 < p, q, r < \infty$ and that $1/p + 1/q = 1/r$. If $S \in S_p(H_1, H_2)$ and $T \in S_q(H_1, H_2)$ then $ST \in S_r(H_1, H_2)$ and*

$$\left(\sum_{j=1}^\infty (s_j(ST))^r \right)^{1/r} \leq \left(\sum_{j=1}^\infty (s_j(S))^p \right)^{1/p} \cdot \left(\sum_{j=1}^\infty (s_j(T))^q \right)^{1/q}.$$

(ii) Suppose that $(X, \|.\|_X)$ is a symmetric Banach sequence space contained in c_0, with associate space $(X', \|.\|_{X'})$ also contained in c_0. If $S \in S_X$ and $T \in S_{X'}$ then $ST \in S_1$ and $\|ST\|_1 \leq \|S\|_X \cdot \|T\|_{X'}$.

15.11 The Hilbert–Schmidt class

There are two particularly important Banach operator ideals, the *trace class* S_1 and the *Hilbert–Schmidt class* S_2. We begin with the Hilbert–Schmidt class.

Theorem 15.11.1 *Suppose that H_1 and H_2 are Hilbert spaces.*

(i) Suppose that $T \in K(H_1, H_2)$. Then the (possibly infinite) sum $\sum_{j=1}^\infty \|T(e_j)\|^2$ is the same for all orthonormal bases (e_j) of H_1. $T \in S_2(H_1, H_2)$ if and only if the sum is finite, and then $\|T\|_2^2 = \sum_{j=1}^\infty \|T(e_j)\|^2$.

(ii) If $S, T \in S_2(H_1, H_2)$ then the series $\sum_{j=1}^\infty \langle S(e_j), T(e_j) \rangle$ is absolutely

convergent for all orthonormal bases (e_j), *and the sum is the same for all orthonormal bases. Let*

$$\langle S, T \rangle = \sum_{j=1}^{\infty} \langle S(e_j), T(e_j) \rangle .$$

Then $\langle S, T \rangle$ *is an inner product on* $S_2(H_1, H_2)$ *for which* $\langle T, T \rangle = \|T\|_2^2$, *for all* $T \in S_2(H_1, H_2)$.

Proof (i) Suppose that (e_j) and (f_k) are orthonormal bases of H_1. Then

$$\sum_{j=1}^{\infty} \|T(e_j)\|^2 = \sum_{j=1}^{\infty} \sum_{k=1}^{\infty} |\langle T(e_j), f_k \rangle|^2$$

$$= \sum_{k=1}^{\infty} \sum_{j=1}^{\infty} |\langle e_j, T^*(f_k) \rangle|^2$$

$$= \sum_{k=1}^{\infty} \|T^*(f_k)\|^2 .$$

Thus the sum does not depend on the choice of orthonormal basis (e_j). Now there exists an orthonormal sequence (x_j) such that $\|T(x_j)\| = s_j(T)$, for all j. Let (z_j) be an orthonormal basis for $(\operatorname{span}(x_j))^{\perp}$, and let (e_j) be an orthonormal basis for H_1 whose terms comprise the x_js and the y_js. Then

$$\sum_{j=1}^{\infty} \|T(e_j)\|^2 = \sum_{j=1}^{\infty} \|T(x_j)\|^2 + \sum_{j=1}^{\infty} \|T(y_j)\|^2 = \sum_{j=1}^{\infty} (s_j(T))^2,$$

so that the sum is finite if and only if $T \in S_2(H_1, H_2)$, and then $\|T\|_2^2 = \sum_{j=1}^{\infty} \|T(e_j)\|^2$.

(ii) This is a simple exercise in polarization. □

The equality in part (i) of this theorem is quite special. For example, let $v_j = (1/\sqrt{j} \log(j+1))$. Then $v = (v_j) \in l^2$; let $w = v/\|v\|_2$. Now let $P_w = w \otimes \bar{w}$ be the one-dimensional orthogonal projection of l_2 onto the span of w. Then $P_w \in S_p$, and $\|P_w\|_p = 1$, for $1 \leq p < \infty$, while

$$\sum_{j=1}^{\infty} \|P_w(e_j)\|^p = \sum_{j=1}^{\infty} \frac{1}{j^{p/2}(\log j)^p \|v\|_2^p} = \infty$$

for $1 \leq p < 2$. This phenomenon is a particular case of the following inequalities.

Proposition 15.11.1 *Suppose that* $T = \sum_{n=1}^{\infty} s_n(T) \langle \cdot, x_n \rangle y_n \in K(H_1, H_2)$ *and that* (e_k) *is an orthonormal basis for* H_1.

(i) *If* $1 \le p < 2$ *then* $\sum_{k=1}^{\infty} \|T(e_k)\|^p \ge \sum_{j=1}^{\infty} (s_j(T))^p$.

(ii) *If* $2 < p < \infty$ *then* $\sum_{k=1}^{\infty} \|T(e_k)\|^p \le \sum_{j=1}^{\infty} (s_j(T))^p$.

Proof (i) We use Hölder's inequality, with exponents $2/p$ and $2/(2-p)$:

$$
\sum_{j=1}^{\infty} (s_j(T))^p = \sum_{j=1}^{\infty} (s_j(T))^p \left(\sum_{k=1}^{\infty} |\langle e_k, x_j \rangle|^2 \right)
$$

$$
= \sum_{k=1}^{\infty} \left(\sum_{j=1}^{\infty} (s_j(T))^p |\langle e_k, x_j \rangle|^2 \right)
$$

$$
= \sum_{k=1}^{\infty} \left(\sum_{j=1}^{\infty} (s_j(T))^p |\langle e_k, x_j \rangle|^p |\langle e_k, x_j \rangle|^{2-p} \right)
$$

$$
\le \sum_{k=1}^{\infty} \left(\left(\sum_{j=1}^{\infty} (s_j(T))^2 |\langle e_k, x_j \rangle|^2 \right)^{p/2} \left(\sum_{j=1}^{\infty} |\langle e_k, x_j \rangle|^2 \right)^{1-p/2} \right)
$$

$$
= \sum_{k=1}^{\infty} \left(\sum_{j=1}^{\infty} (s_j(T))^2 |\langle e_k, x_j \rangle|^2 \right)^{p/2}
$$

$$
= \sum_{k=1}^{\infty} \|T(e_k)\|^p.
$$

(ii) In this case, we use Hölder's inequality with exponents $p/2$ and $p/(p-2)$:

$$
\sum_{k=1}^{\infty} \|T(e_k)\|^p = \sum_{k=1}^{\infty} \left(\sum_{j=1}^{\infty} (s_j(T))^2 |\langle e_k, x_j \rangle|^2 \right)^{p/2}
$$

$$
= \sum_{k=1}^{\infty} \left(\sum_{j=1}^{\infty} (s_j(T))^2 |\langle e_k, x_j \rangle|^{4/p} |\langle e_k, x_j \rangle|^{2-4/p} \right)^{p/2}
$$

$$\leq \sum_{k=1}^{\infty} \left(\left(\sum_{j=1}^{\infty} (s_j(T))^p | \langle e_k, x_j \rangle |^2 \right) \left(\sum_{j=1}^{\infty} | \langle e_k, x_j \rangle |^2 \right)^{(p-2)/2} \right)$$

$$= \sum_{k=1}^{\infty} \sum_{j=1}^{\infty} (s_j(T))^p | \langle e_k, x_j \rangle |^2)$$

$$= \sum_{j=1}^{\infty} (s_j(T))^p \left(\sum_{k=1}^{\infty} | \langle e_k, x_j \rangle |^2 \right) = \sum_{j=1}^{\infty} (s_j(T))^p.$$

$\square$

15.12 The trace class

We now turn to the trace class. First let us note that we can use it to characterize the Muirhead maximal numbers $s_k^\dagger(T)$.

Theorem 15.12.1 *Suppose that $T \in K(H_1, H_2)$. Then*

$$s_k^\dagger(T) = \inf\{ \|R\|_1 / k + \|S\| : \ T = R + S, R \in S_1(H_1, H_2), S \in K(H_1, H_2) \},$$

and the infimum is attained.

Proof First suppose that $T = R + S$, with $R \in S_1(H_1, H_2)$ and $S \in K(H_1, H_2)$. Then by Ky Fan's inequality,

$$s_k^\dagger(T) \leq s_k^\dagger(R) + s_k^\dagger(S) \leq \|R\|_1 / k + \|S\|.$$

On the other hand, if $T = \sum_{n=1}^{\infty} s_n(T) \langle \cdot, x_n \rangle y_n$, let

$$R = \sum_{n=1}^{k} (s_n(T) - s_k(T)) \langle \cdot, x_n \rangle y_n$$

and

$$S = \sum_{n=1}^{k} s_k(T) \langle \cdot, x_n \rangle y_n + \sum_{n=k+1}^{\infty} s_n(T) \langle \cdot, x_n \rangle y_n.$$

Then $T = R + S$ and $\|R\|_1 = k(s_k^\dagger(T) - s_k(T))$ and $\|S\| = s_k(T)$, so that $s_k(T) = \|R\|_1 / k + \|S\|$. $\square$

This enables us to prove an operator version of Calderón's interpolation theorem.

Corollary 15.12.1 *Suppose that Φ is a norm-decreasing linear map of $K(H_1, H_2)$ into $K(H_3, H_4)$, which is also norm-decreasing from $S_1(H_1, H_2)$ into $S_1(H_3, H_4)$. If $T \in K(H_1, H_2)$ then $s_k^{\dagger}(\Phi(T)) \le s_k^{\dagger}(T)$, so that $\|\Phi(T)\|_X \le \|T\|_X$ for any Banach operator ideal J_X.*

The important feature of the trace class $S_1(H)$ is that we can define a special linear functional on it, namely the *trace*.

Theorem 15.12.2 *(i) Suppose that T is a positive compact operator on a Hilbert space H. Then the (possibly infinite) sum $\sum_{j=1}^{\infty} \langle T(e_j), e_j \rangle$ is the same for all orthonormal bases (e_j) of H. $T \in S_1(H_1, H_2)$ if and only if the sum is finite, and then $\|T\|_1 = \sum_{j=1}^{\infty} \langle T(e_j), e_j \rangle$.*

(ii) If $T \in S_1(H)$, then $\sum_{j=1}^{\infty} \langle T(e_j), e_j \rangle$ converges absolutely, and the sum is the same for all orthonormal bases (e_j) of H.

Proof (i) We can write T as $T = \sum_{n=1}^{\infty} s_n(T) \langle \cdot, x_n \rangle x_n$. Let $S = \sum_{n=1}^{\infty} \sqrt{s_n(T)} \langle \cdot, x_n \rangle x_n$. Then S is a positive compact operator, and $T = S^2$. Thus

$$\sum_{j=1}^{\infty} \langle T(e_j), e_j \rangle = \sum_{j=1}^{\infty} \langle S(e_j), S(e_j) \rangle = \sum_{j=1}^{\infty} \|S(e_j)\|^2,$$

and we can apply Theorem 15.11.1. In particular, the sum is finite if and only if $S \in S_2(H)$, and then

$$\sum_{j=1}^{\infty} s_j(T) = \sum_{j=1}^{\infty} (s_j(S))^2 = \sum_{j=1}^{\infty} \langle T(e_j), e_j \rangle.$$

(ii) We can write T as $T = \sum_{n=1}^{\infty} s_n(T) \langle \cdot, x_n \rangle y_n$. Let

$$R = \sum_{n=1}^{\infty} \sqrt{s_n(T)} \langle \cdot, y_n \rangle y_n \quad \text{and} \quad S = \sum_{n=1}^{\infty} \sqrt{s_n(T)} \langle \cdot, x_n \rangle y_n.$$

Then R and S are Hilbert–Schmidt operators, $T = RS$, and if (e_j) is an orthonormal basis then $\langle T(e_j), e_j \rangle = \langle S(e_j), R^*(e_j) \rangle$, so that the result follows from Theorem 15.11.1 (ii). $\qquad \square$

15.13 Lidskii's trace formula

The functional $\text{tr}(T) = \sum_{j=1}^{\infty} \langle T(e_j), e_j \rangle$ is called the *trace* of T. It is a continuous linear functional on $S_1(H)$, which is of norm 1, and which satisfies $\text{tr}(T^*) = \overline{\text{tr}(T)}$. It generalizes the trace of an operator on a finite-dimensional space; can it too be characterized in terms of its eigenvalues?

Theorem 15.13.1 (Lidskii's trace formula) *If $T \in S_1(H)$ then $\sum_{j=1}^{\infty} \lambda_j(T)$ is absolutely convergent, and $tr(T) = \sum_{j=1}^{\infty} \lambda_j(T)$.*

Proof This result had been conjectured for a long time; the first proof was given by Lidskii [Lid 59]: we shall howevever follow the proof given by Leiterer and Pietsch, as described in [Kön 86].

The fact that $\sum_{j=1}^{\infty} \lambda_j(T)$ is absolutely convergent follows immediately from Weyl's inequality. Let us set $\tau(T) = \sum_{j=1}^{\infty} \lambda_j(T)$. If T is of finite rank, then $\tau(T) = tr(T)$. The finite rank operators are dense in $S_1(H)$, and tr is continuous on $S_1(H)$, and so it is enough to show that τ is continuous on $S_1(H)$.

The key idea of the proof is to introduce new parameters which are more useful, in the present circumstances, than the singular numbers. The next lemma gives the details.

Lemma 15.13.1 *Suppose that $S, T \in K(H)$. Let $t_k(T) = (s_k(T))^{1/2}$, $t_k^{\dagger}(T) = (1/k) \sum_{j=1}^{k} t_j(T)$ and $y_k(T) = (t_k^{\dagger}(T))^2$. Then*

(i) $\sum_{k=1}^{l} s_k(T) \le \sum_{k=1}^{l} y_k(T) \le 4 \sum_{k=1}^{l} s_k(T)$;

(ii) $|\lambda_k(T)| \le y_k(T)$;

(iii) $y_{2k}(S+T) \le 2y_k(S) + 2y_k(T)$.

Proof (i) Clearly $s_k(T) \le y_k(T)$; this gives the first inequality. On the other hand, applying the Hardy–Riesz inequality,

$$\sum_{k=1}^{l} y_k(T) = \sum_{k=1}^{l} (t_k^{\dagger}(T))^2 \le 4 \sum_{k=1}^{l} (t_k(T))^2 = 4 \sum_{k=1}^{l} s_k(T).$$

(ii) It follows from Weyl's inequality that

$$|\lambda_k(T)| \le \left(\frac{1}{k} \sum_{j=1}^{k} |\lambda_j(T)|^{1/2}\right)^2 \le \left(\frac{1}{k} \sum_{j=1}^{k} t_k(T)\right)^2 = y_k(T).$$

(iii) Using Proposition 15.7.1, and the inequality $(a+b)^{1/2} \le a^{1/2} + b^{1/2}$ for

$a, b \geq 0$,

$$t_{2k}^\dagger(S+T) \leq \frac{1}{k} \sum_{j=1}^{k} (s_{2j-1}(S+T))^{1/2}$$

$$\leq \frac{1}{k} \sum_{j=1}^{k} (s_j(S) + s_j(T))^{1/2}$$

$$\leq \frac{1}{k} \sum_{j=1}^{k} ((s_j(S))^{1/2} + (s_j(T))^{1/2}) = t_k^\dagger(S) + t_k^\dagger(T);$$

thus

$$y_{2k}(S+T) \leq (t_k^\dagger(S) + t_k^\dagger(T))^2 \leq 2(t_k^\dagger(S))^2 + 2(t_k^\dagger(T))^2 = 2y_k(S) + 2y_k(T).$$

$\square$

Let us now return to the proof of the theorem. Suppose that $T \in S_1(H)$ and that $\epsilon > 0$. $\sum_{j=1}^{\infty} y_j(T) \leq 4 \sum_{j=1}^{\infty} \sigma_j(T) < \infty$, and so there exists J such that $\sum_{j=J+1}^{\infty} |y_j(T)| < \epsilon/24$, and there exists $0 < r < \min(\epsilon/24J, |\lambda_J(T)|)$ such that $T \in O_1^{(r)}$. By Proposition 15.10.1, there exists $0 < \delta < \epsilon/24$ such that if $\|S - T\|_1 < \delta$ then $S \in O_1^{(r)}(H)$, $\|S_{<r} - T_{<r}\|_1 < \epsilon/24$ and $\|S_{>r} - T_{>r}\|_1 < \epsilon/24$. Consequently, for such S,

$$\left| \sum_{|\lambda_j(T)|>r} \lambda_j(T) - \sum_{|\lambda_j(S)|>r} \lambda_j(S) \right| = |\mathrm{tr}(T_{>r}) - \mathrm{tr}(S_{>r})|$$

$$\leq \|T_{>r} - S_{>r}\|_1 < \epsilon/24.$$

On the other hand, using the inequalities of Lemma 15.13.1,

$$\sum_{|\lambda_j(T)|<r} |\lambda_j(T)| \leq \sum_{j=J+1}^{\infty} |y_j(T)| < \epsilon/24,$$

and

$$\sum_{|\lambda_j(S)|<r} |\lambda_j(S)| = \sum_{j=1}^{\infty} |\lambda_j(S_{<r})| \leq \sum_{j=1}^{2J} |\lambda_j(S_{<r})| + \sum_{j=2J+1}^{\infty} y_j(S_{<r})$$

$$\leq 2Jr + \sum_{j=2J+1}^{\infty} y_j(S)$$

$$\leq 2\epsilon/24 + 4 \sum_{j=J+1}^{\infty} y_j(T) + 4 \sum_{j=J+1}^{\infty} y_j(S-T)$$

$$\leq 6\epsilon/24 + 4 \sum_{j=1}^{\infty} y_j(S-T)$$

$$\leq 6\epsilon/24 + 16 \sum_{j=1}^{\infty} s_j(S-T) \leq 22\epsilon/24.$$

Thus $|\tau(T - \tau(S))| < \epsilon$, and τ is continuous. $\qquad\square$

We can now apply Corollary 15.3.1.

Theorem 15.13.2 *If $S \in S_1(H_1)$ and $T \in S_1(H_2)$ are related operators, then $\mathrm{tr}(ST) = \mathrm{tr}(TS)$.*

15.14 Operator ideal duality

We can now establish a duality theory for Banach operator ideals analogous to that for symmetric Banach function spaces. The basic results are summarized in the next theorem; the details are straightforward, and are left to the reader.

Theorem 15.14.1 *Suppose that X is a symmetric Banach sequence space contained in c_0, whose associate space is also contained in c_0. If $S \in J_X(H_1, H_2)$ and $T \in J_{X'}(H_2, H_1)$ then $TS \in S_1(H_1)$ and $ST \in S_1(H_2)$, $\mathrm{tr}(TS) = \mathrm{tr}(ST)$ and $|\mathrm{tr}(TS)| \leq \|S\|_X \cdot \|T\|_{X'}$. Further,*

$$\|S\|_X = \sup\{|\mathrm{tr}(ST)| : T \in J_{X'}(H_2, H_1), \|T\|_{X'} \leq 1\}.$$

The inner product on $S_2(H_1, H_2)$ can also be expressed in terms of the trace: if $S, T \in S_2(H_1, H_2)$, and (E_j) is an orthonormal basis for H_1 then

$$\langle S, T \rangle = \sum_{j=1}^{\infty} \langle S(e_j), T(e_j) \rangle = \sum_{j=1}^{\infty} \langle T^*S(e_j), T(e_j) \rangle = \mathrm{tr}(T^*S).$$

The ideals S_p enjoy the same complex interpolation properties as L^p spaces.

Theorem 15.14.2 *Suppose that $1 \leq p_0, p_1 \leq \infty$, that $0 < \theta < 1$ and that $1/p = (1-\theta)/p_0 + \theta/p_1$. Then $S_p = (S_{p_0}, S_{p_1})_{[\theta]}$ (where $S_\infty = K$).*

Proof The proof is much the same as for the L^p spaces. Suppose that $T = \sum_{n=1}^{\infty} s_n(T) \langle \cdot, x_n \rangle y_n$. Let $u(z) = (1-z)/p_0 + z/p_1$, and let $T(z) = \sum_{n=1}^{\infty} (s_n(T))^{pu(z)} \langle \cdot, x_n \rangle y_n$. Then $\|T(z)\|_{p_j} = \|T\|_p$ for $z \in L_j$ (for $j = 0, 1$) and $T_\theta = T$, so that $\|T\|_{[\theta]} \leq \|T\|_p$. On the other hand, suppose that $F \in \mathcal{F}(S_{p_0}, S_{p_1})$, with $F(\theta) = T$. Let $r_n = (s_n(T))^{p-1}$, and for each N let $R_N = \sum_{n=1}^{N} r_n \langle \cdot, y_n \rangle x_n$, and $G_N(z) = \sum_{n=1}^{N} r_n^{p'v(z)} \langle \cdot, y_n \rangle x_n$, where $v(z) = (1-z)/p_0' + z/p_1'$. Then

$$\sum_{n=1}^{N} (s_n(T))^p = \text{tr}(RT) \leq \max_{j=0,1} \sup_{z \in L_j} |\text{tr} R(z) F(z)|$$

$$\leq \|R\|_{p'} \max_{j=0,1} \sup_{z \in L_j} \|F(z)\|_{p_j} = \left(\sum_{n=1}^{N} (s_n(T))^p \right)^{1/p'} \|T\|_{[\theta]}.$$

Letting $N \to \infty$, we see that $T \in S_p$ and $\|T\|_p \leq \|T\|_{[\theta]}$. $\qquad\square$

15.15 Notes and remarks

Information about the spectrum and resolvent of a bounded linear operator are given in most books on functional analysis, such as [Bol 90], Chapter 12. Accounts of the functional calculus are given in [Dow 78] and [DuS 88].

The study of ideals of operators on a Hilbert space was inaugurated by Schatten [Scha 50], although he expressed his results in terms of tensor products, rather than operators.

Exercises

15.1 Suppose that $T \in L(E)$, where $(E, \|.\|_E)$ is a complex Banach space.
(i) Suppose that $\lambda, \mu \in \rho(T)$. Establish the resolvent equation

$$R_\lambda - R_\mu = -(\lambda - \mu) R_\lambda R_\mu = -(\lambda - \mu) R_\mu R_\lambda.$$

(ii) Suppose that $S, T \in L(E)$, that $\lambda \in \rho(ST)$ and that $\lambda \neq 0$. Show that $\lambda \in \rho(TS)$ and that

$$R_\lambda(TS) = \lambda^{-1}(I - T R_\lambda(ST) S).$$

What happens when $\lambda = 0$?

(iii) Suppose that λ is a boundary point of $\sigma(T)$. Show that λ is an approximate eigenvalue of T: there exists a sequence (x_n) of unit vectors such that $T(x_n) - \lambda x_n \to 0$ as $n \to \infty$. (Use the fact that if $\mu \in \rho(T)$ and $|\nu - \mu| < \|R_\mu\|^{-1}$ then $\nu \in \rho(T)$.) Show that if T is compact and $\lambda \neq 0$ then λ is an eigenvalue of T.

15.2   Show that the functions $\{e^{2\pi int}: n \in \mathbf{Z}\}$ form an orthonormal basis for $L^2(0,1)$. Using this, or otherwise, show that the Fredholm integral operator

$$T(f)(x) = \int_0^x f(t)\, dt$$

is a compact operator on $L^2(0,1)$.

15.3   (i) Suppose that (x_n) is a bounded sequence in a Hilbert space H. Show, by a diagonal argument, that there is a subsequence (x_{n_k}) such that $\langle x_{n_k}, y \rangle$ is convergent for each $y \in H$. (First reduce the problem to the case where H is separable.) Show that there exists $x \in H$ such that $\langle x_{n_k}, y \rangle \to \langle x, y \rangle$ as $n \to \infty$, for each $y \in H$.

(ii) Suppose that $T \in L(H,E)$, where $(E, \|.\|_E)$ is a Banach space. Show that $T(B_H)$ is closed in E.

(iii) Show that $T \in L(H,E)$ is compact if and only if $T(B_H)$ is compact.

(iv) Show that if $T \in K(H,E)$ then there exists $x \in H$ with $\|x\| = 1$ such that $\|T(x)\| = \|T\|$.

(v) Give an example of $T \in L(H)$ for which $\|T(x)\| < 1$ for all $x \in H$ with $\|x\| = 1$.

15.4   Suppose that $T \in K(H_1, H_2)$, where H_1 and H_2 are Hilbert spaces. Suppose that $\|x\| = 1$ and $\|T(x)\| = \|T\|$ (as in the previous question). Show that if $\langle x, y \rangle = 0$ then $\langle T(x), (T(y) \rangle = 0$. Use this to give another proof of Theorem 15.6.1.

15.5   Use the finite-dimensional version of Theorem 15.6.1 to show that an element T of $L(l_2^d)$ with $\|T\| \le 1$ is a convex combination of unitary operators.

15.6   Suppose that $T \in L(H_1, H_2)$.   Show that $T \in K(H_1, H_2)$ if and only if $\|T(e_n)\| \to 0$ as $n \to \infty$ for every orthonormal sequence (e_n) in H_1.

16

Summing operators

16.1 Unconditional convergence

In the previous chapter, we obtained inequalities for operators between Hilbert spaces, and endomorphisms of Hilbert spaces, and considered special spaces of operators, such as the trace class and the space of Hilbert–Schmidt operators. For the rest of the book, we shall investigate inequalities for operators between Banach spaces, and endomorphisms of Banach spaces. Are there spaces of operators that correspond to the trace class and the space of Hilbert–Schmidt operators?

We shall however not approach these problems directly. We begin by considering a problem concerning series in Banach spaces.

Suppose that $\sum_{n=1}^{\infty} x_n$ is a series in a Banach space $(E, \|.\|_E)$. We say that the series is *absolutely convergent* if $\sum_{n=1}^{\infty} \|x_n\|_E < \infty$, and say that it is *unconditionally convergent* if $\sum_{n=1}^{\infty} x_{\sigma(n)}$ is convergent in norm, for each permutation σ of the indices: however we rearrange the terms, the series still converges. An absolutely convergent series is unconditionally convergent, and a standard result of elementary analysis states that the converse holds when E is finite-dimensional. On the other hand, the series $\sum_{n=1}^{\infty} e_n/n$ converges unconditionally in l_2, but does not converge absolutely. What happens in l_1? What happens generally?

Before we go further, let us establish some equivalent characterizations of unconditional convergence.

Proposition 16.1.1 *Suppose that (x_n) is a sequence in a Banach space $(E, \|.\|_E)$. The following are equivalent:*

   (i) The series $\sum_{n=1}^{\infty} x_n$ is unconditionally convergent.
   (ii) If $n_1 < n_2 < \cdots$ then the series $\sum_{n=1}^{\infty} x_{n_i}$ converges.
   (iii) If $\epsilon_n = \pm 1$ then the series $\sum_{n=1}^{\infty} \epsilon_n x_n$ converges.

263

(iv) If $b = (b_n)$ is a bounded sequence then the series $\sum_{n=1}^{\infty} b_n x_n$ converges.

(v) Given $\epsilon > 0$, there exists a finite subset F of $\mathbf{N}$ such that whenever G is a finite subset of $\mathbf{N}$ disjoint from F then $\left\| \sum_{n \in G} x_n \right\|_E < \epsilon$.

Proof It is clear that (ii) and (iii) are equivalent, that (v) implies (i) and (ii) and that (iv) implies (ii). We shall show that each of (ii) and (i) implies (v), and that (v) implies (iv).

Suppose that (v) fails, for some $\epsilon > 0$. Then recursively we can find finite sets F_k such that $\left\| \sum_{n \in F_k} x_n \right\|_E \geq \epsilon$, and with the property that $\min F_k > \sup F_{k-1} = N_{k-1}$, say, for $k > 1$. Thus, setting $N_0 = 0$, $F_k \subseteq J_k$, where $J_k = \{n \colon N_{k-1} < n \leq N_k\}$. We write $\cup_{k=1}^{\infty} F_k$ as $\{n_1 < n_2 < \cdots \}$; then $\sum_{j=1}^{\infty} x_{n_j}$ does not converge. Thus (ii) implies (v). Further there exists a permutation σ of $\mathbf{N}$ such that $\sigma(J_k) = J_k$ for each j and $\sigma(N_{k-1}+i) \in F_k$ for $1 \leq i \leq \#(F_j)$. Then $\sum_{n=1}^{\infty} x_{\sigma(n)}$ does not converge, and so (i) implies (v).

Suppose that (v) holds, and that b is a bounded sequence. Without loss of generality we can suppose that each b_n is real (in the complex case, consider real and imaginary parts) and that $0 \leq b_n < 1$ (scale, and consider positive and negative parts). Suppose that $\epsilon > 0$. Then there exists n_0 such that $\left\| \sum_{n \in G} x_n \right\|_E < \epsilon$ if G is a finite set with $\min G > n_0$. Now suppose that $n_0 < n_1 < n \leq n_2$. Let $b_n = \sum_{k=1}^{\infty} b_{n,k}/2^k$ be the binary expansion of b_n, so that $b_{n,k} = 0$ or 1. Let $B_k = \{n \colon n_1 < n \leq n_2, b_{n,k} = 1\}$. Then

$$\left\| \sum_{n=n_1+1}^{n_2} b_n x_n \right\| = \left\| \sum_{k=1}^{\infty} \left(\frac{1}{2^k} \sum_{n \in B_k} x_n \right) \right\| \leq \sum_{k=1}^{\infty} \frac{1}{2^k} \left\| \sum_{n \in B_k} x_n \right\| < \sum_{k=1}^{\infty} \epsilon/2^k = \epsilon.$$

Thus $\sum_{n=1}^{\infty} b_n x_n$ converges, and (v) implies (iv). $\qquad \square$

Corollary 16.1.1 *Suppose that the series $\sum_{n=1}^{\infty} x_n$ is unconditionally convergent and that σ is a permutation of $\mathbf{N}$. Let $s = \sum_{n=1}^{\infty} x_n$ and $s_\sigma = \sum_{n=1}^{\infty} x_{\sigma(n)}$. Then $s = s_\sigma$.*

Proof Suppose that $\epsilon > 0$. There exists a finite set F satisfying (v). Then if $N > \sup F$, $|\sum_{n=1}^{N} x_n - \sum_{n \in F} x_n| < \epsilon$, and so $|s - \sum_{n \in F} x_n| \leq \epsilon$. Similarly, if $N > \sup\{\sigma^{-1}(n) \colon n \in F\}$, then $|\sum_{n=1}^{N} x_{\sigma(n)} - \sum_{n \in F} x_n| < \epsilon$, and so $|s_\sigma - \sum_{n \in F} x_n| \leq \epsilon$. Thus $|s - s_\sigma| \leq 2\epsilon$. Since this holds for all $\epsilon > 0$, $s = s_\sigma$. $\qquad \square$

Corollary 16.1.2 *If the series $\sum_{n=1}^{\infty} x_n$ is unconditionally convergent and $\phi \in E^*$ then $\sum_{n=1}^{\infty} |\phi(x_n)| < \infty$.*

Proof Let $b_n = \overline{\text{sgn}(\phi(x_n))}$. Then $\sum_{n=1}^{\infty} b_n x_n$ converges, and so therefore does $\sum_{n=1}^{\infty} \phi(b_n x_n) = \sum_{n=1}^{\infty} |\phi(x_n)|$. □

We can measure the size of an unconditionally convergent series.

Proposition 16.1.2 *Suppose that (x_n) is an unconditionally convergent sequence in a Banach space $(E, \|.\|_E)$. Then*

$$M_1 = \sup \left\{ \left\| \sum_{n=1}^{\infty} b_n x_n \right\| : \ b = (b_n) \in l_{\infty}, \|b\|_{\infty} \leq 1 \right\}$$

and $M_2 = \sup \left\{ \sum_{n=1}^{\infty} |\phi(x_n)| : \ \phi \in B_{E^*} \right\}$

are both finite, and equal.

Proof Consider the linear mapping $J : E^* \to l_1$ defined by $J(\phi) = (\phi(x_n))$. This has a closed graph, and is therefore continuous. Thus $M_2 = \|J\|$ is finite.

If $b \in l_{\infty}$ then

$$\left| \phi \left(\sum_{n=1}^{\infty} b_n x_n \right) \right| = \left| \sum_{n=1}^{\infty} b_n \phi(x_n) \right| \leq \sum_{n=1}^{\infty} |\phi(x_n)| \leq M_2.$$

Thus

$$\left\| \sum_{n=1}^{\infty} b_n x_n \right\| = \sup \left\{ \left| \phi \left(\sum_{n=1}^{\infty} b_n x_n \right) \right| : \ \phi \in B_{E^*} \right\} \leq M_2,$$

and $M_1 \leq M_2$. Conversely, suppose that $\phi \in B_{E^*}$. Let $b_n = \overline{\text{sgn}(\phi(x_n))}$. Thus $\sum_{n=1}^{\infty} |\phi(x_n)| = \phi(\sum_{n=1}^{\infty} b_n x_n) \leq M_1 \|\phi\|^*$, so that $M_2 \leq M_1$. □

16.2 Absolutely summing operators

We now linearize and generalize: we say that a linear mapping T from a Banach space $(E, \|.\|_E)$ to a Banach space $(F, \|.\|_F)$ is *absolutely summing* if whenever $\sum_{n=1}^{\infty} x_n$ converges unconditionally in E then $\sum_{n=1}^{\infty} T(x_n)$ converges absolutely in F. Thus every unconditionally convergent series in E is absolutely convergent if and only if the identity mapping on E is absolutely summing.

Theorem 16.2.1 *A linear mapping T from a Banach space $(E, \|.\|_E)$ to a Banach space $(F, \|.\|_F)$ is absolutely summing if and only if there exists a* *constant K such that*

$$\sum_{n=1}^{N} \|T(x_n)\|_F \leq K \sup_{\phi \in B_{E^*}} \sum_{n=1}^{N} |\phi(x_n)|,$$

for all N and all $x_1, \ldots, x_N$ in E.

Proof Suppose first that K exists, and suppose that $\sum_{n=1}^{\infty} x_n$ is unconditionally convergent. Then

$$\sum_{n=1}^{\infty} \|T(x_n)\|_F = \sup_N \sum_{n=1}^{N} \|T(x_n)\|_F \leq K \sup_N \sup_{\phi \in B_{E^*}} \sum_{n=1}^{N} |\phi(x_n)|$$

$$= K \sup_{\phi \in B_{E^*}} \sum_{n=1}^{\infty} |\phi(x_n)| < \infty,$$

so that T is absolutely summing.

Conversely, suppose that K does not exist. Then we can find $0 = N_0 < N_1 < N_2 < \cdots$ and vectors x_n in E such that

$$\sup_{\phi \in B_{E^*}} \left(\sum_{n=N_{k-1}+1}^{N_k} |\phi(x_n)| \right) \leq \frac{1}{2^k} \quad \text{and} \quad \sum_{n=N_{k-1}+1}^{N_k} \|T(x_n)\|_F \geq 1.$$

Then $\sup_{\phi \in B_{E^*}} \sum_{n=1}^{\infty} |\phi(x_n)| \leq 1$, so that $\sum_{n=1}^{\infty} x_n$ is unconditionally convergent. Since $\sum_{n=1}^{\infty} \|T(x_n)\|_F = \infty$, T is not absolutely summing. □

16.3 (p, q)-summing operators

We now generalize again. Suppose that $1 \leq q \leq p < \infty$. We say that a linear mapping T from a Banach space $(E, \|.\|_E)$ to a Banach space $(F, \|.\|_F)$ is (p, q)-summing if there exists a constant K such that

$$\left(\sum_{n=1}^{N} \|T(x_n)\|_F^p \right)^{1/p} \leq K \sup_{\phi \in B_{E^*}} \left(\sum_{n=1}^{N} |\phi(x_n)|^q \right)^{1/q} \qquad (*)$$

for all N and all $x_1, \ldots, x_N$ in E. We denote the smallest such constant K by $\pi_{p,q}(T)$, and denote the set of all (p, q)-summing mappings from E to F by $\Pi_{p,q}(E, F)$. We call a (p, p)-summing mapping a p-summing mapping, and write Π_p for $\Pi_{p,p}$ and π_p for $\pi_{p,p}$. Thus Theorem 16.2.1 states that the absolutely summing mappings are the same as the 1-summing mappings. In fact we shall only be concerned with p-summing operators, for $1 < p < \infty$, and $(p, 2)$ summing operators, for $2 \leq p < \infty$.

We then have the following:

Theorem 16.3.1 *Suppose that* $(E, \|.\|_E)$ *and* $(F, \|.\|_F)$ *are Banach spaces and that* $1 \leq q \leq p < \infty$. *Then* $\Pi_{p,q}(E, F)$ *is a linear subspace of* $L(E, F)$, *and* $\pi_{p,q}$ *is a norm on* $\Pi_{p,q}(E, F)$, *under which* $\Pi_{p,q}(E, F)$ *is a Banach space. If* $T \in \Pi_{p,q}(E, F)$ *then* $\|T\| \leq \pi_{p,q}(T)$, *and if* $R \in L(D, E)$ *and* $S \in L(F, G)$ *then* $STR \in \Pi_{p,q}(D, G)$ *and* $\pi_{p,q}(STR) \leq \|S\| \, \pi_{p,q}(T) \, \|R\|$.

If (∗) *holds for all* $x_1, \ldots, x_N$ *in a dense subset of* E *then* $T \in \Pi_{p,q}(E, F)$, *and* $\pi_{p,q}(T)$ *is the smallest constant* K.

Proof We outline the steps that need to be taken, and leave the details to the reader. First, $\|T\| \leq \pi_{p,q}(T)$: consider a sequence of length 1. Next, $\pi_{p,q}(\lambda T) = |\lambda| \pi_{p,q}(T)$ (trivial) and $\pi_{p,q}(S + T) \leq \pi_{p,q}(S) + \pi_{p,q}(T)$ (use Minkowski's inequality on the left-hand side of (∗)), so that $\Pi_{p,q}(E, F)$ is a linear subspace of $L(E, F)$, and $\pi_{p,q}$ is a norm on $\Pi_{p,q}(E, F)$. If (T_n) is a $\pi_{p,q}$-Cauchy sequence, then it is a $\|.\|$-Cauchy sequence, and so converges in the operator norm, to T, say. Then $T \in \Pi_{p,q}$ and $\pi_{p,q}(T_n - T) \to 0$ (using (∗)), so that $\Pi_{p,q}(E, F)$ is a Banach space. The remaining results are even more straightforward. □

Recall that if $1 \leq r < s < \infty$ then $l_r \subseteq l_s$, and the inclusion is norm-decreasing. From this it follows that if $1 \leq q_1 \leq q_0 \leq p_0 \leq p_1 < \infty$ and $T \in \Pi_{p_0, q_0}(E, F)$ then $T \in \Pi_{p_1, q_1}(E, F)$ and $\pi_{p_1, q_1}(T) \leq \pi_{p_0, q_0}(T)$. We can however say more.

Proposition 16.3.1 *Suppose that* $1 \leq q_0 \leq p_0 < \infty$, *that* $1 \leq q_1 \leq p_1 < \infty$ *and that* $1/p_0 - 1/p_1 = 1/q_0 - 1/q_1 > 0$. *If* $T \in \Pi_{p_0, q_0}(E, F)$ *then* $T \in \Pi_{p_1, q_1}(E, F)$ *and* $\pi_{p_1, q_1}(T) \leq \pi_{p_0, q_0}(T)$.

In particular, if $1 \leq p_0 < p_1$ *and* $T \in \Pi_{p_0}(E, F)$ *then* $T \in \Pi_{p_1}(E, F)$ *and* $\pi_{p_1}(T) \leq \pi_{p_0}(T)$.

Proof Let $r = p_1/p_0$ and $s = q_1/q_0$. If $x_1, \ldots, x_N \in E$, then using Hölder's inequality with exponents s' and s,

$$\left(\sum_{n=1}^{N} \|T(x_n)\|^{p_1} \right)^{1/p_0}$$

$$= \left(\sum_{n=1}^{N} \left\| T(\|T(x_n)\|^{r-1} x_n) \right\|^{p_0} \right)^{1/p_0}$$

$$\leq \pi_{p_0,q_0}(T) \sup_{\|\phi\|^* \leq 1} \left(\sum_{n=1}^{N} |\phi(\|T(x_n)\|^{r-1} x_n)|^{q_0} \right)^{1/q_0}$$

$$= \pi_{p_0,q_0}(T) \sup_{\|\phi\|^* \leq 1} \left(\sum_{n=1}^{N} \|T(x_n)\|^{(r-1)q_0} |\phi(x_n)|^{q_0} \right)^{1/q_0}$$

$$\leq \pi_{p_0,q_0}(T) \left(\sum_{n=1}^{N} \|T(x_n)\|^{(r-1)q_0 s} \right)^{1/s'q_0} \sup_{\|\phi\|^* \leq 1} \left(\sum_{n=1}^{N} |\phi(x_n)|^{sq_0} \right)^{1/sq_0}$$

$$= \pi_{p_0,q_0}(T) \left(\sum_{n=1}^{N} \|T(x_n)\|^{p_1} \right)^{1/p_0 - 1/p_1} \sup_{\|\phi\|^* \leq 1} \left(\sum_{n=1}^{N} |\phi(x_n)|^{q_1} \right)^{1/q_1},$$

since $(r-1)q_0 s' = p_1$ and $1/s'q_0 = 1/p_0 - 1/p_1$. Dividing, we obtain the desired result. $\qquad\square$

The following easy proposition provides a useful characterization of (p,q)-summing operators.

Proposition 16.3.2 *Suppose that $(E, \|.\|_E)$ and $(F, \|.\|_F)$ are Banach spaces, that $T \in L(E, F)$, that $1 \leq q \leq p < \infty$ and that $K > 0$. Then $T \in \Pi_{p,q}$ and $\pi_{p,q} \leq K$ if and only if for each N and each $S \in L(l_{q'}^N, E)$*

$$\left(\sum_{n=1}^{N} \|TS(e_i)\|^p \right)^{1/p} \leq K \|S\|.$$

Proof Suppose first that $T \in \Pi_{p,q}$ and $S \in L(l_{q'}^N, E)$. Let $x_n = S(e_n)$. If $\phi \in B_{E^*}$ then

$$\sum |\phi(x_n)|^q = \sum |(S^*\phi)(e_n)|^q = \|S^*(\phi)\|^q \leq \|S^*\|^q = \|S\|^q,$$

so that

$$\left(\sum_{n=1}^{N} \|TS(e_n)\|^p \right)^{1/p} = \left(\sum_{n=1}^{N} \|T(x_n)\|^p \right)^{1/p} \leq \pi_{pq}(T) \|S\| \leq K \|S\|.$$

Conversely, suppose that the condition is satisfied. If $x_1, \ldots, x_N \in E$, define $S: l_{q'}^N \to E$ by setting $T(\alpha_1, \ldots, \alpha_N) = \alpha_1 x_1 + \cdots + \alpha_N x_N$. Then

$$\|S\| = \|S^*\| = \sup_{\phi \in B_{E^*}} \left(\sum_{n=1}^N |S^*(\phi)(e_n)|^q \right)^{1/q} = \sup_{\phi \in B_{E^*}} \left(\sum_{n=1}^N |\phi(x_n)|^q \right)^{1/q},$$

so that

$$\left(\sum_{n=1}^N \|T(x_n)\|^p \right)^{1/p} \leq K \sup_{\phi \in B_{E^*}} \left(\sum_{n=1}^N |\phi(x_n)|^q \right)^{1/q}.$$

$\square$

Corollary 16.3.1 *Suppose that $1 \leq q \leq p_1 \leq p_2$ and that $T \in \Pi_{p_1,q}$. Then*
$$\pi_{p_2,q}(T) \leq \|T\|^{1-p_1/p_2} (\pi_{p_1,q}(T))^{p_1/p_2}.$$

Proof For

$$\left(\sum_{n=1}^N \|TS(e_n)\|^{p_2} \right)^{1/p_2} \leq \left(\sup_{n=1}^N \|TS(e_n)\| \right)^{1-p_1/p_2} \left(\sum_{n=1}^N \|TS(e_n)\|^{p_1} \right)^{1/p_2}$$

$$\leq (\|T\| \cdot \|S\|)^{1-p_1/p_2} \pi_{p_1,q}(T)^{p_1/p_2} \|S\|^{p_1/p_2}$$

$$= \|T\|^{1-p_1/p_2} \pi_{p_1,q}(T)^{p_1/p_2} \|S\|.$$

$\square$

16.4 Examples of p-summing operators

One of the reasons why p-summing operators are important is that they occur naturally in various situations. Let us give some examples. First, let us introduce some notation that we shall use from now on. Suppose that K is a compact Hausdorff space and that μ is a probability measure on the Baire subsets of K. We denote the natural mapping from $C(K)$ to $L^p(\mu)$, sending f to its equivalence class in L^p, by j_p.

Proposition 16.4.1 *Suppose that K is a compact Hausdorff space and that μ is a probability measure on the Baire subsets of K. If $1 \leq p < \infty$ then j_p is p-summing, and $\pi_p(j_p) = 1$.*

Proof Suppose that $f_1, \ldots, f_N \in C(K)$. If $x \in K$, the mapping $f \to f(x)$ is a continuous linear functional of norm 1 on $C(K)$, and so

$$\sum_{n=1}^{N} \|j_p(f_n)\|_p^p = \sum_{n=1}^{N} \int_K |f_n(x)|^p \, d\mu(x)$$

$$= \int_K \sum_{n=1}^{N} |f_n(x)|^p \, d\mu(x)$$

$$\leq \sup \left\{ \sum_{n=1}^{N} |\phi(f_n)|^p : \ \phi \in C(K)^*, \|\phi\|^* \leq 1 \right\}.$$

Thus j_p is p-summing, and $\pi_p(j_p) \leq 1$. But also $\pi_p(j_p) \geq \|j_p\| = 1$. $\qquad \square$

Proposition 16.4.2 *Suppose that (Ω, Σ, μ) is a measure space, that $1 \leq p < \infty$ and that $f \in L^p(\Omega, \Sigma, \mu)$. Let $M_f(g) = fg$, for $g \in L^\infty$. Then $M_f \in \Pi_p(L^\infty, L^p)$ and $\pi_p(M_f) = \|M_f\| = \|f\|_p$.*

Proof We use Proposition 16.3.2. Suppose first that $p > 1$. Suppose that $S \in L(l_{p'}^N, L^\infty)$. Let $g_n = S(e_n)$. If $\alpha_1, \ldots, \alpha_N$ are rational and $\|(\alpha_1, \ldots, \alpha_n)\|_{p'} \leq 1$ then $|\sum_{n=1}^{N} \alpha_n g_n(\omega)| \leq \|S\|$, for almost all ω. Taking the supremum over the countable collection of all such $\alpha_1, \ldots, \alpha_N$, we see that $\|(g_1(\omega), \ldots, g_n(\omega))\|_p \leq \|S\|$, for almost all ω. Then

$$\sum_{n=1}^{N} \|M_f S(e_n)\|_p^p = \sum_{n=1}^{N} \|fg_n\|_p^p = \sum_{n=1}^{N} \int |fg_n|^p \, d\mu$$

$$= \int |f|^p (\sum_{n=1}^{N} |g_n|^p) \, d\mu \leq \|S\|^p \|f\|_p^p.$$

Thus it follows from Proposition 16.3.2 that M_f is p-summing, and $\pi_p(M_f) \leq \|f\|_p$. But $\pi_p(M_f) \geq \|M_f\| = \|f\|_p$.

If $p = 1$ and $S \in L(l_\infty^N, L^\infty)$ then for each ω

$$\sum_{n=1}^{N} |S(e_n)(\omega)| = S\left(\sum_{n=1}^{N} \alpha_n e_n\right)(\omega)$$

for some $\alpha = (\alpha_n)$ with $\|\alpha\|_\infty = 1$. Thus $\left\|\sum_{n=1}^{N} |S(e_n)|\right\|_\infty \leq \|S\|$, and so

$$\sum_{n=1}^{N} \|M_f S(e_n)\|_1 \leq \left\|\sum_{n=1}^{N} |S(e_n)|\right\|_\infty \|f\|_1 \leq \|S\| \|f\|_1. \qquad \square$$

Proposition 16.4.3 *Suppose that (Ω, Σ, μ) is a measure space, and that $\phi \in L^p(E^*)$, where E is a Banach space and $1 \le p < \infty$. Then the mapping $I_\phi : x \to \int_\Omega \phi(\omega)(x) \, d\mu(\omega)$ from E to $L^p(\Omega, \Sigma, \mu)$ is p-summing, and $\pi_p(I_\phi) \le \|\phi\|_p$.*

Proof Suppose that $x_1, \dots, x_N \in E$. Let $A = \{\omega : \phi(\omega) \neq 0\}$. Then

$$\sum_{n=1}^{N} \|I_\phi(x_n)\|_p^p = \int_A \sum_{n=1}^{N} |\phi(\omega)(x_n)|^p \, d\mu(\omega)$$

$$= \int_A \sum_{n=1}^{N} |(\phi(\omega)/\|\phi(\omega)\|)(x_n)|^p \, \|\phi(\omega)\|^p \, d\mu(\omega)$$

$$\le \left(\sup_{\|\psi\|^* \le 1} \sum_{n=1}^{N} |\psi(x_n)|^p \right) \int_A \|\phi(\omega)\|^p \, d\mu(\omega).$$

$\square$

We wish to apply this when E is an L^q space. Suppose that K is a measurable function on $(\Omega_1, \Sigma_1, \mu_1) \times (\Omega_2, \Sigma_2, \mu_2)$ for which

$$\int_{\Omega_1} \left(\int_{\Omega_2} |K(x, y)|^{q'} \, d\mu_2(y) \right)^{p/q'} d\mu_1(x) < \infty,$$

where $1 \le p < \infty$ and $1 < q \le \infty$. We can consider K as an element of $L^p(L^{q'}) = L^p((L^q)')$; then I_K is the integral operator

$$I_K(f)(x) = \int_{\Omega_2} K(x, y) f(y) \, d\mu_2(y).$$

The proposition then states that I_K is p-summing from $L^q(\Omega_2, \Sigma_2, \mu_2)$ to $L^p(\Omega_1, \Sigma_1, \mu_1)$, and

$$\pi_p(I_K) \le \left(\int_{\Omega_1} \left(\int_{\Omega_2} |K(x, y)|^{q'} \, d\mu_2(y) \right)^{p/q'} d\mu_1(x) \right)^{1/p}.$$

16.5 (p, 2)-summing operators between Hilbert spaces

How do these ideas work when we consider linear operators between Hilbert spaces? Do they relate to the ideas of the previous chapter?

Proposition 16.5.1 *Suppose that H_1 and H_2 are Hilbert spaces and that $2 \le p < \infty$. Then $\Pi_{p,2}(H_1, H_2) = S_p(H_1, H_2)$, and if $T \in S_p(H_1, H_2)$ then $\pi_{p,2}(T) = \|T\|_p$.*

Proof Suppose that $T \in \Pi_{p,2}(H_1, H_2)$. If (e_n) is an orthonormal sequence in H_1 and $y \in H_1$, then $\sum_{n=1}^{N} |\langle e_n, y \rangle|^2 \leq \|y\|^2$, and so $\sum_{n=1}^{N} \|T(e_n)\|^p \leq (\pi_{p,2}(T))^p$. Consequently, $\sum_{n=1}^{\infty} \|T(e_n)\|^p \leq (\pi_{p,2}(T))^p$, and in particular $\|T(e_n)\| \to 0$ as $n \to \infty$. Thus T is compact (Exercise 15.7). Suppose that $T = \sum_{n=1}^{\infty} s_n(T) \langle \cdot, x_n \rangle y_n$. Then

$$\sum_{j=1}^{\infty} (s_j(T))^p = \sum_{j=1}^{\infty} \|T(x_j)\|^p \leq (\pi_{p,2}(T))^p,$$

so that $T \in S_p(H_1, H_2)$, and $\|T\|_p \leq \pi_{p,2}(T)$.

Conversely, if $T \in S_p(H_1, H_2)$ and $S \in L(l_2^N, H_1)$, then $(\sum_{n=1}^{N} \|TS(e_n)\|^p)^{1/p} \leq \|TS\|_p \leq \|S\| \|T\|_p$, by Proposition 15.11.1 (ii). By Proposition 16.3.2, $T \in \Pi_{p,2}(H_1, H_2)$ and $\pi_{p,2}(T) \leq \|T\|_p$. $\square$

In particular, $\Pi_2(H_1, H_2) = S_2(H_1, H_2)$. Let us interpret this when H_1 and H_2 are L^2 spaces.

Theorem 16.5.1 *Suppose that $H_1 = L^2(\Omega_1, \Sigma_1, \mu_1)$ and $H_2 = L^2(\Omega_2, \Sigma_2, \mu_2)$, and that $T \in L(H_2, H_1)$. Then $T \in S_2(H_2, H_1)$ if and only if there exists $K \in L^2(\Omega_1 \times \Omega_2)$ such that $T = I_K$. If so, and if $T = \sum_{j=1}^{\infty} s_j \langle \cdot, g_j \rangle f_j$, then*

$$K(x, y) = \sum_{j=1}^{\infty} s_j f_j(x) \overline{g_j(y)},$$

the sum converging in norm in $L^2(\Omega_1 \times \Omega_2)$, and $\|K\|_2 = \|T\|_2$.

Proof If $T = I_K$, then $T \in \Pi_2(H_2, H_1)$, by Proposition 16.4.3, and $\|T\|_2 = \|K\|_2$. Conversely, suppose that $T = \sum_{j=1}^{\infty} s_j \langle \cdot, g_j \rangle f_j \in \Pi_2(H_2, H_1)$. Let $h_j(x, y) = f_j(x) \overline{g_j(y)}$. Then (h_j) is an orthonormal sequence in $L^2(\Omega_1 \times \Omega_2)$, and so the sum $\sum_{j=1}^{\infty} s_j h_j$ converges in L^2 norm, to K, say. Let $K_n = \sum_{j=1}^{n} s_j h_j$. If $f \in L^2(\Omega_2)$ then

$$T(f) = \lim_{n \to \infty} \sum_{j=1}^{n} s_j \langle f, g_j \rangle f_j = \lim_{n \to \infty} I_{K_n}(f) = I_K(f)$$

since

$$\|I_K(f) - I_{K_n}(f)\|_2 \leq \|I_{K-K_n}\| \|f\|_2 \leq \|I_{K-K_n}\|_2 \|f\|_2,$$

and $\|I_{K-K_n}\|_2 \to 0$ as $n \to \infty$. $\square$

16.6 Positive operators on L^1

The identification of 2-summing mappings with Hilbert–Schmidt mappings, together with the results of the previous section, lead to some strong conclusions.

Let us introduce some more notation that we shall use from now on. Suppose that $(\Omega, \Sigma, \mathbf{P})$ is a probability space. Then if $1 \leq p < q \leq \infty$ we denote the inclusion mapping $L^q \to L^p$ by $I_{q,p}$.

Theorem 16.6.1 *Suppose that $(\Omega, \Sigma, \mathbf{P})$ is a probability space. Suppose that $T \in L(L^1, L^\infty)$ and that $\int T(f)\bar{f}\,d\mathbf{P} \geq 0$ for $f \in L^1$. Let $T_1 = I_{\infty,1}T$. Then T_1 is a Riesz operator on L^1, every non-zero eigenvalue λ_j is positive, the corresponding generalized eigenvector is an eigenvector, and $\sum_{j=1}^{\infty} \lambda_j \leq \|T\|$. The corresponding eigenvectors f_j are in L^∞ and can be chosen to be orthonormal in L^2. The series*

$$\sum_{j=1}^{\infty} \lambda_j \overline{f_j(y)} f_j(x)$$

then converges in $L^2(\Omega \times \Omega)$ norm to a function $K \in L^\infty(\Omega \times \Omega)$ and if $f \in L^1$ then $T(f)(x) = \int_\Omega K(x,y)f(y)\,d\mathbf{P}(y)$.

Proof Let $T_2 = I_{\infty,2}TI_{2,1} : L^2 \to L^2$. Then T_2 is a positive Hermitian operator on L^2. Since, by Proposition 16.4.1, $I_{2,\infty}$ is 2-summing, with $\pi_2(I_{\infty,2}) = 1$, T_2 is also a 2-summing operator, with $\pi_2(T_2) \leq \|T\|$. Thus T_2 is a positive Hilbert–Schmidt operator, and we can write $T_2 = \sum_{j=1}^{\infty} \lambda_j \langle \cdot, f_j \rangle f_j$, where $(\lambda_j) = (\sigma_j(T_2))$ is a decreasing sequence of non-negative numbers in l_2. Now $T_1^2 = I_{2,1}T_2I_{\infty,2}T$, so that T_1^2 is compact, and T_1 is a Riesz operator. Since $T_1 = I_{2,1}I_{\infty,2}T$, the operators T_1 and T_2 are related, and (λ_j) is the sequence of eigenvalues of T_1, repeated according to their multiplicity, and each principal vector is in fact an eigenvector. Since $T_2(f_j) = \lambda_j I_{2,\infty}TI_{2,1}(f_j)$, $f_j \in L^\infty$.

Now let $S = \sum_{j=1}^{\infty} \sqrt{\lambda_j} \langle \cdot, f_j \rangle f_j$, so that $S^2 = T_2$. If $f \in L^2$ then

$$\|S(f)\|_2^2 = \langle S(f), S(f) \rangle = \langle T_2(f), f \rangle$$
$$= \int_\Omega T(f)\bar{f}\,d\mathbf{P} \leq \|T(f)\|_\infty \|f\|_1 \leq \|T\| \|f\|_1^2.$$

Thus S extends to a bounded linear mapping $S_1 : L^1 \to L^2$ with $\|S_1\| \leq \|T\|^{1/2}$. Then $S_1^* \in L(L^2, L^\infty)$, with $\|S_1^*\| \leq \|T\|^{1/2}$. Since S is self-adjoint, $S = I_{\infty,2}S_1^*$, and so S is 2-summing, by Proposition 16.4.1, with $\pi_2(S) \leq \|T\|^{1/2}$.

But $\pi_2(S) = (\sum_{j=1}^{\infty}(\sqrt{\lambda_j})^2)^{1/2} = (\sum_{j=1}^{\infty}\lambda_j)^{1/2}$, and so $\sum_{j=1}^{\infty}\lambda_j \leq \|T\|$. Thus T_2 is a trace class operator.

Now let $W_n = \sum_{j=1}^{n}\lambda_j(T_2)\langle\cdot, f_j\rangle f_j$ and let $K_n(x,y) = \sum_{j=1}^{n}\lambda_j(T_2)\overline{f_j(y)} f_j(x)$. Then

$$\langle W_n(f), f\rangle = \sum_{j=1}^{n}\lambda_j(T_2)|\langle f, f_j\rangle|^2 \leq \sum_{j=1}^{\infty}\lambda_j(T_2)|\langle f, f_j\rangle|^2 = \langle T(f), f\rangle,$$

and $|\langle W_n(f), g\rangle|^2 \leq \langle W_n(f), f\rangle \langle W_n(g), g\rangle$, so that

$$\left|\int_{A\times B} K_n(x,y)\,d\mathbf{P}(x)d\mathbf{P}(y)\right|^2 = |\langle W_n(I_A), I_B\rangle|^2$$
$$\leq \langle W_n(I_A), I_A\rangle \langle W_n(I_B), I_B\rangle$$
$$\leq \langle T(I_A), I_A\rangle \langle T(I_B), I_B\rangle$$
$$\leq \|T\|^2 (\mathbf{P}(A))^2(\mathbf{P}(B))^2,$$

so that $|K_n(x,y)| \leq \|T\|$ almost everywhere. Since $K_n \to K$ in $L^2(\Omega\times\Omega)$, it follows that $|K(x,y)| \leq \|T\|$ almost everywhere. Thus I_K defines an element T_K of $L(L^1, L^\infty)$. But $I_K = T_2$ on L^2, and L^2 is dense in L^1, and so $T = T_K$.

$\square$

16.7 Mercer's theorem

Theorem 16.6.1 involved a bounded kernel K. If we consider a continuous positive-definite kernel on $X \times X$, where (X, τ) is a compact Hausdorff space, we obtain even stronger results.

Theorem 16.7.1 (Mercer's theorem) *Suppose that $\mathbf{P}$ is a probability measure on the Baire sets of a compact Hausdorff space (X, τ), with the property that if U is a non-empty open Baire set then $\mathbf{P}(U) > 0$, and that K is a continuous function on $X \times X$ such that*

$$\int_{X\times X} K(x,y)\overline{f(x)}f(y) \geq 0 \quad \text{for } f \in L^1(\mathbf{P}).$$

Then $T = I_K$ satisfies the conditions and conclusions of Theorem 16.6.1. With the notation of Theorem 16.6.1, the eigenvectors f_j are continuous, and the series $\sum_{j=1}^{\infty}\lambda_j\overline{f_j(x)}f_j(y)$ converges absolutely to $K(x,y)$, uniformly in x and y. T is a compact operator from $L^1(\mathbf{P})$ to $C(X)$, and $\sum_{j=1}^{\infty}\lambda_j = \int_X K(x,x)\,d\mathbf{P}(x)$.

Proof If $x \in X$ and $\epsilon > 0$ then there exists a neighbourhood U of x such that $|K(x', y) - K(x, y)| < \epsilon$ for $x' \in U$ and all $y \in X$. Then $|T(f)(x') - T(f)(x)| \leq \epsilon \|f\|_1$ for $x' \in U$, and so T is a bounded linear mapping from $L^1(\mathbf{P})$ into $C(X)$, which we can identify with a closed linear subspace of $L^\infty(\mathbf{P})$. Then T satisfies the conditions of Theorem 16.6.1. If λ_j is a non-zero eigenvalue, then $T(f_j) = \lambda_j f_j \in C(X)$, and so f_j is continuous.

Now let $W_n = \sum_{j=1}^n \lambda_j \langle \cdot, f_j \rangle f_j$; let $R_n = T - W_n$ and $L_n = K - K_n$, so that $R_n = I_{L_n} = \sum_{j=n+1}^\infty \lambda_j \langle \cdot, f_j \rangle f_j$. Thus $L_n(x, y) = \sum_{j=n+1}^\infty \lambda_j \overline{f_j(x)} f_j(y)$, the sum converging in norm in $L^2(\mathbf{P} \times \mathbf{P})$. Consequently, $L_n(x, y) = \overline{L_n(y, x)}$, almost everywhere. But L_n is continuous, and so $L_n(x, y) = \overline{L_n(y, x)}$ for all (x, y). In particular, $L_n(x, x)$ is real, for all x. If $x_0 \in X$ and U is an open Baire neighbourhood of x_0 then

$$\int_{U \times U} L_n(x, y)\, d\mathbf{P}(x) d\mathbf{P}(y) = \langle R_n(I_U), I_U \rangle = \sum_{j=n+1}^\infty \lambda_j |\int_U f_j\, d\mathbf{P}|^2 \geq 0,$$

and so it follows from the continuity of L_n that $L_n(x_0, x_0) \geq 0$, for all $x_0 \in X$. Thus

$$K_n(x, x) = \sum_{j=1}^n \lambda_j |f_j(x)|^2 \leq K(x, x) \text{ for all } x \in X,$$

and so $\sum_{j=1}^\infty \lambda_j |f_j(x)|^2$ converges to a sum $Q(x)$, say, with $Q(x) \leq K(x, x)$, for all $x \in X$.

Suppose now that $x \in X$ and that $\epsilon > 0$. There exists n_0 such that $\sum_{j=n+1}^m \lambda_j |f_j(x)|^2 < \epsilon^2$, for $m > n \geq n_0$. But if $y \in X$ then

$$\sum_{j=n+1}^m \lambda_j |f_j(x) f_j(y)| \leq \left(\sum_{j=n+1}^m \lambda_j |f_j(x)|^2 \right)^{1/2} \left(\sum_{j=n+1}^m \lambda_j |f_j(y)|^2 \right)^{1/2}$$

$$\leq \epsilon (K(y, y))^{1/2} \leq \epsilon \|K\|_\infty^{1/2} \qquad (\dagger)$$

by the Cauchy–Schwartz inequality, so that $\sum_{j=1}^\infty \lambda_j f_j(x) f_j(y)$ converges absolutely, uniformly in y, to $B(x, y)$, say. Similarly, for fixed y, the series converges absolutely, uniformly in x. Thus $B(x, y)$ is a separately continuous function on $X \times X$. We want to show that $B = K$. Let $D = K - B$. Since $\sum_{j=1}^\infty \lambda_j f_j(x) f_j(y)$ converges to K in norm in $L^2(\mathbf{P} \times \mathbf{P})$, it follows that $D = 0$ $\mathbf{P} \times \mathbf{P}$-almost everywhere. Let $G = \{x \colon D(x, y) = 0 \text{ for all } y\}$. For almost all x, $D(x, y) = 0$ for almost all y. But $D(x, y)$ is a continuous function of y, and so $x \in G$ for almost all x. Suppose that $D(x, y) \neq 0$. Then there exists a Baire open neighbourhood U of x such that $D(z, y) \neq 0$, for $z \in U$. Thus $U \cap G = \emptyset$. But this implies that $\mathbf{P}(U) = 0$, giving a contradiction. Thus $B = K$.

In particular, $Q(x) = K(x, x)$ for all x, and $\sum_{j=1}^{\infty} \lambda_j |f_j(x)|^2 = K(x, x)$. Since the summands are positive and continuous and K is continuous, it follows from Dini's Theorem (see Exercise 16.3) that the convergence is uniform in x. Using the inequality (†) again, it follows that $\sum_{j=1}^{\infty} \lambda_j f_j(x) f_j(y)$ converges absolutely to $K(x, y)$, uniformly in (x, y). Thus $I_{K_n} \to I_K = T$ in operator norm. Since I_{K_n} is a finite-rank operator, T is compact. Finally,

$$\sum_{j=1}^{\infty} \lambda_j = \sum_{j=1}^{\infty} \lambda_j \int_X |f_j|^2 \, d\mathbf{P} = \int_X \sum_{j=1}^{\infty} \lambda_j |f_j|^2 \, d\mathbf{P} = \int_X K(x, x) \, d\mathbf{P}(x).$$

$\square$

It is not possible to replace the condition that K is continuous by the condition that $T \in L(L^1, C(K))$ (see Exercise 16.4).

16.8 p-summing operators between Hilbert spaces $(1 \le p \le 2)$

We know that the 2-summing operators between Hilbert spaces are simply the Hilbert–Schmidt operators, and the π_2 norm is the same as the Hilbert–Schmidt norm. What about p-summing operators between Hilbert spaces, for other values of p? Here the results are rather surprising. First we establish a result of interest in its own right, and a precursor of stronger results yet to come.

Proposition 16.8.1 *The inclusion mapping $i_{1,2} : l_1 \to l_2$ is 1-summing, and $\pi_1(i_{1,2}) = \sqrt{2}$.*

Proof The proof uses the Kahane–Khintchine inequality for complex numbers. Suppose that $x^{(1)}, \ldots, x^{(N)} \in l_1$. Suppose that $K \in \mathbf{N}$, and let $\epsilon_1, \ldots, \epsilon_K$ be Bernoulli random variables on D_2^K. Then, by Theorem 13.3.1,

$$\sum_{n=1}^{N} \left(\sum_{k=1}^{K} |x_k^{(n)}|^2 \right)^{1/2} \le \sqrt{2} \sum_{n=1}^{N} \left(\mathbf{E} \left| \sum_{k=1}^{K} \epsilon_k(\omega) x_k^{(n)} \right| \right)$$

$$= \sqrt{2} \mathbf{E} \left(\sum_{n=1}^{N} \left| \sum_{k=1}^{K} \epsilon_k(\omega) x_k^{(n)} \right| \right)$$

$$\le \sqrt{2} \sup \left\{ \sum_{n=1}^{N} \left| \sum_{k=1}^{\infty} \phi_k x_k^{(n)} \right| : |\phi_k| \le 1 \text{ for all } k \right\}.$$

Thus

$$\sum_{n=1}^{N}\left\|x^{(n)}\right\|_2 \le \sqrt{2}\sup\{\sum_{n=1}^{N}|\phi(x^{(n)})|: \ \phi \in (l_1)^* = l_\infty, \|\phi\|^* \le 1\},$$

so that $i_{1,2}$ is 1-summing, and $\pi_1(i_{1,2}) \le \sqrt{2}$. To show that $\sqrt{2}$ is the best possible constant, consider $x^{(1)} = (1/2, 1/2, 0, 0, \ldots)$, $x^{(2)} = (1/2, -1/2, 0, 0, \ldots)$. $\qquad\square$

Theorem 16.8.1 *If* $T = \sum_{j=1}^{\infty} s_j(T) y_j \otimes \bar{x}_j \in S_2(H_1, H_2)$ *then* $T \in \Pi_1(H_1, H_2)$ *and* $\pi_1(T) \le \sqrt{2}\|T\|_2$.

Proof If $x \in H_1$, let $S(x) = (s_j(T)\langle x, x_j\rangle)$. Applying the Cauchy–Schwartz inequality,

$$\sum_{j=1}^{\infty}|S(x)_j| \le (\sum_{j=1}^{\infty}(s_j(T))^2)^{1/2}(\sum_{j=1}^{\infty}|\langle x, x_j\rangle|^2)^{1/2} \le \|T\|_2 \|x\|,$$

so that $S \in L(H_1, l_1)$ and $\|S\| \le \|T\|_2$. If $\alpha \in l_2$ let $R(\alpha) = \sum_{j=1}^{\infty} \alpha_j y_j$. Clearly $R \in L(l_2, H_2)$ and $\|R\| = 1$. Since $T = Ri_{1,2}S$, the result follows from Proposition 16.8.1. $\qquad\square$

Corollary 16.8.1 $S_2(H_1, H_2) = \Pi_p(H_1, H_2)$, *for* $1 \le p \le 2$.

We shall consider the case $2 < p < \infty$ later, after we have developed the general theory further.

16.9 Pietsch's domination theorem

We now establish a fundamental theorem, whose proof uses the Hahn–Banach separation theorem in a beautiful way. First we make two remarks. If $(E, \|.\|_E)$ is a Banach space, there is an isometric embedding i of E into $C(K)$, for some compact Hausdorff space K: for example, we can take K to be the unit ball of E^*, with the weak* topology, and let $i(x)(\phi) = \phi(x)$. Second, the Riesz representation theorem states that if ϕ is a continuous linear functional on $C(K)$ then there exists a probability measure μ in $P(K)$, the set of probability measures on the Baire subsets of K, and a measurable function h with $|h(k)| = \|\phi\|^*$ for all $k \in K$ such that $\phi(f) = \int_X fh \, d\mu$ for all $f \in C(K)$. We write $\phi = h \, d\mu$.

Theorem 16.9.1 (Pietsch's domination theorem) *Suppose that* $(E, \|.\|_E)$ *and* $(F, \|.\|_F)$ *are Banach spaces and that* $T \in L(E, F)$. *Suppose* *that $i : E \to C(K)$ is an isometric embedding, and that $1 \le p < \infty$. Then $T \in \Pi_p(E, F)$ if and only if there exists $\mu \in P(K)$ and a constant M such that $\|T(x)\| \le M(\int |i(x)|^p \, d\mu)^{1/p}$ for each $x \in E$. If so, then $M \ge \pi_p(T)$, and we can choose μ so that $M = \pi_p(T)$.*

Proof If such μ and M exist, and $x_1, \ldots, x_N \in E$ then, since for each $k \in K$ the mapping $x \to i(x)(k)$ is a continuous linear functional of norm at most 1 on E,

$$\sum_{n=1}^{N} \|T(x_n)\|_F \le M^p \int_K \sum_{n=1}^{N} |i(x_n)(k)|^p \, d\mu(k)$$

$$\le M^p \sup \left\{ \sum_{n=1}^{N} |\phi(x_n)|^p : \phi \in E^*, \|\phi\|^* \le 1 \right\},$$

and so $T \in \Pi_p(E, F)$ and $\pi_p(T) \le M$.

Conversely, suppose that $T \in \Pi_p(E, F)$; by scaling, we can suppose that $\pi_p(T) = 1$. For $S = (x_1, \ldots, x_N)$ a finite sequence in E and $k \in K$, set

$$g_S(k) = \sum_{n=1}^{N} |i(x_n)(k)|^p \quad \text{and} \quad l_S(k) = \sum_{n=1}^{N} \|T(x_n)\|_F^p - g_S(k).$$

Then $g_S \in C_{\mathbf{R}}(K)$. Since K is compact, g_S attains its supremum G_S at a point k_S of K. Now if $\phi \in E^*$ then by the Hahn–Banach extension theorem there exists $h \, d\mu \in C_{\mathbf{R}}(K)^*$ with $\|h \, d\mu\| = \|\phi\|^*$ such that $\phi(x) = \int_K i(x) h \, d\mu$, and so

$$\sum_{n=1}^{N} \|T(x_n)\|_F^p \le \sup \left\{ \sum_{n=1}^{N} |\phi(x_n)|^p : \phi \in E^*, \|\phi\|^* \le 1 \right\}$$

$$= \sup \left\{ \sum_{n=1}^{N} |\int i(x_n) h \, d\mu|^p : h \, d\mu \in C(K)^*, \|h \, d\mu\|^* \le 1 \right\}$$

$$\le \sup \left\{ \sum_{n=1}^{N} \int |i(x_n)|^p \, d\mu : \mu \in P(K) \right\}$$

$$= \sup \left\{ \int \sum_{n=1}^{N} |i(x_n)|^p \, d\mu : \mu \in P(K) \right\} \le G_S.$$

Thus $l_S(k_S) \le 0$. Now let

$$L = \{l_S : S = (x_1, \ldots, x_N) \text{ a finite sequence in } E\},$$

and let

$$U = \{f \in C_{\mathbf{R}}(K) \colon\ f(k) > 0 \text{ for all } k \in K\}.$$

Then L and U are disjoint, and U is convex and open. L is also convex: for if $S = (x_1, \ldots, x_N)$ and $S' = (x'_1, \ldots, x'_{N'})$ are finite sets in E and $0 < \lambda < 1$ then $(1 - \lambda)h_S + \lambda h_{S'} = h_{S''}$, where

$$S'' = ((1 - \lambda)^{1/p} x_1, \ldots, (1 - \lambda)^{1/p} x_N, \lambda^{1/p} x'_1, \ldots, \lambda^{1/p} x'_{N'}).$$

Thus by the Hahn–Banach separation theorem (Theorem 4.6.2), there exist $h\,d\mu \in C_{\mathbf{R}}(K)^*$ and $\lambda \in \mathbf{R}$ such that $\int fh\,d\mu > \lambda$ for $f \in U$ and $\int l_S h\,d\mu \le \lambda$ for $l_S \in L$. Since $0 \in L$, $\lambda \ge 0$. If $f \in U$ and $\epsilon > 0$ then $\epsilon f \in U$, and so $\epsilon \int fh\,d\nu > \lambda$. Since this holds for all $\epsilon > 0$, it follows that $\lambda = 0$. Thus $\int fh\,d\mu > 0$ if $f \in U$, and so $h(k) = \|h\,d\mu\|^*$ μ-almost everywhere. Thus $\int l_S\,d\mu \le 0$ for $l_S \in L$. Applying this to a one-term sequence $S = (x)$, this says that $\|T(x)\|_F^p \le \int_K |i(x)(k)|^p\,d\mu(k)$. Thus the required inequality holds with $M = 1 = \pi_p(T)$. $\qquad\square$

16.10 Pietsch's factorization theorem

Proposition 16.4.1 shows that if μ is a probability measure on the Baire sets of a compact Hausdorff space, and if $1 \le p < \infty$, then the natural map $j_p : C(K) \to L^p(\mu)$ is p-summing, and $\pi_p(j_p) = 1$. We can also interpret Pietsch's domination theorem as a factorization theorem, which shows that j_p is the archetypical p-summing operator.

Theorem 16.10.1 (The Pietsch factorization theorem) *Suppose that $(E, \|.\|_E)$ and $(F, \|.\|_F)$ are Banach spaces and that $T \in L(E, F)$. Suppose that $i : E \to C(K)$ is an isometric embedding, and that $1 \le p < \infty$. Then $T \in \Pi_p(E, F)$ if and only if there exists $\mu \in P(K)$ and a continuous linear mapping $R : \overline{j_p i(E)} \to F$ (where $\overline{j_p i(E)}$ is the closure of $j_p i(E)$ in $L^p(\mu)$, and is given the L^p norm) such that $T = Rj_p i$. If so, then we can find a factorization such that $\|R\| = \pi_p(T)$.*

Proof If $T = Rj_p i$, then since j_p is p-summing, so is T, and $\pi_p(T) \le \|R\|\,\pi_p(j_p)\,\|i\| = \|R\|$. Conversely, suppose that $T \in \Pi_p(E, F)$. Let μ be a probability measure satisfying the conclusions of Theorem 16.9.1. If $f = j_p i(x) = j_p i(y) \in j_p i(E)$ then $\|T(x) - T(y)\|_F \le \pi_p(T)\,\|j_p i(x) - j_p i(y)\|_p = 0$, so that $T(x) = T(y)$. We can therefore define $R(f) = T(x)$ without ambiguity, and then $\|R(f)\|_F \le \pi_p(T)\,\|f\|_p$. Finally, we extend R to $\overline{j_p i(E)}$, by continuity. $\qquad\square$

We therefore have the following diagram:

$$
\begin{array}{ccc}
E & \xrightarrow{\;\;T\;\;} & F \\[4pt]
\Big\downarrow{\scriptstyle i} & & \Big\uparrow{\scriptstyle R} \\[4pt]
i(E) & \xrightarrow{\;\;j_p\;\;} & \overline{j_p i(E)} \\[4pt]
\subseteq\Big\downarrow & & \Big\downarrow\subseteq \\[4pt]
C(K) & \xrightarrow{\;\;j_p\;\;} & L^p(\mu)
\end{array}
$$

In general, we cannot extend R to $L^p(\mu)$, but there are two special cases when we can. First, if $p = 2$ we can compose R with the orthogonal projection of $L^2(\mu)$ onto $\overline{j_2 i(E)}$. We therefore have the following.

Corollary 16.10.1 *Suppose that $(E, \|.\|_E)$ and $(F, \|.\|_F)$ are Banach spaces and that $T \in L(E, F)$. Suppose that $i : E \to C(K)$ is an isometric embedding. Then $T \in \Pi_2(E, F)$ if and only if there exists $\mu \in P(K)$ and a continuous linear mapping $R : L^2(\mu) \to F$ such that $T = Rj_2i$. If so, we can find a factorization such that $\|R\| = \pi_2(T)$.*

$$
\begin{array}{ccc}
E & \xrightarrow{\;\;T\;\;} & F \\[4pt]
\Big\downarrow{\scriptstyle i} & & \Big\uparrow{\scriptstyle R} \\[4pt]
C(K) & \xrightarrow{\;\;j_2\;\;} & L^2(\mu)
\end{array}
$$

Second, suppose that $E = C(K)$, where K is a compact Hausdorff space. In this case, $j_p(E)$ is dense in $L^p(\mu)$, so that $R \in L(L^p(\mu), F)$. Thus we have the following.

Corollary 16.10.2 *Suppose that K is a compact Hausdorff space, that $(F, \|.\|_F)$ is a Banach space and that $T \in L(C(K), F)$. Then $T \in \Pi_p(C(K), F)$ if and only if there exists $\mu \in P(K)$ and a continuous linear mapping $R : L^p(\mu) \to F$ such that $T = Rj_p$. If so, then we can find a factorization such that $\|R\| = \pi_p(T)$.*

This corollary has the following useful consequence.

Proposition 16.10.1 *Suppose that K is a compact Hausdorff space, that $(F, \|.\|_F)$ is a Banach space and that $T \in \Pi_p(C(K), F)$. If $p < q < \infty$ then $\pi_q(T) \le \|T\|^{1-p/q} (\pi_p(T))^{p/q}$.*

Proof Let $T = Rj_p$ be a factorization with $\|R\| = \pi_p(T)$. Let $j_q : C(K) \to L^q(\mu)$ be the natural map, and let $I_{q,p} : L^q(\mu) \to L^p(\mu)$ be the inclusion map. If $\phi \in F^*$ then $g_\phi = R^*(\phi) \in (L^p(\mu))^* = L^{p'}(\mu)$. By Littlewood's inequality, $\|g_\phi\|_{q'} \le \|g_\phi\|_1^{1-p/q} \|g_\phi\|_{p'}^{p/q}$, and

$$\|g_\phi\|_1 = \left\|j_p^*(g_\phi)\right\|^* = \left\|j_p^* R^*(\phi)\right\|^* = \|T^*(\phi)\|^* \le \|T^*\| \cdot \|\phi\|^* = \|T\| \cdot \|\phi\|^*.$$

Thus

$$\begin{aligned}
\pi_q(T) = \pi_q(RI_{q,p}j_q) &\le \|RI_{q,p}\| \, \pi_q(j_q) \\
&= \|RI_{q,p}\| = \left\|I_{q,p}^* R^*\right\| \\
&= \sup\left\{\left\|I_{q,p}^* R^*(\phi)\right\| : \|\phi\|^* \le 1\right\} = \sup\left\{\|g_\phi\|_{q'} : \|\phi\|^* \le 1\right\} \\
&\le \sup\left\{\|g_\phi\|_1^{1-p/q} : \|\phi\|^* \le 1\right\} \sup\left\{\|g_\phi\|_{p'}^{p/q} : \|\phi\|^* \le 1\right\} \\
&\le \|T\|^{1-p/q} \|R\|^{p/q} = \|T\|^{1-p/q} (\pi_p(T))^{p/q}.
\end{aligned}$$

$\square$

16.11 *p*-summing operators between Hilbert spaces $(2 \le p \le \infty)$

Pietsch's theorems have many applications. First let us complete the results on operators between Hilbert spaces.

Theorem 16.11.1 *Suppose that H_1 and H_2 are Hilbert spaces and that $2 \le p < \infty$. Then $T \in \Pi_p(H_1, H_2)$ if and only if $T \in S_2(H_1, H_2)$.*

Proof If $T \in S_2(H_1, H_2)$ then $T \in \Pi_2(H_1, H_2)$, and so $T \in \Pi_p(H_1, H_2)$. Conversely, if $T \in \Pi_p(H_1, H_2)$ then $T \in \Pi_{p,2}(H_1, H_2)$, and so $T \in S_p(H_1, H_2)$. Thus T is compact, and we can write $T = \sum_{j=1}^\infty s_j(T) \langle \cdot, x_j \rangle y_j$. Let B_1 be the unit ball of H_1, with the weak topology. By Pietsch's domination theorem, there exists $\mu \in P(B_1)$ such that $\|T(x)\|^p \le (\pi_p(T))^p \int_{B_1} |\langle x, y \rangle|^p \, d\mu(y)$ for all $x \in H_1$. Once again, we make use of the Kahane–Khintchine inequality. Let $\epsilon_1, \ldots, \epsilon_J$ be Bernoulli random variables on D_2^J, and let $x(\omega) = \sum_{j=1}^J \epsilon_j(\omega)x_j$. Then $T(x(\omega)) = \sum_{j=1}^J \epsilon_j(\omega)s_j(T)y_j$, so that $\|T(x(\omega))\| =$

$(\sum_{j=1}^{J}(s_j(T))^2)^{1/2}$, for each ω. Thus

$$\sum_{j=1}^{J}(s_j(T))^2)^{p/2} \le (\pi_p(T))^p \int_{B_1} |\langle x(\omega), y \rangle|^p \, d\mu(y).$$

Integrating over D_2^J, changing the order of integration, and using the Kahane–Khintchine inequality, we see that

$$\left(\sum_{j=1}^{J}(s_j(T))^2 \right)^{p/2} \le (\pi_p(T))^p \int_{D_2^J} \left(\int_{B_1} |\langle x(\omega), y \rangle|^p \, d\mu(y) \right) d\mathbf{P}(\omega)$$

$$= (\pi_p(T))^p \int_{B_1} \left(\int_{D_2^J} \left| \sum_{j=1}^{J} \epsilon_j(\omega) \langle x_j, y \rangle \right|^p \, d\mathbf{P}(\omega) \right) d\mu(y)$$

$$\le (\pi_p(T))^p B_p^p \int_{B_1} \left(\sum_{j=1}^{J} |\langle x_j, y \rangle|^2 \right)^{p/2} d\mu(y),$$

where B_p is the constant in the Kahane–Khintchine inequality. But $\sum_{j=1}^{J}|\langle x_j, y \rangle|^2 \le \|y\|^2 \le 1$ for $y \in B_1$, and so $\|T\|_2 = \|(S_j(T))\|_2 \le B_p \pi_p(T)$. $\qquad\Box$

16.12 The Dvoretzky–Rogers theorem

Pietsch's factorization theorem enables us to prove the following.

Theorem 16.12.1 *Suppose that $S \in \Pi_2(E, F)$ and $T \in \Pi_2(F, G)$. Then TS is 1-summing, and compact.*

Proof Let i_E be an isometry of E into $C(K_E)$ and let i_F be an isometry of F into $C(K_F)$. We can write $S = \tilde{S}j_2 i_E$ and $T = \tilde{T}j'_2 i_F$:

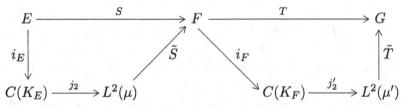

Then $j'_2 i_F \tilde{S}$ is 2-summing, and therefore is a Hilbert–Schmidt operator. Thus it is 1-summing, and compact, and so therefore is $TS = \tilde{T}(j'_2 i_F \tilde{S})j_2 i_E$. $\qquad\Box$

We can now answer the question that was raised at the beginning of the chapter.

Theorem 16.12.2 (The Dvoretzky–Rogers theorem) *If $(E, \|.\|_E)$ is a Banach space in which every unconditionally convergent series is absolutely convergent, then E is finite-dimensional.*

Proof For the identity mapping I_E is 1-summing, and therefore 2-summing, and so $I_E = I_E^2$ is compact. □

Since $\pi_1(T) \geq \pi_2(T)$, the next result can be thought of as a finite-dimensional metric version of the Dvoretzky–Rogers theorem.

Theorem 16.12.3 *If $(E, \|.\|_E)$ is a n-dimensional normed space, then $\pi_2(E) = \sqrt{n}$.*

Proof Let I_E be the identity mapping on E. We can factorize $I_E = Rj_2i$, with $\|R\| = \pi_2(I_E)$. Let $H_n = j_2i(E)$. Then $\dim H_n = n$ and j_2iR is the identity mapping on H_n. Thus

$$\sqrt{n} = \pi_2(I_{H_n}) \leq \pi_2(j_2) \|i\| . \|R\| = \|R\| = \pi_2(I_E).$$

For the converse, we use Proposition 16.3.2. Let $S \in L(l_2^J, E)$, let K be the null-space of S, and let Q be the orthogonal projection of l_2^J onto $K^\perp$. Then $\dim K^\perp \leq n$, and $I_E S = S = S I_{K^\perp} Q$, so that $\pi_2(S) \leq \|S\| \pi_2(I_{K^\perp}) \leq \sqrt{n} \|S\|$. Thus $(\sum_{j=1}^{J} \|I_E S(e_j)\|^2)^{1/2} \leq \sqrt{n} \|S\|$, and so $\pi_2(I_E) \leq \sqrt{n}$. □

This result is due to Garling and Gordon [GaG 71], but this elegant proof is due to Kwapień. It has three immediate consequences.

Corollary 16.12.1 *Suppose that $(E, \|.\|_E)$ is an n-dimensional normed space. Then there exists an invertible linear mapping $T : E \to l_2^n$ with $\|T\| = 1$ and $\|T^{-1}\| \leq \sqrt{n}$.*

Proof Let $U : l_2^n \to H_n$ be an isometry, and take $T = U^{-1}j_2i$, so that $T^{-1} = RU$, and $\|T^{-1}\| = \|R\| = \sqrt{n}$. □

Corollary 16.12.2 *Suppose that E_n is an n-dimensional subspace of a normed space $(E, \|.\|_E)$. Then there exists a projection P of E onto E_n with $\|P\| \leq \sqrt{n}$.*

Proof Let i be an isometric embedding of E into $C(K)$, for some compact Hausdorff space K, and let $I_{E_n} = Rj_2i_{|E_n}$ be a factorization with $\|R\| = \sqrt{n}$. Then $P = Rj_2i$ is a suitable projection. $\qquad\square$

Corollary 16.12.3 *Suppose that $(E, \|.\|_E)$ is an n-dimensional normed space and that $2 < p < \infty$. Then $\pi_{p,2}(I_E) \le n^{1/p}$.*

Proof By Corollary 16.3.1, $\pi_{p,2}(I_E) \le \|I_E\|^{1-2/p} (\pi_2(i_E))^{2/p} = n^{1/p}$. $\qquad\square$

We shall obtain a lower bound for $\pi_{p,2}(I_E)$ later (Corollary 17.4.2).

16.13 Operators that factor through a Hilbert space

Corollary 16.10.1 raises the problem: when does $T \in L(E, F)$ factor through a Hilbert space? We say that $T \in \Gamma_2 = \Gamma_2(E, F)$ if there exist a Hilbert space H and $A \in L(H, F)$, $B \in E, H$ such that $T = AB$. If so, we set $\gamma_2(T) = \inf\{\|A\| \|B\| : T = AB\}$.

To help us solve the problem, we introduce the following notation: if $x = (x_1, \ldots, x_m)$ and $y = (y_1, \ldots, y_n)$ are finite sequences in a Banach space $(E, \|.\|_E)$ we write $x \prec\prec y$ if $\sum_{i=1}^m |\phi(x_i)|^2 \le \sum_{j=1}^n |\phi(y_j)|^2$ for all $\phi \in E^*$.

Theorem 16.13.1 *Suppose that $T \in L(E, F)$. Then $T \in \Gamma_2$ if and only if there exists $C \ge 0$ such that whenever $x \prec\prec y$ then $\sum_{i=1}^m \|T(x_i)\|^2 \le C^2 \sum_{j=1}^n \|y_j\|^2$. If so, then γ_2 is the infimum of the C for which the condition holds.*

Proof Suppose first that $T \in \Gamma_2$ and that $C > \gamma_2(T)$. Then there is a factorization $T = AB$ with $\|B\| = 1$ and $\|A\| < C$. Suppose that $x \prec\prec y$. Let $(e_1, \ldots, e_l)$ be an orthonormal basis for span $(B(x_1), \ldots, B(x_m))$, and let $\phi_k = B^*(e_k)$ for $1 \le k \le l$. Then

$$\sum_{i=1}^m \|T(x_i)\|^2 \le C^2 \sum_{i=1}^m \|B(x_i)\|^2$$

$$= C^2 \sum_{i=1}^m \sum_{k=1}^l |\langle B(x_i), e_k \rangle|^2$$

$$= C^2 \sum_{k=1}^l \sum_{i=1}^m |\phi_k(x_i)|^2$$

$$\le C^2 \sum_{k=1}^{l} \sum_{j=1}^{n} |\phi_k(y_j)|^2$$

$$= C^2 \sum_{j=1}^{n} \sum_{k=1}^{l} |\langle B(y_j), e_k \rangle|^2$$

$$\le C^2 \sum_{j=1}^{n} \|B(y_j)\|^2 \le C^2 \sum_{j=1}^{n} \|y_j\|^2.$$

Thus the condition is necessary.

Second, suppose that the condition is satisfied. First we consider the case where E is finite-dimensional. Let K be the unit sphere of E^*: K is compact. If $x \in E$ and $k \in K$, let $\hat{x}(k) = k(x)$. Then $\hat{x} \in C(K)$. Now let

$$S = \left\{ (x, y): \ \sum_{i=1}^{m} \|T(x_i)\|^2 > C^2 \sum_{j=1}^{n} \|y_j\|^2 \right\},$$

and let

$$D = \left\{ \sum_{j=1}^{n} |\hat{y}_j|^2 - \sum_{i=1}^{m} |\hat{x}_i|^2: \ (x, y) \in S \right\}.$$

Then D is a convex subset of $C(K)$, and the condition ensures that D is disjoint from the convex open set $U = \{f: \ f(k) > 0 \text{ for all } k \in K\}$. By the Hahn–Banach theorem, there exists a probability measure $\mathbf{P}$ on K so that $\int g \, d\mathbf{P} \le 0$ for all $g \in D$. Then it follows by considering sequences of length 1 that if $\|T(x)\| > C \|y\|$ then $\int |\hat{x}|^2 \, d\mathbf{P} \ge \int |\hat{y}|^2 \, d\mathbf{P}$. Let $a = \sup\{\int |\hat{x}|^2 \, d\mathbf{P}: \ \|x\| = 1\}$. Then $a \le 1$, and it is easy to see that $a > 0$ (why?). Let $\mu = a\mathbf{P}$, and let $B(x) = j_2(\hat{x})$, where j_2 is the natural map from $C(K) \to L^2(\mu)$, and let $H = B(E)$. Then $\|B\| = 1$, and it follows that if $\|B(x)\| < \|B(y)\|$ then $\|T(x)\| \le C \|y\|$. Choose y so that $\|B(y)\| = \|y\| = 1$. Thus if $\|B(x)\| < 1$ then $\|T(x)\| \le C$. This implies that $\|T(x)\| \le C \|B(x)\|$ for all $x \in E$, so that if $B(x) = B(z)$ then $T(x) = T(z)$. We can therefore define $A \in L(H, F)$ such that $T = AB$ and $\|A\| \le C$.

We now consider the case where E is infinite-dimensional. First suppose that E is separable, so that there is an increasing sequence (E_i) of finite-dimensional subspaces whose union E_∞ is dense in E. For each i there is a factorization $T_{|E_i} = A_i B_i$, with $\|A_i\| \le C$ and $\|B_i\| = 1$. For $x, y \in E_i$ let $\langle x, y \rangle_i = \langle B_i(x), B_i(y) \rangle$. Then a standard approximation and diagonalization argument shows that there is a subsequence (i_k) such that if $x, y \in E_\infty$ then $\langle x, y \rangle_{i_k}$ converges, to $\langle x, y \rangle_\infty$, say. $\langle x, y \rangle_\infty$ is a pre-inner product; it satisfies all the conditions of an inner product except that

$N = \{x: \langle x, y \rangle_\infty = 0 \text{ for all } y \in E_\infty\}$ may be a non-trivial linear subspace of E_∞. But then we can consider E/N, define an inner product on it, and complete it, to obtain a Hilbert space H. Having done this, it is then straightforward to obtain a factorization of T; the details are left to the reader. If E is non-separable, a more sophisticated transfinite induction is needed; an elegant way to provide this is to consider a free ultrafilter defined on the set of finite-dimensional subspaces of E. □

Let us now consider the relation $x \prec\prec y$ further.

Proposition 16.13.1 *Suppose that $x = (x_1, \ldots, x_m)$ and $y = (y_1, \ldots, y_n)$ are finite sequences in a Banach space $(E, \|.\|_E)$. Then $x \prec\prec y$ if and only if there exists $A = (a_{ij}) \in L(l_2^m, l_2^n)$ with $\|A\| \leq 1$ such that $x_i = \sum_{j=1}^n a_{ij} y_j$ for $1 \leq i \leq m$.*

Proof Suppose that $x \prec\prec y$. Consider the subspace $V = \{(\phi(x_i))_{i=1}^m : \phi \in E^*\}$ of l_2^m. If $v = (\phi(x_i))_{i=1}^m \in V$, let $A_0(v) = (\phi(y_j))_{j=1}^n \in l_2^n$. Then A_0 is well-defined, and $\|A_0\| \leq 1$. Let $A = A_0 P$, where P is the orthogonal projection of l_2^m onto V. Then A has the required properties.

Conversely, if the condition is satisfied and $\phi \in E^*$ then

$$\sum_{i=1}^m |\phi(x_i)|^2 = \sum_{i=1}^m \left| \sum_{j=1}^n a_{ij} \phi(y_j) \right|^2 \leq \sum_{j=1}^m |\phi(y_j)|^2.$$

□

In Theorem 16.13.1, we can clearly restrict attention to sequences x and y of equal length. Combining Theorem 16.13.1 with this proposition, and with Exercise 16.6, we obtain the following.

Theorem 16.13.2 *Suppose that $T \in L(E, F)$. Then the following are equivalent:*

(i) $T \in \Gamma_2$;

(ii) there exists $C \geq 0$ such that if $y_1, \ldots, y_n \in X$ and $A \in L(l_2^n, l_2^n)$ then

$$\sum_{i=1}^n \left\| T\left(\sum_{j=1}^n u_{ij} y_j \right) \right\|^2 \leq C^2 \|A\|^2 \sum_{i=1}^n \|T(y_j)\|^2;$$

(iii) there exists $C \geq 0$ such that if $y_1, \ldots, y_n \in X$ and $U = (u_{ij})$ is an $n \times n$ unitary matrix then

$$\sum_{i=1}^{n} \left\| T\left(\sum_{j=1}^{n} u_{ij} y_j\right) \right\|^2 \leq C^2 \sum_{i=1}^{n} \|T(y_j)\|^2.$$

If so, then γ_2 is the infimum of the C for which the conditions hold.

16.14 Notes and remarks

Absolutely summing operators were introduced by Grothendieck [Grot 53] as *applications semi-intégrales à droite* and many of the results of the rest of the book have their origin in this fundamental work. It was however written in a very compressed style, and most of the results were expressed in terms of tensor products, rather than linear operators, and so it remained impenetrable until the magnificent paper of Lindenstrauss and Pełczyński [LiP 68] appeared. This explained Grothendieck's work clearly in terms of linear operators, presented many new results, and ended with a large number of problems that needed to be resolved.

Theorem 16.8.1 was first proved by Grothendieck [Grot 53]. The proof given here is due to Pietsch [Pie 67], who extended the result to p-summing operators, for $1 \leq p \leq 2$. Theorem 16.11.1 was proved by Pełczyński [Pel 67]. Grothendieck proved his result by calculating the 1-summing norm of a Hilbert–Schmidt operator directly. Garling [Gar 70] did the same for the p-summing norms, thus giving a proof that does not make use of the Kahane–Khintchine inequality.

If $(E, \|.\|_E)$ and $(F, \|.\|_F)$ are finite-dimensional spaces of the same dimension, the *Banach–Mazur distance* $d(E, F)$ is defined as

$$\inf\{\|T\| \|T^{-1}\| : \ T \text{ a linear isomorphism of } E \text{ onto } F\}.$$

This is a basic concept in the local theory of Banach spaces, and the geometry of finite-dimensional normed spaces. Corollary 16.12.1 was originally proved by John [Joh 48], by considering the ellipsoid of maximal volume contained in the unit ball of E. This more geometric approach has led to many interesting results about finite-dimensional normed spaces. For this, see [Tom 89] and [Pis 89].

Mercer was a near contemporary of Littlewood at Trinity College, Cambridge (they were bracketed as Senior Wrangler in 1905): he proved his theorem in 1909 [Mer 09] for functions on $[a, b] \times [a, b]$. His proof was classical: a good account is given in [Smi 62].

Exercises

16.1   Prove Proposition 16.1.2 without appealing to the closed graph theorem.

16.2   Why do we not consider (p, q)-summing operators with $p < q$?

16.3   Suppose that (f_n) is a sequence in $C(K)$, where K is a compact Hausdorff space, which increases pointwise to a continuous function f. Show that the convergence is uniform (Dini's theorem). [Hint: consider $A_{n,\epsilon} = \{k: f_n(k) \geq f(k) - \epsilon\}$.]

16.4   Give an example where P is a probability measure on the Baire sets of a compact Hausdorff space K, and $T \in L(L^1, C(K))$ satisfies the conditions of Theorem 16.6.1, but where the conclusions of Mercer's theorem do not hold.

16.5   (i) Suppose that $\mathbf{P}$ is a probability measure on the unit sphere K of l_2^d. Show that there exists $x \in l_2^d$ with $\|x\| = 1$ and $\int_K |\langle x, k \rangle|^2 \, d\mathbf{P}(k) \geq 1/d$.

(ii) Give an example of a probability measure $\mathbf{P}$ on the unit sphere K of l_2^d for which $\int_K |\langle x, k \rangle|^2 \, d\mathbf{P}(k) \leq \|x\|^2 /d$ for all x.

(iii) Use Corollary 16.12.1 to obtain a lower bound for a in Theorem 16.13.1.

16.6   Suppose that $\sum_{i=1}^{\infty} f_i$ is an unconditionally convergent series in $L_{\mathbf{R}}^1(\Omega, \Sigma, \mu)$. Show that

$$\left(\sum_{i=1}^{m} \|f_i\|_1^2 \right)^{1/2} \leq \left\| \left(\sum_{i=1}^{m} f_i^2 \right)^{1/2} \right\|_1 \leq \sqrt{2} \mathbf{E} \left(\left\| \sum_{i=1}^{m} \epsilon_i f_i \right\|_1 \right),$$

where (ϵ_i) is a sequence of Bernoulli random variables. Deduce that $\sum_{i=1}^{\infty} \|f_i\|_1^2 < \infty$ (Orlicz' theorem).

What happens if L^1 is replaced by L^p, for $1 < p \leq 2$, and for $2 < p < \infty$?

16.7   Prove the following extension of Theorem 16.13.1.

Suppose that G is a linear subspace of E and that $T \in L(G, F)$. Suppose that there exists $C \geq 0$ such that if $x \in G$, $y \in E$ and $x \prec \prec y$ then $\sum_{i=1}^{m} \|T(x_i)\|^2 \leq C^2 \sum_{j=1}^{n} \|y_j\|^2$. Show that there exists a Hilbert space H and $B \in L(E, H)$, $A \in L(H, F)$ with $\|A\| \leq C$, $\|B\| \leq 1$ such that $T(x) = AB(x)$ for $x \in G$.

Show that there exists $\tilde{T} \in \Gamma_2(E, F)$ such that $\tilde{T}(x) = T(x)$ for $x \in G$, with $\gamma_2(\tilde{T}) \leq C$.

16.8   Show that $\Gamma_2(E, F)$ is a vector space and that γ_2 is a norm on it. Show that $(\Gamma_2(E, F), \gamma_2)$ is complete.

Approximation numbers and eigenvalues

17.1 The approximation, Gelfand and Weyl numbers

We have identified the p-summing operators between Hilbert spaces H_1 and H_2 with the Hilbert–Schmidt operators $S_2(H_1, H_2)$, and the $(p, 2)$-summing operators with $S_p(H_1, H_2)$. These spaces were defined using singular numbers: are there corresponding numbers for operators between Banach spaces? In fact there are many analogues of the singular numbers, and we shall mention three. Suppose that $T \in L(E, F)$, where E and F are Banach spaces.

- The *n-th approximation number* $a_n(T)$ is defined as

$$a_n(T) = \inf\{\|T - R\| : R \in L(E, F), \ \mathrm{rank}(R) < n\}.$$

- The *n-th Gelfand number* $c_n(T)$ is defined as

$$c_n(T) = \inf\{\|T_{|G}\| : G \text{ a closed subspace of } E \text{ of codimension less than } n\}.$$

- The *n-th Weyl number* $x_n(T)$ is defined as

$$x_n(T) = \sup\{c_n(TS) : S \in L(l_2, E), \|S\| \leq 1\}.$$

The approximation numbers, Gelfand numbers and Weyl numbers are closely related to singular numbers, as the next proposition shows. The Weyl numbers were introduced by Pietsch; they are technically useful, since they enable us to exploit the strong geometric properties of Hilbert space.

Proposition 17.1.1 *Suppose that $T \in L(E, F)$, where E and F are Banach spaces. Then $x_n(T) \leq c_n(T) \leq a_n(T)$, and if E is a Hilbert space, they are all equal.*

$$x_n(T) = \sup\{a_n(TS) : S \in L(l_2, E), \|S\| \leq 1\}.$$

If E and F are Hilbert spaces and T is compact then $a_n(T) = c_n(T) = x_n(T) = s_n(T)$.

Proof If $S \in L(l_2, E)$ and G is a subspace of E with codim $G < n$ then codim $S^{-1}(G) < n$, so that $c_n(TS) \leq c_n(T) \|S\|$, and $x_n(T) \leq c_n(T)$. If $R \in L(E, F)$ and rank $R < n$ then the null-space N of R has codimension less than n, and $\|T_{|N}\| \leq \|T - R\|$; thus $c_n(T) \leq a_n(T)$. If E is a Hilbert space then clearly $x_n(T) = c_n(T)$; if G is a closed subspace of E of codimension less than n, and P is the orthogonal projection onto $G^{\perp}$ then rank$(TP) < n$ and $\|T - TP\| = \|T_{|G}\|$, so that $c_n(T) = a_n(T)$. Consequently

$$x_n(T) = \sup\{a_n(TS) : S \in L(l_2, E) \; \|S\| \leq 1\}.$$

Finally, the Rayleigh–Ritz minimax formula (Theorem 15.7.1) states if $T \in K(H_1, H_2)$ then $s_n(T) = c_n(T)$. $\qquad \square$

In general, the inequalities can be strict: if J is the identity map from $l_1^3(\mathbf{R})$ to $l_2^3(\mathbf{R})$, then $a_2(J) = 1/\sqrt{2} < \sqrt{2/3} = c_2(T)$; if I is the identity map on $l_1^2(\mathbf{R})$ then $x_2(I) = 1/\sqrt{2} < 1 = c_2(I)$.

It is clear that if $T \in L(E, F)$ then T can be approximated in operator norm by a finite rank operator if and only if $a_n(T) \to 0$ as $n \to \infty$. In particular, if $a_n(T) \to 0$ as $n \to \infty$ then T is compact. It is however a deep and difficult result that not every compact operator between Banach spaces can be approximated by finite rank operators. This illuminates the importance of the following result.

Theorem 17.1.1 *If $T \in L(E, F)$ then T is compact if and only if $c_n(T) \to 0$ as $n \to \infty$.*

Proof First, suppose that T is compact, and that $\epsilon > 0$. There exist $y_1, \ldots, y_n$ in the unit ball B_F of F such that $T(B_E) \subseteq \cup_{i=1}^n (y_i + \epsilon B_F)$. By the Hahn–Banach theorem, for each i there exists $\phi_i \in F^*$ with $\|\phi_i\|^* = 1$ and $\phi_i(y_i) = \|y_i\|$. Let $G = \{x \in E : \phi_i(T(x)) = 0 \text{ for } 1 \leq i \leq n\}$. G has codimension less than $n + 1$. Suppose that $x \in B_E \cap G$. Then there exists i such that $\|T(x) - y_i\| < \epsilon$. Then $\|y_i\| = \phi(y_i) = \phi_i(y_i - T(x)) < \epsilon$, and so $\|T(x)\| < 2\epsilon$. Thus $c_{n+1} < 2\epsilon$, and so $c_n \to 0$ as $n \to \infty$.

Conversely, suppose that $T \in L(E, F)$, that $\|T\| = 1$ and that $c_n(T) \to 0$ as $n \to \infty$. Suppose that $0 < \epsilon < 1$ and that G is a finite-codimensional subspace such that $\|T_{|G}\| < \epsilon$. Since $\|T_{|\bar{G}}\| = \|T_{|G}\| < \epsilon$, we can suppose that G is closed, and so there is a continuous projection P_G of E onto G. Let $P_K = I - P_G$, and let $K = P_K(E)$. Since K is finite-dimensional, P_K is compact, and there exist $x_1, \ldots, x_n$ in B_E such that $P_K(B_E) \subseteq \cup_{i=1}^{n}(P_K(x_i) + \epsilon B_E)$. If $x \in B_E$ there exists i such that $\|P_K(x - x_i)\| \leq \epsilon$; then

$$\|P_G(x - x_i)\| \leq \|x - x_i\| + \|P_K(x - x_i)\| < \|x\| + \|x_i\| + \epsilon \leq 2 + \epsilon.$$

Consequently

$$\|T(x) - T(x_i)\| \leq \|T(P_G(x - x_i))\| + \|T(P_K(x - x_i))\| \leq \epsilon(2 + \epsilon) + \epsilon < 4\epsilon.$$

Thus T is compact. $\qquad\square$

17.2 Subadditive and submultiplicative properties

The approximation numbers, Gelfand numbers and Weyl numbers enjoy subadditive properties. These lead to inequalities which correspond to the Ky Fan inequalities.

Proposition 17.2.1 *Let σ_n denote one of a_n, c_n or x_n. If $S, T \in L(E, F)$ and $m, n, J \in \mathbf{N}$ then $\sigma_{m+n-1}(S + T) \leq \sigma_m(S) + \sigma_n(T)$, and*

$$\sum_{j=1}^{2J} \sigma_j(S + T) \leq 2 \left(\sum_{j=1}^{J} \sigma_j(S) + \sum_{j=1}^{J} \sigma_j(T) \right)$$

$$\sum_{j=1}^{2J-1} \sigma_j(S + T) \leq 2 \left(\sum_{j=1}^{J-1} \sigma_n(S) + \sum_{j=1}^{J-1} \sigma_n(T) \right) + \sigma_J(S) + \sigma_J(T).$$

If $(X, \|.\|_X)$ is a symmetric Banach sequence space and $(\sigma_n(S))$ and $(\sigma_n(T))$ are both in X then $(\sigma_n(S + T)) \in X$ and

$$\|(\sigma_n(S + T))\|_X \leq 2 \|(\sigma_n(S) + \sigma_n(T))\|_X \leq 2(\|(\sigma_n(S))\|_X + \|(\sigma_n(T))\|_X).$$

Proof The first set of inequalities follow easily from the definitions, and the next two follow from the fact that

$$\sigma_{2j}(S + T) \leq \sigma_{2j-1}(S + T) \leq \sigma_j(S) + \sigma_j(T).$$

Let $u_{2n-1} = u_{2n} = \sigma_n(S) + \sigma_n(T)$. Then $(\sigma_n(S + T)) \prec_w (u_n)$, and so

$$\|(\sigma_n(S + T))\|_X \leq \|(u_n)\|_X \leq 2 \|(\sigma_n(S) + \sigma_n(T))\|_X,$$

by Corollary 7.4.1. $\qquad\square$

The approximation numbers, Gelfand numbers and Weyl numbers also enjoy submultiplicative properties. These lead to inequalities which correspond to the Horn inequalities.

Proposition 17.2.2 *Let σ_n denote one of a_n, c_n or x_n. If $S \in L(E,F)$ and $T \in L(F,G)$ and $m,n,J \in \mathbf{N}$ then $\sigma_{m+n-1}(TS) \le \sigma_n(T).\sigma_m(S)$, and*

$$\prod_{j=1}^{2J} \sigma_j(TS) \le \left(\prod_{j=1}^{J} \sigma_j(T).\sigma_j(S) \right)^2$$

$$\prod_{j=1}^{2J-1} \sigma_j(TS) \le \left(\prod_{j=1}^{J-1} \sigma_j(T).\sigma_j(S) \right)^2 \sigma_J(T)\sigma_J(S).$$

Suppose that ϕ is an increasing function on $[0,\infty)$ and that $\phi(e^t)$ is a convex function of t. Then

$$\sum_{j=1}^{2J} \phi(\sigma_j(TS)) \le 2 \sum_{j=1}^{J} \phi(\sigma_j(T).\sigma_j(S)), \quad \text{for each } J.$$

In particular,

$$\sum_{j=1}^{2J} |\sigma_j(TS)|^p \le 2 \sum_{j=1}^{J} (\sigma_j(T).\sigma_j(S))^p, \quad \text{for } 0 < p < \infty, \text{ for each } J.$$

Suppose that $(X, \|.\|_X)$ is a symmetric Banach sequence space. If $(\sigma_j(T))$ and $(\sigma_j(S))$ are both in X then $(\sigma_j(TS)) \in X$ and $\|(\sigma_j(TS))\|_X \le 2\|(\sigma_j(T) \cdot \sigma_j(S))\|_X$.

Proof For (a_n) and (c_n), the first inequality follows easily from the definitions. Let us prove it for (x_n). Suppose that $R \in L(l_2, E)$, that $\|R\| \le 1$, and that $\epsilon > 0$. Then there exists $A_m \in L(l_2, F)$ with $\text{rank}(A_m) < m$ and

$$\|SR - A_m\| < a_m(SR) + \epsilon \le x_m(S) + \epsilon.$$

There also exists $B_n \in L(l_2, G)$ with $\text{rank}(B_n) < n$ and

$$\|T(SR - A_m) - B_n\| \le a_n(T(SR - A_m)) + \epsilon$$
$$\le x_n(T) \|SR - A_m\| + \epsilon.$$

Then rank $(TA_m + B_n) < m + n - 1$, and so

$$a_{m+n-1}(TSR) \le \|T(SR - A_m) - B_n\|$$
$$\le x_n(T) \|SR - A_m\| + \epsilon \le x_n(T)(x_m(S) + \epsilon) + \epsilon.$$

Taking the supremum as R varies over the unit ball of $L(l_2, E)$,

$$x_{m+n-1}(TS) \leq x_n(T)(x_m(S) + \epsilon) + \epsilon;$$

this holds for all $\epsilon > 0$, and so the inequality follows.

The next two inequalities then follow from the fact that

$$\sigma_{2j}(TS) \leq \sigma_{2j-1}(TS) \leq \sigma_j(T)\sigma_j(S).$$

Thus if we set $v_{2j-1} = v_{2j} = \sigma_j(T)\sigma_j(S)$ then $\prod_{j=1}^{J} \sigma_j(TS) \leq \prod_{j=1}^{J} v_j$, and the remaining results follow from Proposition 7.6.3. $\quad\square$

We next consider the Gelfand and Weyl numbers of $(p, 2)$-summing operators. For this, we need the following elementary result.

Proposition 17.2.3 *Suppose that $T \in L(H, F)$, where H is a Hilbert space, and that $0 < \epsilon_n < 1$, for $n \in \mathbf{N}$. Then there exists an orthonormal sequence (e_n) in H such that $\|T(e_n)\| \geq (1 - \epsilon)c_n(T)$ for each n.*

Proof This follows from an easy recursion argument. Choose a unit vector E_1 such that $\|T(e_1)\| > (1 - \epsilon_1) \|T\| = (1 - \epsilon_1)c_1(T)$. Suppose that we have found $e_1, \ldots, e_n$. If $G = \{e_1, \ldots, e_n\}^{\perp}$, then codim $G = n$, so that there exists a unit vector e_{n+1} in G with $\|T(e_{n+1})\| > (1 - \epsilon_{n+1})c_{n+1}(T)$. $\quad\square$

Corollary 17.2.1 *If $T \in \Pi_{p,2}(H, F)$, where $2 \leq p < \infty$, then*

$$\left(\sum_{n=1}^{\infty} (c_n(T))^p\right)^{1/p} \leq \pi_{p,2}(T).$$

Proof Suppose that $\epsilon > 0$. Let (e_n) satisfy the conclusions of the proposition. If $N \in \mathbf{N}$ then

$$(1 - \epsilon)\left(\sum_{n=1}^{N} (c_n(T))^p\right)^{1/p} \leq \left(\sum_{n=1}^{N} \|T(e_n)\|^p\right)^{1/p}$$

$$\leq \pi_{p,2}(T) \sup\left\{\left(\sum_{n=1}^{N} |\langle e_n, y\rangle|^2\right)^{1/2} : \|y\| \leq 1\right\}$$

$$\leq \pi_{p,2}(T).$$

Since ϵ and N are arbitrary, the inequality follows. $\quad\square$

Corollary 17.2.2 *If $T \in \Pi_{p,2}(E, F)$, where E and F are Banach spaces and $2 \leq p < \infty$, then $x_n(T) \leq \pi_{p,2}(T)/n^{1/p}$.*

Proof Suppose that $S \in L(l_2, E)$ and that $\|S\| \leq 1$. Then $\pi_{p,2}(TS) \leq \pi_{p,2}(T)$, and so

$$c_n(TS) \leq \left(\frac{1}{n} \sum_{i=1}^{n} c_i(TS)^p \right)^{1/p} \leq \frac{\pi_{p,2}(TS)}{n^{1/p}} \leq \frac{\pi_{p,2}(T)}{n^{1/p}}.$$

The result follows on taking the supremum over all S in the unit ball of $L(l_2, E)$.   □

17.3 Pietsch's inequality

We are now in a position to prove a fundamental inequality, which is the Banach space equivalent of Weyl's inequality.

Theorem 17.3.1 (Pietsch's inequality) *Suppose that T is a Riesz operator on a Banach space $(E, \|.\|_E)$. Then*

$$\prod_{j=1}^{2n} |\lambda_j(T)| \leq (2e)^n \left(\prod_{j=1}^{n} x_j(T) \right)^2,$$

$$\prod_{j=1}^{2n+1} |\lambda_j(T)| \leq (2e)^{n+1/2} \left(\prod_{j=1}^{n} x_j(T) \right)^2 \ldots x_{n+1}(T).$$

Proof We shall prove this for $2n$; the proof for $2n + 1$ is very similar. As in Sections 15.1 and 15.2, there exists a T-invariant $2n$-dimensional subspace E_{2n} of E for which $T_{2n} = T_{|E_{2n}}$ has eigenvalues $\lambda_1(T), \ldots, \lambda_{2n}(T)$. Note that $x_j(T_{2n}) \leq x_j(T)$ for $1 \leq j \leq 2n$. Since $\pi_2(I_{E_{2n}}) = \sqrt{2n}$, the Pietsch factorization theorem tells us that there exists an isomorphism S of E_{2n} onto l_2^{2n} with $\pi_2(S) = \sqrt{2n}$ and $\|S^{-1}\| = 1$. Let $R = ST_{2n}S^{-1} : l_2^{2n} \to l_2^{2n}$. Then R and T_{2n} are related operators, and so R has the same eigenvalues as T. Using Weyl's inequality and Proposition 17.2.1,

$$\prod_{j=1}^{2n} |\lambda_j(T)| = \prod_{j=1}^{2n} |\lambda_j(R)| \leq \prod_{j=1}^{2n} s_j(R) \leq \left(\prod_{j=1}^{n} s_{2j-1}(R) \right)^2$$

$$= \left(\prod_{j=1}^{n} x_{2j-1}(ST) \right)^2 \leq \left(\prod_{j=1}^{n} x_j(S)x_j(T) \right)^2.$$

Now $x_j(S) \leq \pi_2(S)/\sqrt{j} = (2n/j)^{1/2}$, by Corollary 17.2.2, and $\prod_{j=1}^{n}(2n/j) = 2^n n^n/n! \leq (2e)^n$, since $n^n \leq e^n n!$ (Exercise 3.5), so that

$$\prod_{j=1}^{2n} |\lambda_j(T)| \leq (2e)^n \left(\prod_{j=1}^{n} x_j(T) \right)^2 .$$

$\square$

Corollary 17.3.1 *(i) Suppose that ϕ is an increasing function on $[0, \infty)$ and that $\phi(e^t)$ is a convex function of t. Then*

$$\sum_{j=1}^{2J} \phi(|\lambda_j(T)|) \leq 2 \sum_{j=1}^{J} \phi(\sqrt{2e}x_j(T)), \quad \textit{for each } J.$$

In particular,

$$\sum_{j=1}^{2J} |\lambda_j(T)|^p \leq 2(2e)^{p/2} \sum_{j=1}^{J} (x_j(T))^p, \quad \textit{for } 0 < p < \infty, \textit{ for each } J.$$

Suppose that $(X, \|.\|_X)$ is a symmetric Banach sequence space. If $(x_j(T)) \in X$ then $(\lambda_j(T)) \in X$ and $\|(\lambda_j(T))\|_X \leq 2\sqrt{2e}\,\|(x_j(T))\|_X$.

Proof Let $y_{2j-1}(T) = y_{2j}(T) = \sqrt{2e}x_j(T)$. Then $\prod_{j=1}^{J} |\lambda_j(T)| \leq \prod_{j=1}^{J} y_j(T)$, for each J, and the result follows from Proposition 7.6.3. $\square$

We use Weyl's inequality to establish the following inequality.

Theorem 17.3.2 *If $T \in L(E, F)$ then*

$$\prod_{j=1}^{2n} c_j(T) \leq (4en)^n \left(\prod_{j=1}^{n} x_j(T) \right)^2 .$$

Proof Suppose that $0 < \epsilon < 1$. A straightforward recursion argument shows that there exist unit vectors z_j in E and ϕ_j in F^* such that $\phi_j(z_i) = 0$ for $i < j$ and $|\phi_j(T(x_j))| \geq (1-\epsilon)c_j(T)$. Let $A: l_2^{2n} \to E$ be defined by $A(e_j) = z_j$, let $B: F \to l_\infty^{2n}$ be defined by $(B(y))_j = \phi_j(y)$, let $I_{\infty,2}^{(2n)}: l_\infty^{2n} \to l_2^{2n}$ be the identity map and let $S_{2n} = I_{\infty,2}^{(2n)}BTA$. Then $\|A\| \leq \sqrt{2n}$, since

$$\|A(\alpha)\| \leq \sum_{j=1}^{2n} |\alpha_j|.\,\|z_j\| \leq \sqrt{2n}\,\|\alpha\|,$$

by the Cauchy–Schwarz inequality. Further, $\|B\| \le 1$ and $\pi_2(I_{\infty,2}^{(2n)}) = \sqrt{2n}$, so that $x_j(I_{\infty,2}^{(2n)}B) \le \sqrt{2n/j}$, for $1 \le j \le 2n$, by Corollary 17.2.2.

Now S_{2n} is represented by a lower triangular matrix with diagonal entries $\phi_j(T(x_j))$, and so

$$(1-\epsilon)^{2n} \prod_{j=1}^{2n} c_j(T) \le \prod_{j=1}^{2n} s_j(S_{2n}) \le \left(\prod_{j=1}^{n} s_{2j-1}(S_{2n})\right)^2,$$

by Weyl's inequality. But, arguing as in the proof of Pietsch's inequality,

$$s_{2j-1}(S_{2n}) \le \|A\|\, x_{2j-1}(I_{\infty,2}^{(2n)}BT) \le \sqrt{2n}x_j(I_{\infty,2}^{(2n)}B)x_j(T) \le (2n/\sqrt{j})x_j(T),$$

so that

$$(1-\epsilon)^{2n} \prod_{j=1}^{2n} c_j(T) \le \left(\frac{(2n)^{2n}}{n!}\right)\left(\prod_{j=1}^{n} x_j(T)\right)^2 \le (4en)^n \left(\prod_{j=1}^{n} x_j(T)\right)^2.$$

Since ϵ is arbitrary, the result follows. $\qquad\square$

Since $(2n)^{2n} \le e^{2n}.(2n)!$ we have the following corollary.

Corollary 17.3.2 $\prod_{j=1}^{2n}(c_j(T)/\sqrt{j}) \le 2^n e^{2n}(\prod_{j=1}^{n} x_j(T))^2$

Applying Proposition 7.6.3, we deduce this corollary.

Corollary 17.3.3 $\sum_{j=1}^{\infty}(c_j(T))^2/j \le 2e^2 \sum_{j=1}^{\infty}(x_j(T))^2$.

Corollary 17.3.4 *If* $\sum_{j=1}^{\infty}(x_j(T))^2 < \infty$ *then* T *is compact.*

Proof For then $\sum_{j=1}^{\infty}(c_j(T))^2/j < \infty$, so that $c_j(T) \to 0$, and the result follows from Theorem 17.1.1. $\qquad\square$

17.4 Eigenvalues of p-summing and $(p,2)$-summing endomorphisms

We now use these results to obtain information about the eigenvalues of p-summing and $(p,2)$-summing endomorphisms of a complex Banach space.

Theorem 17.4.1 *If* $(E, \|.\|_E)$ *is a complex Banach space and* $T \in \pi_2(E)$, *then* T^2 *is compact, so that* T *is a Riesz operator. Further,* $(\sum_{j=1}^{\infty}|\lambda_j(T)|^2)^{1/2} \le \pi_2(T)$.

Proof Let $T = Rj_2i$ be a factorization, with $\|R\| = \pi_2(T)$, and let $S = j_2iR$. Then T and S are related operators, and S is a Hilbert–Schmidt operator with $\|S\|_2 \le \pi_2(T)$. As $T^2 = RSj_2i$, T^2 is compact, and so T is a Riesz operator. Since T and S are related,

$$\left(\sum_{j=1}^{\infty}|\lambda_j(T)|^2\right) = \left(\sum_{j=1}^{\infty}|\lambda_j(S)|^2\right) \le \|S\|_2 \le \pi_2(T).$$

$\square$

Theorem 17.4.2 *If $T \in \Pi_{p,2}(E)$ and $m > p$ then T^m is compact, and so T is a Riesz operator.*

Proof Using submultiplicity, and applying Corollary 17.2.2,

$$x_{mn-1}(T^m) \le (x_n(T))^m \le (\pi_{p,2}(T))^m/n^{m/p},$$

and so $\sum_{j=1}^{\infty}(x_j(T^m))^2 < \infty$. The result follows from Corollary 17.3.4. $\square$

Corollary 17.4.1 *Suppose that $T \in \Pi_{p,2}(E)$. Then*

$$n^{1/p}|\lambda_n(T)| \le n^{1/p}\lambda_n^{\dagger}(T) \le 2p'\sqrt{2e}\pi_{p,2}(T).$$

Proof

$$\begin{aligned}
n^{1/p}|\lambda_n| &\le n^{1/p}\lambda_n^{\dagger}(T) \le \|(\lambda(T))\|_{p,\infty}^{\dagger} \\
&\le 2\sqrt{2e}\,\|(x(T))\|_{p,\infty}^{\dagger} \quad \text{(by Corollary 17.3.1)} \\
&\le 2p'\sqrt{2e}\,\|(x(T))\|_{p,\infty}^{*} \quad \text{(by Proposition 10.2.1)} \\
&= 2p'\sqrt{2e}\sup_{j} j^{1/p}x_j(T) \\
&\le 2p'\sqrt{2e}\pi_{p,2}(T) \quad \text{(by Corollary 17.2.2)}.
\end{aligned}$$

$\square$

Applying this to the identity mapping on a finite-dimensional space, we have the following, which complements Corollary 16.12.3.

Corollary 17.4.2 *If $(E, \|.\|_E)$ is an n-dimensional normed space, then $\pi_{p,2}(E) \ge n^{1/p}/(2p'\sqrt{2e})$.*

If $T \in \Pi_p(E)$ for some $1 \le p \le 2$, then $T \in \Pi_2(E)$, and T is a Riesz operator with $(\sum_{j=1}^{\infty}|\lambda_j(T)|^2)^{1/2} \le \pi_2(T) \le \pi_p(T)$ (Theorem 17.4.1). What happens when $2 < p < \infty$?

Theorem 17.4.3 *If $T \in \Pi_p(E)$ for some $2 < p < \infty$, then T is a Riesz operator and $(\sum_{j=1}^{\infty} |\lambda_j(T)|^p)^{1/p} \leq \pi_p(T)$.*

Proof Since $T \in \Pi_{p,2}(E)$, T is a Riesz operator. Suppose that $p < r < \infty$. Then, by Corollary 17.4.1,

$$|\lambda_j(T)|^r \leq (2p'\sqrt{2e}\pi_{p,2}(T))^r/j^{r/p} \leq (2p'\sqrt{2e}\pi_p(T))^r/j^{r/p},$$

so that

$$\sum_{j=1}^{\infty} |\lambda_j(T)|^r \leq C_r \pi_p(T)^r, \quad \text{where } C_r = (2p'\sqrt{2e})^r p/(r-p).$$

Note that $C_r \to \infty$ as $r \searrow p$: this seems to be an unpromising approach. But let us set

$$D_r = \inf \left\{ C : \sum_{j=1}^{\infty} |\lambda_j(T)|^r \leq C(\pi_p(T))^r, E \quad \text{a Banach space}, T \in \Pi_p(E) \right\}.$$

Then $1 \leq D_r \leq C_r$: we shall show that $D_r = 1$. Then

$$\left(\sum_{j=1}^{\infty} |\lambda_j(T)|^p \right)^{1/p} = \lim_{r \searrow p} \left(\sum_{j=1}^{\infty} |\lambda_j(T)|^r \right)^{1/r} \leq \pi_p(T).$$

In order to show that $D_r = 1$, we consider tensor products. Suppose that E and F are Banach spaces. Then an element $t = \sum_{j=1}^{n} x_j \otimes y_j$ of $E \otimes F$ defines an element T_t of $L(E^*, F)$: $T_t(\phi) = \sum_{j=1}^{n} \phi(x_j)y_j$. We give t the corresponding operator norm:

$$\|t\|_\epsilon = \|T_t\| = \sup \left\{ \left\| \sum_{j=1}^{n} \phi(x_j)y_j \right\|_F : \|\phi\|_{E^*} \leq 1 \right\}$$

$$= \sup \left\{ \left| \sum_{j=1}^{n} \phi(x_j)\psi(y_j) \right| : \|\phi\|_{E^*} \leq 1, \|\psi\|_{F^*} \leq 1 \right\}.$$

This is the *injective* norm on $E \otimes F$. We denote the completion of $E \otimes F$ under this norm by $E \hat{\otimes}_\epsilon F$. If $S \in L(E_1, E_2)$ and $T \in L(F_1, F_2)$ and $t = \sum_{j=1}^{n} x_j \otimes y_j$ we set $(S \otimes T)(t) = \sum_{j=1}^{n} S(x_j) \otimes T(y_j)$. Then it follows from the definition that $\|(S \otimes T)(t)\|_\epsilon \leq \|S\| \|T\| \|t\|_\epsilon$.

Proposition 17.4.1 *Suppose that $i_1 : E_1 \to C(K_1)$ and $i_1 : E_2 \to C(K_2)$ are isometries. If $t = \sum_{j=1}^{n} x_j \otimes y_j \in E_1 \otimes E_2$, let $I(t)(k_1, k_2) =$*

$\sum_{j=1}^{n} i_1(x_j)(k_1) \otimes i_2(y_j)(k_2) \in C(K_1 \times K_2)$. *Then* $\|I(t)\| = \|t\|_{\epsilon}$, *so that* I *extends to an isometry of* $E_1 \otimes_{\epsilon} E_2$ *into* $C(K_1 \otimes K_2)$.

Proof Let $f_j = i_1(x_j)$, $g_j = i_2(y_j)$. Since

$$|I(t)(k_1, k_2)| = \left| \sum_{j=1}^{n} \delta_{k_1}(f_j)\delta_{k_2}(g_j) \right| \leq \|t\|_{\epsilon}, \quad \|I(t)\| \leq \|t\|_{\epsilon}.$$

If, for $k = 1, 2$, $\phi_k \in E_k^*$ and $\|\phi_k\|_{E_k^*} = 1$, then by the Hahn–Banach theorem, ϕ extends, without increase of norm, to a continuous linear functional on $C(K_k)$, and by the Riesz representation theorem this is given by $h_k \, d\mu_k$, where μ_k is a Baire probability measure and $|h_k| = 1$. Thus

$$\left| \sum_{j=1}^{n} \phi_1(x_j)\phi_2(y_j) \right|$$

$$= \left| \int_{K_1} \left(\int_{K_2} \sum_{j=1}^{n} f_j(k_1)g_j(k_2)h_2(k_2) \, d\mu_2 \right) h_1(k_1) \, d\mu_1 \right|$$

$$= \left| \int_{K_1} \left(\int_{K_2} I(t)h_2(k_2) \, d\mu_2 \right) h_1(k_1) \, d\mu_1 \right|$$

$$\leq \int_{K_1} \left(\int_{K_2} |I(t)| \, d\mu_2 \right) d\mu_1 \leq \|I(t)\|.$$

Consequently $\|t\|_{\epsilon} \leq \|I(t)\|$.   □

Theorem 17.4.4 *Suppose that* $1 \leq p < \infty$ *and that* $T_1 \in \Pi_p(E_1, F_1)$, $T_2 \in \Pi_p(E_2, F_2)$. *Then* $T_1 \otimes T_2 \in \Pi_p(E_1 \hat{\otimes}_{\epsilon} F_1, E_2 \hat{\otimes}_{\epsilon} F_2)$ *and*

$$\pi_p(T_1 \otimes T_2) \leq \pi_p(T_1)\pi_p(T_2).$$

Proof Let $i_1 \colon E_1 \to C(K_1)$ and $i_2 \colon E_2 \to C(K_2)$ be isometric embeddings, and let $I \colon E_1 \hat{\otimes}_{\epsilon} E_2 \to C(K_1 \times K_2)$ be the corresponding embedding. By Pietsch's domination theorem, there exist, for $k = 1, 2$, probability measures μ_k on the Baire sets of K_k such that

$$\|T_k(x)\| \leq \pi_p(T_k) \left(\int_{K_k} |i_k(x)|^p \, d\mu_k \right)^{1/p}.$$

Now let $\mu = \mu_1 \times \mu_2$ be the product measure on $K_1 \times K_2$. Suppose that $t = \sum_{j=1}^{n} x_i \otimes y_i$ and $\phi \in B_{F_1^*}$, $\psi \in B_{F_2^*}$. Let $f_j = i_1(x_j)$, $g_j = i_2(y_j)$. Then

$$\left| \sum_{j=1}^{n} \phi(T_1(x_j)) \psi(T_2(y_j)) \right|$$

$$= \left| \phi \left(T_1 \left(\sum_{j=1}^{n} \psi(T_2(y_j)) x_j \right) \right) \right|$$

$$\leq \left\| T_1(\sum_{j=1}^{n} \psi(T_2(y_j)) x_j) \right\|$$

$$\leq \pi_p(T_1) \left(\int_{K_1} \left| \sum_{j=1}^{n} \psi(T_2(y_j)) f_j(k_1) \right|^p d\mu_1(k_1) \right)^{1/p}$$

$$\leq \pi_p(T_1) \left(\int_{K_1} \left\| T_2 \left(\sum_{j=1}^{n} f_j(k_1) y_j \right) \right\|^p d\mu_1(k_1) \right)^{1/p}$$

$$\leq \pi_p(T_1) \pi_p(T_2) \left(\int_{K_1} \int_{K_2} \left| \sum_{j=1}^{n} f_j(k_1) g_j(k_2) \right|^p d\mu_1(k_1) \, d\mu_2(k_2) \right)^{1/p}$$

$$= \pi_p(T_1) \pi_p(T_2) \left(\int_{K_1 \times K_2} |I(t)|^p d\mu \right)^{1/p}.$$

Thus $\|(T_1 \otimes T_2)(t)\|_\epsilon \leq \pi_p(T_1) \pi_p(T_2) (\int_{K_1 \times K_2} |I(t)|^p \, d\mu)^{1/p}$, and this inequality extends by continuity to any $t \in E_1 \hat{\otimes}_\epsilon F_1$. □

We now complete the proof of Theorem 17.4.3. We consider $T \otimes T$. If λ_1, λ_2 are eigenvalues of T then $\lambda_1 \lambda_2$ is an eigenvalue of $T \otimes T$, whose generalized eigenspace contains

$$\oplus \{ G_\alpha \otimes G_\beta : \alpha, \beta \text{ eigenvalues of } T, \alpha\beta = \lambda_1 \lambda_2 \}$$

and so

$$\left(\sum_{j=1}^{\infty} |\lambda_j(T)|^r \right)^2 \leq \sum_{j=1}^{\infty} |\lambda_j(T \otimes T)|^r \leq D_r \pi_p(T \otimes T)^r = D_r(\pi_p(T))^{2r}.$$

Thus $D_r \leq D_r^{1/2}$, and $D_r = 1$. □

17.5 Notes and remarks

Detailed accounts of the distribution of eigenvalues are given in [Kön 86] and [Pie 87]; the latter also contains a fascinating historical survey.

Theorem 17.1.1 was proved by Lacey [Lac 63]. Enflo [Enf 73] gave the first example of a compact operator which could not be approximated in norm by operators of finite rank; this was a problem which went back to Banach.

Exercises

17.1   Verify the calculations that follow Proposition 17.1.1.

17.2   Suppose that (Ω, Σ, μ) is a measure space, and that $1 < p < \infty$. Suppose that K is a measurable kernel such that

$$K_p = \left(\int_\Omega \left(\int_\Omega |K(\omega_1, \omega_2)|^{p'} \, d\mu(\omega_2) \right)^{p/p'} \, d\mu(\omega_1) \right)^{1/p} < \infty.$$

Show that K defines an operator T_K in $L(L^p(\Omega, \Sigma, \mu))$ with $\|T_K\| \leq K_p$. Show that T_K is a Riesz operator, and that if $1 < p \leq 2$ then $\sum_{k=1}^\infty |\lambda_k(T_K)|^2 \leq K_p^2$, while if $2 < p < \infty$ then $\sum_{k=1}^\infty |\lambda_k(T_K)|^p \leq K_p^p$.

17.3   Let (Ω, Σ, μ) be **T**, with Haar measure. Suppose that $2 < p < \infty$ and that $f \in L^{p'}$. Let $K(s, t) = f(s - t)$. Show that K satisfies the conditions of the preceding exercise. What are the eigenvectors and eigenvalues of T_K? What conclusion do you draw from the preceding exercise?

18

Grothendieck's inequality, type and cotype

18.1 Littlewood's 4/3 inequality

In the previous chapter, we saw that p-summing and $(p, 2)$-summing properties of a linear operator can give useful information about its structure. Pietsch's factorization theorem shows that if μ is a probability measure on the Baire sets of a compact Hausdorff space and $1 \leq p < \infty$ then the natural mapping $j_p : C(K) \to L^p(\mu)$ is p-summing. This implies that $C(K)$ and $L^p(\mu)$ are very different. In this chapter, we shall explore this idea further, and obtain more examples of p-summing and $(p, 2)$-summing mappings.

We consider inequalities between norms on the space $M_{m,n} = M_{m,n}(\mathbf{R})$ or $M_{m,n}(\mathbf{C})$ of real or complex $m \times n$ matrices. Suppose that $A = (a_{ij}) \in M_{m,n}$. Our main object of study will be the norm

$$\|A\| = \sup \left\{ \sum_{i=1}^m \left| \sum_{j=1}^n a_{ij} t_j \right| : |t_j| \leq 1 \right\}$$

$$= \sup \left\{ \left| \sum_{i=1}^m \sum_{j=1}^n a_{ij} s_i t_j \right| : |s_i| \leq 1, |t_j| \leq 1 \right\}.$$

$\|A\|$ is simply the operator norm of the operator $T_A : l_\infty^n \to l_1^m$ defined by $T_A(t) = (\sum_{j=1}^n a_{ij} t_j)_{i=1}^m$, for $t = (t_1, \ldots, t_n) \in l_\infty^n$. In this section, we restrict attention to the real case, where

$$\|A\| = \sup \left\{ \sum_{i=1}^m \left| \sum_{j=1}^n a_{ij} t_j \right| : t_j = \pm 1 \right\}$$

$$= \sup \left\{ \left| \sum_{i=1}^m \sum_{j=1}^n a_{ij} s_i t_j \right| : s_i = \pm 1, t_j = \pm 1 \right\}.$$

We set $a_i = (a_{ij})_{j=1}^n$, so that $a_i \in \mathbf{R}^n$. The following inequalities are due to Littlewood and Orlicz.

Proposition 18.1.1 *If $A \in M_{m,n}(\mathbf{R})$ then $\sum_{i=1}^m \|a_i\|_2 \leq \sqrt{2}\,\|A\|$ (Littlewood) and $(\sum_{i=1}^m \|a_i\|_1^2)^{1/2} \leq \sqrt{2}\,\|A\|$ (Orlicz).*

Proof Using Khintchine's inequality,

$$
\sum_{i=1}^m \|a_i\|_2 = \sum_{i=1}^m \left(\sum_{j=1}^n |a_{ij}|^2 \right)^{1/2}
$$

$$
\leq \sqrt{2} \sum_{i=1}^m \mathbf{E}(|\sum_{j=1}^n \epsilon_j a_{ij}|)
$$

$$
= \sqrt{2}\mathbf{E}\left(\sum_{i=1}^m |\sum_{j=1}^n \epsilon_j a_{ij}| \right) \leq \sqrt{2}\,\|A\|.
$$

Similarly $\sum_{j=1}^n (\sum_{i=1}^m |a_{ij}|^2)^{1/2} \leq \sqrt{2}\,\|A\|$. Orlicz's inequality now follows by applying Corollary 5.4.2. $\qquad\square$

As a corollary, we obtain Littlewood's 4/3 inequality; it was for this that he proved Khintchine's inequality.

Corollary 18.1.1 (Littlewood's 4/3 inequality) *If $A \in M_{m,n}(\mathbf{R})$ then $(\sum_{i,j} |a_{ij}|^{4/3})^{3/4} \leq \sqrt{2}\,\|A\|$.*

Proof We use Hölder's inequality twice.

$$
\sum_{i,j} |a_{ij}|^{4/3} = \sum_i \left(\sum_j |a_{ij}|^{2/3} |a_{ij}|^{2/3} \right)
$$

$$
\leq \sum_i \left((\sum_j |a_{ij}|^2)^{1/3} (\sum_j |a_{ij}|)^{2/3} \right)
$$

$$
\leq \left(\sum_i (\sum_j |a_{ij}|^2)^{1/2} \right)^{2/3} \left(\sum_i (\sum_j |a_{ij}|)^2 \right)^{1/3}
$$

$$
= \left(\sum_i \|a_i\|_2 \right)^{2/3} \left(\sum_i \|a_i\|_1^2 \right)^{1/3} \leq \left(\sqrt{2}\,\|A\| \right)^{4/3}.
$$

$\qquad\square$

The exponent $4/3$ is best possible. To see this, let A be an $n \times n$ Hadamard matrix. Then $(\sum_{i,j} |a_{ij}|^p)^{1/p} = n^{2/p}$, while if $\|t\|_\infty = 1$ then, since the a_i are orthogonal,

$$\sum_i \left| \sum_j a_{ij} t_j \right| \leq \sqrt{n} \left(\sum_i (\sum_j a_{ij} t_j)^2 \right)^{1/2}$$

$$= \sqrt{n} \left(\sum_i \langle a_i, t \rangle^2 \right)^{1/2}$$

$$= n \|t\|_2 \leq n^{3/2}.$$

18.2 Grothendieck's inequality

We now come to Grothendieck's inequality. We set

$$g(A) = \sup \left\{ \sum_{i=1}^m \left\| \sum_{j=1}^n a_{ij} k_j \right\|_H : k_j \in H, \|k_j\| \leq 1 \right\}$$

$$= \sup \left\{ \left| \sum_{i=1}^m \sum_{j=1}^n a_{ij} \langle h_i, k_j \rangle \right| : h_i, k_j \in H, \|h_i\| \leq 1, \|k_j\| \leq 1 \right\},$$

where H is a real or complex Hilbert space. $g(A)$ is the operator norm of the operator $T_A : l_\infty^n(H) \to l_1^m(H)$ defined by $T_A(k) = (\sum_{j=1}^n a_{ij} k_j)_{i=1}^m$ for $k = (k_1, \ldots, k_n) \in l_\infty^n(H)$.

Theorem 18.2.1 (Grothendieck's inequality) *There exists a constant C, independent of m and n, such that if $A \in M_{m,n}$ then $g(A) \leq C \|A\|$.*

The smallest value of the constant C is denoted by $K_G = K_G(\mathbf{R})$ or $K_G(\mathbf{C})$, and is called *Grothendieck's constant*. The exact values are not known, but it is known that $1.338 \leq K_G(\mathbf{C}) \leq 1.405$ and that $\pi/2 = 1.571 \leq K_G(\mathbf{R}) \leq 1.782 = \pi/(2 \sinh^{-1}(1))$.

Proof There are several proofs of this inequality. We shall give two, neither of which is the proof given by Grothendieck, and neither of which gives good values for the constants.

We begin by giving what is probably the shortest and easiest proof. Let

$$K_{m,n} = \sup\{g(A) : A \in M_{m,n}, \|A\| \leq 1\}.$$

If $\|A\| \leq 1$ then $\sum_{i=1}^{m} |a_{ij}| \leq 1$, and so $g(A) \leq n$; we need to show that there is a constant C, independent of m and n, such that $K_{m,n} \leq C$.

We can suppose that H is an infinite-dimensional separable Hilbert space. Since all such spaces are isometrically isomorphic, we can suppose that H is a Gaussian Hilbert space, a subspace of $L^2(\Omega, \Sigma, \mathbf{P})$. (Recall that H is a closed linear subspace of $L^2(\Omega, \Sigma, \mathbf{P})$ with the property that if $h \in H$ then h has a normal, or Gaussian, distribution with mean 0 and variance $\|h\|_2^2$; such a space can be obtained by taking the closed linear span of a sequence of independent standard Gaussian random variables.) The random variables h_i and k_j are then unbounded random variables; the idea of the proof is to truncate them at a judiciously chosen level. Suppose that $0 < \delta < 1/2$. There exists M such that if $h \in H$ and $\|h\| = 1$ then $\int_{|h|>M} |h|^2 \, d\mathbf{P} = \delta^2$. If $h \in H$, let $h^M = hI_{(|h| \leq M\|h\|)}$. Then $\|h - \tilde{h}^M\| = \delta \|h\|$.

If $\|A\| \leq 1$ and $\|h_i\|_H \leq 1$, $\|k_j\|_H \leq 1$ then

$$\left| \sum_{i=1}^{m} \sum_{j=1}^{n} a_{ij} \langle h_i, k_j \rangle \right| \leq \left| \sum_{i=1}^{m} \sum_{j=1}^{n} a_{ij} \langle h_i^M, k_j^M \rangle \right|$$
$$+ \left| \sum_{i=1}^{m} \sum_{j=1}^{n} a_{ij} \langle h_i - h_i^M, k_j^M \rangle \right|$$
$$+ \left| \sum_{i=1}^{m} \sum_{j=1}^{n} a_{ij} \langle h_i, k_j - k_j^M \rangle \right|.$$

Now

$$\left| \sum_{i=1}^{m} \sum_{j=1}^{n} a_{ij} \langle h_i^M, k_j^M \rangle \right| = \left| \int_{\Omega} \sum_{i=1}^{m} \sum_{j=1}^{n} a_{ij} h_i^M(\omega) \overline{k_j^M(\omega)} \, d\mathbf{P}(\omega) \right| \leq M^2,$$

while

$$\left| \sum_{i=1}^{m} \sum_{j=1}^{n} a_{ij} \langle h_i - h_i^M, k_j^M \rangle \right| \leq \delta K_{m,n}$$

and

$$\left| \sum_{i=1}^{m} \sum_{j=1}^{n} a_{ij} \langle h_i, k_j - k_j^M \rangle \right| \leq \delta K_{m,n},$$

so that

$$K_{m,n} \leq M^2 + 2\delta K_{m,n}, \text{ and } K_{m,n} \leq M^2/(1 - 2\delta).$$

□

For example, in the real case if $M = 3$ then $\delta = 0.16$ and $K_G \leq 13.5$.

18.3 Grothendieck's theorem

The following theorem is the first and most important consequence of Grothendieck's inequality.

Theorem 18.3.1 (Grothendieck's theorem) *If $T \in L(L^1(\Omega, \Sigma, \mu), H)$, where H is a Hilbert space, then T is absolutely summing and $\pi_1(T) \leq K_G \|T\|$.*

Proof By Theorem 16.3.1, it is enough to consider simple functions $f_1, \ldots, f_n$ with

$$\sup \left\{ \left\| \sum_{j=1}^{n} b_j f_j \right\|_1 : |b_j| \leq 1 \right\} \leq 1.$$

We can write

$$f_j = \sum_{i=1}^{m} c_{ij} I_{A_i} = \sum_{i=1}^{m} a_{ij} g_i,$$

where $A_1, \ldots, A_m$ are disjoint sets of positive measure, and where $g_i = I_{A_i}/\mu(A_i)$, so that $\|g_i\|_1 = 1$. Let $h_i = T(g_i)$, so that $\|h_i\|_H \leq \|T\|$. Then

$$\sum_{j=1}^{n} \|T(f_j)\|_H = \sum_{j=1}^{n} \left\| \sum_{i=1}^{m} a_{ij} h_i \right\|_H \leq g(A) \|T\| \leq K_G \|A\| \|T\|,$$

where A is the matrix (a_{ij}). But if $|t_j| \leq 1$ for $1 \leq j \leq n$ then

$$\sum_{i=1}^{m} \left| \sum_{j=1}^{n} a_{ij} t_j \right| = \left\| \sum_{j=1}^{n} t_j f_j \right\|_1 \leq 1,$$

so that $\|A\| \leq 1$.

□

Grothendieck's theorem is essentially equivalent to Grothendieck's inequality. For suppose that we know that $\pi_1(S) \leq K \|S\|$ for each $S \in L(l_1, H)$, and suppose that $A \in M_{m,n}$. If $h_1, \ldots, h_m$ are in the unit ball of

H, let $S: l_1 \to H$ be defined by $S(z) = \sum_{i=1}^{m} z_i h_i$. Then $\|S\| \le 1$, so that $\pi_1(ST_A) \le \pi_1(S)\|T_A\| \le K\|A\|$. But then

$$\sum_{j=1}^{n}\left\|\sum_{i=1}^{m} a_{ij} h_i\right\| = \sum_{j=1}^{n} \|ST_A(e_j)\|$$

$$\le \pi_1(ST_A) \sup\left\{\left\|\sum_{j=1}^{n} b_j e_j\right\|_\infty : |b_j| \le 1 \text{ for } 1 \le j \le n\right\}$$

$$\le K\|A\|.$$

18.4 Another proof, using Paley's inequality

It is of interest to give a direct proof of Grothendieck's Theorem for operators in $L(l_1, H)$, and this was done by Pełczyński and Wojtaszczyk [Pel 77]. It is essentially a complex proof, but the real version then follows from it. It uses an interesting inequality of Paley.

Recall that if $1 \le p < \infty$ then

$$H^p = \left\{ f: f \text{ analytic on } \mathbf{D}, \|f\|_p = \sup_{0 \le r < 1}\left(\frac{1}{2\pi}\int_0^{2\pi} |f(re^{i\theta})|^p \, d\theta\right)^{1/p} < \infty \right\},$$

and that

$$A(D) = \{f \in C(\bar{\mathbf{D}}): f \text{ analytic on } \mathbf{D}\}.$$

We give $A(D)$ the supremum norm. If $f \in H^p$ or $A(D)$ we can write $f(z) = \sum_{n=0}^{\infty} \hat{f}_n z^n$, for $z \in D$. If $f \in H^2$, then $\|f\|_{H^2} = (\sum_{n=0}^{\infty} |\hat{f}_n|^2)^{1/2}$.

Theorem 18.4.1 (Paley's inequality) *If $f \in H^1$ then $(\sum_{k=0}^{\infty} |\hat{f}_{2^k - 1}|^2)^{1/2}$*
$\le 2\|f\|_1$.

Proof We use the fact that if $f \in H^1$ then we can write $f = bg$, where b is a Blaschke product (a bounded function on $\mathbf{D}$ for which $\lim_{r \nearrow 1} |f(re^{i\theta})| = 1$ for almost all θ), and g is a function in H^1 with no zeros in $\mathbf{D}$. From this it follows that g has a square root in H^2: there exists $h \in H^2$ with $h^2 = g$. Thus, setting $k = bh$, we can write $f = hk$, where $h, k \in H^2$ and $\|f\|_1 = \|h\|_2 \|k\|_2$. For all this, see [Dur 70].

Thus $\hat{f}_n = \sum_{j=0}^{n} \hat{h}_j \hat{k}_{n-j}$, and so

$$
\sum_{k=0}^{\infty} |\hat{f}_{2^k-1}|^2 \leq \sum_{k=0}^{\infty} \left(\sum_{j=0}^{2^k-1} |\hat{h}_j||\hat{k}_{2^k-1-j}| \right)^2
$$

$$
= \sum_{k=0}^{\infty} \left(\sum_{j=0}^{2^{k-1}-1} |\hat{h}_j||\hat{k}_{2^k-1-j}| + \sum_{j=0}^{2^{k-1}-1} |\hat{h}_{2^k-1-j}||\hat{k}_j| \right)^2
$$

$$
\leq 2 \sum_{k=0}^{\infty} \left(\left(\sum_{j=0}^{2^{k-1}-1} |\hat{h}_j||\hat{k}_{2^k-1-j}| \right)^2 + \left(\sum_{j=0}^{2^{k-1}-1} |\hat{h}_{2^k-1-j}||\hat{k}_j| \right)^2 \right).
$$

By the Cauchy–Schwarz inequality,

$$
\left(\sum_{j=0}^{2^{k-1}-1} |\hat{h}_j||\hat{k}_{2^k-1-j}| \right)^2 \leq \left(\sum_{j=0}^{2^{k-1}-1} |\hat{h}_j|^2 \right) \left(\sum_{j=2^{k-1}}^{2^k-1} |\hat{k}_j|^2 \right)
$$

$$
\leq \|h\|_2^2 \left(\sum_{j=2^{k-1}}^{2^k-1} |\hat{k}_j|^2 \right),
$$

so that

$$
\sum_{k=0}^{\infty} \left(\sum_{j=0}^{2^{k-1}-1} |\hat{h}_j||\hat{k}_{2^k-1-j}| \right)^2 \leq \|h\|_2^2 \|k\|_2^2 ;
$$

similarly

$$
\sum_{k=0}^{\infty} \left(\sum_{j=0}^{2^{k-1}-1} |\hat{h}_{2^k-1-j}||\hat{k}_j| \right)^2 \leq \|h\|_2^2 \|k\|_2^2,
$$

and so

$$
\sum_{k=0}^{\infty} |\hat{f}_{2^k-1}|^2 \leq 4 \|h\|_2^2 \|k\|_2^2.
$$

$\square$

We also need the following surjection theorem.

Theorem 18.4.2 *If $y \in l_2$, there exists $f \in A(\mathbf{D})$ with $\|f\|_\infty \leq \sqrt{e}\,\|y\|_2$ such that $\hat{f}_{2^k-1} = y_k$ for $k = 0, 1, \ldots$.*

Proof We follow the proof of Fournier [Fou 74]. By homogeneity, it is enough to prove the result when $\|y\|_2 = 1$. Note that

$$\log\left(\prod_{k=0}^{\infty}(1+|y_k|^2)\right) = \sum_{k=0}^{\infty}\log(1+|y_k|^2) < \sum_{k=0}^{\infty}|y_k|^2 = 1,$$

so that $\prod_{k=0}^{\infty}(1+|y_k|^2) < e$.

First we consider sequences of finite support. We show that if $y_k = 0$ for $k \geq K$ then there exists $f(z) = \sum_{j=0}^{2^K-1}\hat{f}_jz^j$ with $\hat{f}_{2^k-1} = y_k$ for $k = 0, 1, \ldots, K$ and $\|f\|_{\infty}^2 \leq \prod_{k=0}^{K}(1+|y_k|^2)$. Let us set $f^{(0)}(z) = y_0$ and $g^{(0)}(z) = 1$, and define $f^{(1)}, \ldots, f^{(K)}$ and $g^{(1)}, \ldots, g^{(K)}$ recursively by setting

$$\begin{bmatrix} f^{(k)}(z) \\ g^{(k)}(z) \end{bmatrix} = \begin{bmatrix} 1 & y_kz^{2^k-1} \\ -y_kz^{-(2^k-1)} & 1 \end{bmatrix}\begin{bmatrix} f^{(k-1)}(z) \\ g^{(k-1)}(z) \end{bmatrix} = M_k\begin{bmatrix} f^{(k-1)}(z) \\ g^{(k-1)}(z) \end{bmatrix},$$

for $z \neq 0$.

Now if $|z| = 1$ then $M_kM_k^* = (1+|y_k|^2)I_2$, so that

$$|f^{(k)}(z)|^2 + |g^{(k)}(z)|^2 = (1+|y_k|^2)(|f^{(k-1)}(z)|^2 + |g^{(k-1)}(z)|)^2 = \prod_{j=0}^{k}(1+|y_j|^2).$$

It also follows inductively that $f^{(k)}$ is a polynomial of degree $2^k - 1$ in z, and $g^{(k)}$ is a polynomial of degree $2^k - 1$ in z^{-1}. Thus $f^{(k)} \in A(\mathbf{D})$ and $\|f^{(k)}\|_{\infty}^2 \leq \prod_{j=0}^{k}(1+|y_j|^2)$. Further, $f^{(k)} = f^{(k-1)} + y_kz^{2^k-1}g^{(k-1)}$, and $y_kz^{2^k-1}g^{(k-1)}$ is a polynomial in z whose non-zero coefficients lie in the range $[2^{k-1}, 2^k - 1]$. Thus there is no cancellation of coefficients in the iteration, and so $\widehat{(f^{(k)})}_{2^j-1} = y_j$ for $0 \leq j \leq k$. Thus the result is established for sequences of finite support.

Now suppose that $y \in l_2$ and that $\|y\| = 1$. Let $\prod_{k=0}^{\infty}(1+|y_k|^2) = \alpha^2e$, so that $0 < \alpha < 1$. There exists an increasing sequence $(k_j)_{j=0}^{\infty}$ of indices such that $\sum_{n=k_j+1}^{\infty}|y_n|^2 < (1-\alpha)^2/4^{j+1}$. Let

$$a^{(0)} = \sum_{i=0}^{k_0}y_ie_i \text{ and } a^{(j)} = \sum_{i=k_{j-1}+1}^{k_j}y_ie_i \text{ for } j > 0.$$

Then there exist polynomials f_j with $(\hat{f}_j)_{2^k-1} = a_k^{(j)}$ for all k, and with

$$\|f_0\|_{\infty} \leq (\prod_{k=0}^{k_0}(1+|y_k|^2))^{1/2} \leq \alpha\sqrt{e},$$

$$\|f_j\|_{\infty} \leq (1-\alpha)\sqrt{e}/2^j \text{ for } j > 0.$$

Then $\sum_{j=0}^{\infty} f_j$ converges in norm in $A(D)$ to f say, with $\|f\|_{\infty} \leq \sqrt{e}$, and $\hat{f}_{2^k-1} = y_k$ for $0 \leq k < \infty$. $\qquad \square$

We combine these results to prove Grothendieck's theorem for l_1.

Theorem 18.4.3 *If $T \in L(l_1, l_2)$ then T is absolutely summing and $\pi_1(T) \leq 2\sqrt{e}\,\|T\|$.*

Proof Let $T(e_i) = h^{(i)}$. For each i, there exists $f^{(i)} \in A(D)$ with $\left\|f^{(i)}\right\|_{\infty} \leq \sqrt{e}\,\left\|h^{(i)}\right\| \leq \sqrt{e}\,\|T\|$ such that $\widehat{(f^{(i)})}_{2^k-1} = h_k^{(i)}$, for each k. Let $S : l_1 \to A(\mathbf{D})$ be defined by $S(x) = \sum_{i=0}^{\infty} x_i f^{(i)}$, let J be the inclusion $A(\mathbf{D}) \to H^1$, and let $P : H^1 \to l_2$ be defined by $P(f)_k = \hat{f}_{2^k-1}$, so that $T = PJS$. Then $\|S\| \leq \sqrt{e}\,\|T\|$, $\pi_1(J) = 1$, by Pietsch's domination theorem, and $\|P\| \leq 2$, by Paley's inequality. Thus $T = PJS$ is absolutely summing, and $\pi_1(T) \leq \|P\|\,\pi_1(J)\,\|S\| \leq 2\sqrt{e}\,\|T\|$. $\qquad \square$

18.5 The little Grothendieck theorem

We can extend Grothendieck's theorem to spaces of measures. We need the following elementary result.

Lemma 18.5.1 *Suppose that K is a compact Hausdorff space and that $\phi_1, \ldots, \phi_n \in C(K)^*$. Then there exists a probability measure $\mathbf{P}$ on the Baire sets of K and $f_1, \ldots, f_n$ in $L^1(\mathbf{P})$ such that $\phi_j = f_j d\mathbf{P}$ for each j.*

Proof By the Riesz representation theorem, for each j there exists a probability measure $\mathbf{P}_j$ on the Baire sets of K and a measurable h_j with $|h_j| = \|\phi_j\|^*$ everywhere, such that $\Phi_j = h_j\, d\mathbf{P}_j$. Let $\mathbf{P} = (1/n)\sum_{j=1}^{n} \mathbf{P}_j$. Then $\mathbf{P}$ is a probability measure P_j on the Baire sets of K, and each P_j is absolutely continuous with respect to $\mathbf{P}$. Thus for each j there exists $g_j \geq 0$ with $\int_K g_j\, d\mathbf{P} = 1$ such that $\mathbf{P}_j = g_j\, d\mathbf{P}$. Take $f_j = h_j g_j$. $\qquad \square$

Theorem 18.5.1 *Suppose that K is a compact Hausdorff space. If $T \in L(C(K)^*, H)$, where H is a Hilbert space, then T is absolutely summing and $\pi_1(T) \leq K_G\,\|T\|$.*

Proof Suppose that $\phi_i, \ldots, \phi_n \in C(K)^*$. By the lemma, there exist a probability measure $\mathbf{P}$ and $f_1, \ldots, f_n \in L^1(\mathbf{P})$ such that $\phi_j = f_j d\mathbf{P}$ for $1 \leq j \leq n$. We can consider $L^1(\mathbf{P})$ as a subspace of $C(K)^*$. T maps $L^1(\mathbf{P})$

into H, and

$$\sum_{j=1}^{n} \|T(\phi_j)\| \leq K_G \sup\left\{ \left\| \sum_{j=1}^{n} b_j f_j \right\|_1 : |b_j| \leq 1 \right\}$$

$$= K_G \sup\left\{ \left\| \sum_{j=1}^{n} b_j \phi_j \right\|^* : |b_j| \leq 1 \right\}.$$

$\square$

Corollary 18.5.1 (The little Grothendieck theorem) *If $T \in L(C(K), H)$, where K is a compact Hausdorff space and H is a Hilbert space, then $T \in \Pi_2(C(K), H)$ and $\pi_2(T) \leq K_G \|T\|$.*

Proof We use Proposition 16.3.2. Suppose that $S \in L(l_2^N, C(K))$. Then $S^* \in L(C(K)^*, l_2^N)$. Thus $\pi_1(S^*) \leq K_G \|S^*\|$, and so $\pi_2(S^*T^*) \leq \pi_1(S^*T^*) \leq K_G \|S^*\| \|T^*\|$. But $\pi_2(S^*T^*)$ is the Hilbert–Schmidt norm of S^*T^*, and so $\pi_2(S^*T^*) = \pi_2(TS)$. Thus $(\sum_{n=1}^{N} \|TS(e_n)\|^2)^{1/2} \leq K_G^2 \|T\| \|S\|$, so that $T \in \Pi_2(C(K), H)$ and $\pi_2(T) \leq K_G \|T\|$. $\square$

We also have a dual version of the little Grothendieck theorem.

Theorem 18.5.2 *It $T \in L(L^1(\Omega, \Sigma, \mu), H)$, where H is a Hilbert space, then T is 2-summing, and $\pi_2(T) \leq K_G \|T\|$.*

Proof By Theorem 16.3.1, it is enough to consider simple functions in $L^1(\Omega, \Sigma, \mu)$, and so it is enough to consider $T \in L(l_1^d, H)$. We use Proposition 16.3.2. Suppose that $S \in L(l_2^N, l_1^d)$. Then $S^* \in L(l_\infty^d, l_2^N)$, and so $\pi_2(S^*) \leq K_G \|S^*\|$, by the little Grothendieck theorem. Then $\pi_2(S^*T^*) \leq K_G \|S^*\| \|T^*\|$. But $\pi_2(S^*T^*)$ is the Hilbert–Schmidt norm of S^*T^*, and so $\pi_2(S^*T^*) = \pi_2(TS)$. Thus

$$\left(\sum_{n=1}^{N} \|TS(e_n)\|^2 \right)^{1/2} \leq K_G \|S\| \|T\| \sup\left\{ \left(\sum_{n=1}^{N} |\langle e_n, h \rangle|^2 \right)^{1/2} : \|h\| \leq 1 \right\}$$

$$= K_G \|S\| \|T\|.$$

Thus T is 2-summing, and $\pi_2(T) \leq K_G \|T\|$. $\square$

18.6 Type and cotype

In fact, we can obtain a better constant in the little Grothendieck theorem, and can extend to the result to more general operators. In order to do this, we introduce the notions of type and cotype. These involve Bernoulli sequences of random variables: for the rest of this chapter, (ϵ_n) will denote such a sequence.

Let us begin, by considering the parallelogram law. This says that if $x_1, \ldots, x_n$ are vectors in a Hilbert space H then

$$\mathbf{E}\left(\left\|\sum_{j=1}^{n} \epsilon_j x_j\right\|^2\right) = \sum_{j=1}^{n} \|x_j\|_H^2 .$$

We deconstruct this equation; we split it into two inequalities, we change an index, we introduce constants, and we consider linear operators.

Suppose that $(E, \|.\|_E)$ and $(F, \|.\|_F)$ are Banach spaces, that $T \in L(E, F)$ and that $1 \leq p < \infty$. We say that T is of *type* p if there is a constant C such that if $x_1, \ldots, x_n$ are vectors in E then

$$\left(\mathbf{E}\left(\left\|\sum_{j=1}^{n} \epsilon_j T(x_j)\right\|_F^2\right)\right)^{1/2} \leq C \left(\sum_{j=1}^{n} \|x_j\|_E^p\right)^{1/p}.$$

The smallest possible constant C is denoted by $T_p(T)$, and is called the *type p constant* of T. Similarly, we say that T is of *cotype* p if there is a constant C such that if $x_1, \ldots, x_n$ are vectors in E then

$$\left(\sum_{j=1}^{n} \|T(x_j)\|_F^p\right)^{1/p} \leq C \left(\mathbf{E}\left(\left\|\sum_{j=1}^{n} \epsilon_j(x_j)\right\|_E^2\right)\right)^{1/2}.$$

The smallest possible constant C is denoted by $C_p(T)$, and is called the *cotype p constant* of T.

It follows from the parallelogram law that if T is of type p, for $p > 2$, or cotype p, for $p < 2$, then $T = 0$. If T is of type p then T is of type q, for $1 \leq q < p$, and $T_q(T) \leq T_p(T)$; if T is of cotype p then T is of cotype q, for $p < q < \infty$, and $C_q(T) \leq C_p(T)$. Every Banach space is of type 1. By the Kahane inequalities, we can replace

$$\left(\mathbf{E}\left(\left\|\sum_{j=1}^{n} \epsilon_j T(x_j)\right\|_F^2\right)\right)^{1/2} \quad \text{by} \quad \left(\mathbf{E}\left(\left\|\sum_{j=1}^{n} \epsilon_j T(x_j)\right\|_F^q\right)\right)^{1/q}$$

in the definition, for any $1 < q < \infty$, with a corresponding change of constant.

Proposition 18.6.1 *If $T \in L(E, F)$ and T is of type p, then $T^* \in L(F^*, E^*)$ is of cotype p', and $C_{p'}(T^*) \le T_p(T)$.*

Proof Suppose that $\phi_1, \ldots, \phi_n$ are vectors in F^* and $x_1, \ldots, x_n$ are vectors in E. Then

$$|\sum_{j=1}^{n} T^*(\phi_j)(x_j)| = |\sum_{j=1}^{n} \phi_j(T(x_j))|$$

$$= |\mathbf{E}\left(\left(\sum_{j=1}^{n} \epsilon_j \phi_j\right)\left(\sum_{j=1}^{n} \epsilon_j T(x_j)\right)\right)|$$

$$\le \left(\mathbf{E}\left(\left\|\sum_{j=1}^{n} \epsilon_j \phi_j\right\|^2\right)\right)^{1/2}\left(\mathbf{E}\left(\left\|\sum_{j=1}^{n} \epsilon_j T(x_j)\right\|^2\right)\right)^{1/2}$$

$$\le \left(\mathbf{E}\left(\left\|\sum_{j=1}^{n} \epsilon_j \phi_j\right\|^2\right)\right)^{1/2} T_p(T)(\sum_{j=1}^{n} \|x_j\|^p)^{1/p}.$$

But

$$\left(\sum_{j=1}^{n} \|T^*(\phi_j)\|^{p'}\right)^{1/p'} = \sup\left\{|\sum_{j=1}^{n} T^*(\phi_j)(x_j)|: \left(\sum_{j=1}^{n} \|x_j\|^p\right)^{1/p} \le 1\right\},$$

and so

$$\left(\sum_{j=1}^{n} \|T^*(\phi_j)\|^{p'}\right)^{1/p'} \le T_p(T)\left(\mathbf{E}\left(\left\|\sum_{j=1}^{n} \epsilon_j \phi_j\right\|^2\right)\right)^{1/2}.$$

$\square$

Corollary 18.6.1 *If $T \in L(E, F)$ and T^* is of type p, then T is of cotype p', and $C_{p'}(T) \le T_p(T^*)$.*

The converse of this proposition is not true (Exercise 18.3).

An important special case occurs when we consider the identity operator I_E on a Banach space E. If I_E is of type p (cotype p), we say that E is of type p (cotype p), and we write $T_p(E)$ $(C_p(E))$ for $T_p(I_E)$ $(C_p(I_E))$, and call it the *type p constant (cotype p constant)* of E. Thus the parallelogram law states that a Hilbert space H is of type 2 and cotype 2, and $T_2(H) = C_2(H) = 1$.

18.7 Gaussian type and cotype

It is sometimes helpful to work with sequences of Gaussian random variables, rather than with Bernoulli sequences. Recall that a standard Gaussian random variable is, in the real case, a real-valued Gaussian random variable with mean 0 and variance 1, so that its density function on the real line is $(1/\sqrt{2\pi})e^{-x^2/2}$, and in the complex case is a rotationally invariant, complex-valued Gaussian random variable with mean 0 and variance 1, so that its density function on the complex plane is $(1/\pi)e^{-|z|^2}$. For the rest of this chapter, (g_n) will denote an independent sequence of standard Gaussian random variables, real or complex. The theories are essentially the same in the real and complex cases, but with different constants. For example, for $0 < p < \infty$ we define $\gamma_p = \|g\|_p$, where g is a standard Gaussian random variable. Then in the real case, $\gamma_1 = \sqrt{2/\pi}$, $\gamma_2 = 1$ and $\gamma_4 = 3^{1/4}$, while, in the complex case, $\gamma_1 = \sqrt{\pi}/2, \gamma_2 = 1$ and $\gamma_4 = 2^{1/4}$.

If in the definitions of type and cotype we replace the Bernoulli sequence (ϵ_n) by (g_n), we obtain the definitions of Gaussian type and cotype. We denote the corresponding constants by T_p^γ and C_p^γ.

Proposition 18.7.1 *If $T \in L(E, F)$ is of type 2 (cotype 2) then it is of Gaussian type 2 (Gaussian cotype 2), and $T_2^\gamma(T) \le T_2(T)$ $(C_2^\gamma(T) \le C_2(T))$.*

Proof Let us prove this for cotype: the proof for type is just the same. Let $x_1, \ldots, x_n$ be vectors in E. Suppose that the sequence (g_n) is defined on Ω and the sequence (ϵ_n) on Ω'. Then for fixed $\omega \in \Omega$,

$$\sum_{j=1}^n |g_j(\omega)|^2 \, \|T(x_j)\|_F^2 \le C_2(T) \mathbf{E}_{\Omega'} \left(\left\| \sum_{j=1}^n \epsilon_j g_j(\omega) x_j \right\|_E^2 \right).$$

Taking expectations over Ω, and using the symmetry of the Gaussian sequence, we find that

$$\sum_{j=1}^{n}\|T(x_j)\|_F^2 \leq C_2(T)\mathbf{E}_\Omega\left(\mathbf{E}_{\Omega'}\left(\left\|\sum_{j=1}^{n}\epsilon_j g_j x_j\right\|_E^2\right)\right)$$

$$= C_2(T)\mathbf{E}_\Omega\left(\left\|\sum_{j=1}^{n}g_j x_j\right\|_E^2\right).$$

$\square$

The next theorem shows the virtue of considering Gaussian random variables.

Theorem 18.7.1 (Kwapień's theorem) *Suppose that $T \in L(E,F)$ and $S \in L(F,G)$. If T is of Gaussian type 2 and S is of Gaussian cotype 2 then $ST \in \Gamma_2(E,F)$, and $\gamma_2(ST) \leq T_2^\gamma(T)C_2^\gamma(S)$.*

Proof We use Theorem 16.13.2. Suppose that $y_1,\ldots,y_n \in E$ and that $U = (u_{ij})$ is unitary (or orthogonal, in the real case). Let $h_j = \sum_{i=1}^{n}g_i u_{ij}$. Then $h_1,\ldots,h_n$ are independent standard Gaussian random variables. Thus

$$\left(\sum_{i=1}^{n}\left\|ST(\sum_{j=1}^{n}u_{ij}y_j)\right\|^2\right)^{1/2} \leq C_2^\gamma(S)\left(\mathbf{E}\left(\left\|\sum_{i=1}^{n}g_i(\sum_{j=1}^{n}u_{ij}T(y_j))\right\|^2\right)\right)^{1/2}$$

$$= C_2^\gamma(S)\left(\mathbf{E}\left(\left\|\sum_{j=1}^{n}h_j T(y_j)\right\|^2\right)\right)^{1/2}$$

$$\leq T_2^\gamma(T)C_2^\gamma(S)\left(\sum_{j=1}^{n}\|y_j\|^2\right)^{1/2}.$$

$\square$

Corollary 18.7.1 *A Banach space $(E,\|.\|_E)$ is isomorphic to a Hilbert space if and only if it is of type 2 and cotype 2, and if and only if it is of Gaussian type 2 and Gaussian cotype 2.*

18.8 Type and cotype of L^p spaces

Let us give some examples.

Theorem 18.8.1 *Suppose that (Ω, Σ, μ) is a measure space.*
(i) If $1 \leq p \leq 2$ then $L^p(\Omega, \Sigma, \mu)$ is of type p and cotype 2.
(ii) If $2 \leq p < \infty$ then $L^p(\Omega, \Sigma, \mu)$ is of type 2 and cotype p.

Proof (i) Suppose that $f_1, \ldots, f_n$ are in $L^p(\Omega, \Sigma, \mu)$. To prove the cotype inequality, we use Khintchine's inequality and Corollary 5.4.2.

$$
\left(\mathbf{E} \left(\left\| \sum_{j=1}^n \epsilon_j f_j \right\|_p^2 \right) \right)^{1/2} \geq \left(\mathbf{E} \left(\left\| \sum_{j=1}^n \epsilon_j f_j \right\|_p^p \right) \right)^{1/p}
$$

$$
= \left(\mathbf{E} \left(\int_\Omega | \sum_{j=1}^n \epsilon_j f_j(\omega) |^p \, d\mu(\omega) \right) \right)^{1/p}
$$

$$
= \left(\int_\Omega \mathbf{E} \left(| \sum_{j=1}^n \epsilon_j f_j(\omega) |^p \right) d\mu(\omega) \right)^{1/p}
$$

$$
\geq A_p^{-1} \left(\int_\Omega (\sum_{j=1}^n |f_j(\omega)|^2)^{p/2} \, d\mu(\omega) \right)^{1/p}
$$

$$
\geq A_p^{-1} \left(\sum_{j=1}^n \left(\int_\Omega (|f_j(\omega)|^p) \, d\mu(\omega) \right)^{2/p} \right)^{1/2}
$$

$$
= A_p^{-1} \left(\sum_{j=1}^n \|f_j\|_p^2 \right)^{1/2}.
$$

Thus $L^p(\Omega, \Sigma, \mu)$ is of cotype 2.

To prove the type inequality, we use the Kahane inequality.

$$
\left(\mathbf{E} \left(\left\| \sum_{j=1}^n \epsilon_j f_j \right\|_p^2 \right) \right)^{1/2} \leq K_{p,2} \left(\mathbf{E} \left(\left\| \sum_{j=1}^n \epsilon_j f_j \right\|_p^p \right) \right)^{1/p}
$$

$$
= K_{p,2} \left(\mathbf{E} \left(\int_\Omega \left| \sum_{j=1}^n \epsilon_j f_j(\omega) \right|^p d\mu(\omega) \right) \right)^{1/p}
$$

$$= K_{p,2} \left(\int_\Omega \mathbf{E} \left(\left| \sum_{j=1}^n \epsilon_j f_j(\omega) \right|^p \right) d\mu(\omega) \right)^{1/p}$$

$$\leq K_{p,2} \left(\int_\Omega \left(\sum_{j=1}^n |f_j(\omega)|^2 \right)^{p/2} d\mu(\omega) \right)^{1/p}$$

$$\leq K_{p,2} \left(\sum_{j=1}^n \left(\int_\Omega |f_j(\omega)|^p \, d\mu(\omega) \right) \right)^{1/p}$$

$$= K_{p,2} \left(\sum_{j=1}^n \|f_j\|^p \right)^{1/p}_p .$$

Thus $L^p(\Omega, \Sigma, \mu)$ is of type p.

(ii) Since $L^{p'}(\Omega, \Sigma, \mu)$ is of type p', $L^p(\Omega, \Sigma, \mu)$ is of type p, by Proposition 18.6.1. Suppose that $f_1, \ldots, f_n$ are in $L^p(\Omega, \Sigma, \mu)$. To prove the type inequality, we use Khintchine's inequality and Corollary 5.4.2.

$$\left(\mathbf{E} \left(\left\| \sum_{j=1}^n \epsilon_j f_j \right\|_p^2 \right) \right)^{1/2} \leq \left(\mathbf{E} \left(\left\| \sum_{j=1}^n \epsilon_j f_j \right\|_p^p \right) \right)^{1/p}$$

$$= \left(\mathbf{E} \left(\int_\Omega \left| \sum_{j=1}^n \epsilon_j f_j(\omega) \right|^p d\mu(\omega) \right) \right)^{1/p}$$

$$= \left(\int_\Omega \mathbf{E} \left(\left| \sum_{j=1}^n \epsilon_j f_j(\omega) \right|^p \right) d\mu(\omega) \right)^{1/p}$$

$$\leq B_p \left(\int_\Omega \left(\sum_{j=1}^n |f_j(\omega)|^2 \right)^{p/2} d\mu(\omega) \right)^{1/p}$$

$$\leq B_p \left(\sum_{j=1}^n \left(\int_\Omega |f_j(\omega)|^p \, d\mu(\omega) \right)^{2/p} \right)^{1/2}$$

$$= B_p \left(\sum_{j=1}^n \|f_j\|_p^2 \right)^{1/2} .$$

Thus $L^p(\Omega, \Sigma, \mu)$ is of type 2. $\qquad\square$

18.9 The little Grothendieck theorem revisited

We now give the first generalization of the little Grothendieck theorem.

Theorem 18.9.1 *Suppose that* $(E, \|.\|_E)$ *is a Banach space whose dual* E^* *is of Gaussian type* p*, where* $1 < p \le 2$*. If* $T \in L(C(K), E)$*, then* $T \in \Pi_{p',2}(C(K), E)$*, and* $\pi_{p',2}(T) \le \gamma_1^{-1} T_p^\gamma(E^*) \|T\|$.

Proof Suppose that $f_1, \ldots, f_n, \in C(K)$. We must show that

$$\left(\sum_{j=1}^n \|T(f_j)\|^{p'} \right)^{1/p'} \le C \sup_{k \in K} \left(\sum_{j=1}^n |f_j(k)|^2 \right)^{1/2},$$

where $C = \gamma_1^{-1} T_p^\gamma(E^*)$.

For $f = (f_1, \ldots, f_n) \in C(K; l_2^n)$, let $R(f) = (T(f_j))_{j=1}^n \in l_n^{p'}(E)$. Then we need to show that $\|R\| \le C \|T\|$. To do this, let us consider the dual mapping $R^*: l_n^p(E^*) \to C(K; l_2^n)^*$. If $\Phi = (\phi_j)_{j=1}^n \in C(K; l_2^n)^*$, then $R^*(\Phi) = (T^*(\phi_1), \ldots, T^*(\phi_n))$. By Lemma 18.5.1, there exist a Baire probability measure $\mathbf{P}$ on K and $w_1, \ldots, w_n \in L^1(\mathbf{P})$ such that $T^*(\phi_j) = w_j \, d\mathbf{P}$ for $1 \le j \le n$. Then

$$\|R^*(\Phi)\|_{M(K:l_2^n)} = \int_K \left(\sum_{j=1}^n |w_j(k)|^2 \right)^{1/2} d\mathbf{P}(k)$$

$$= \int_K \left(\mathbf{E} \left(\left| \sum_{j=1}^n g_j w_j(k) \right|^2 \right) \right)^{1/2} d\mathbf{P}(k)$$

$$= \gamma_1^{-1} \int_K \mathbf{E} \left(\left| \sum_{j=1}^n g_j w_j(k) \right| \right) d\mathbf{P}(k)$$

$$= \gamma_1^{-1} \mathbf{E} \left(\int_K \left| \sum_{j=1}^n g_j w_j(k) \right| d\mathbf{P}(k) \right)$$

$$= \gamma_1^{-1} \mathbf{E} \left(\left\| T^* (\sum_{j=1}^n g_j \phi_j) \right\|_{E^*} \right)$$

$$\leq \gamma_1^{-1} \|T^*\| \, \mathbf{E}\left(\left\| \sum_{j=1}^n g_j \phi_j \right\|_{E^*} \right)$$

$$\leq \gamma_1^{-1} \|T^*\| \left(\mathbf{E}\left(\left\| \sum_{j=1}^n g_j \phi_j \right\|_{E^*}^2 \right) \right)^{1/2}$$

$$\leq \gamma_1^{-1} \|T^*\| \, T_p^\gamma(E^*) \left(\sum_{j=1}^n \|\phi_j\|_{E^*}^p \right)^{1/p}$$

$$\leq \gamma_1^{-1} \|T^*\| \, T_p^\gamma(E^*) \, \|\Phi\|_{l_n^p(E^*)}.$$

$\square$

This gives the best constant in the little Grothendieck theorem.

Proposition 18.9.1 *The best constant in the little Grothendieck theorem is γ_1^{-1} ($\sqrt{\pi/2}$ in the real case, $2/\sqrt{\pi}$ in the complex case).*

Proof Theorem 18.9.1 shows that γ_1^{-1} is a suitable upper bound. Let $\mathbf{P}$ be standard Gaussian measure on $\mathbf{R}^d$ (or $\mathbf{C}^d$), so that if we set $g_j(x) = x_j$ then $g_1, \ldots, g_d$ are independent standard Gaussian random variables. Let K be the one-point compactification of $\mathbf{R}^d$ (or $\mathbf{C}^d$), and extend $\mathbf{P}$ to a probability measure on K by setting $\mathbf{P}(\{\infty\}) = 0$.

Now let $G\colon C(K) \to l_2^d$ be defined by $G(f) = (\mathbf{E}(f\bar{g}_j))_{j=1}^d$. Then

$$\|G(f)\| = \left(\sum_{j=1}^d |\mathbf{E}(f\bar{g}_j)|^2 \right)^{1/2}$$

$$= \sup\left\{ \left| \mathbf{E}\left(f\left(\sum_{j=1}^d \alpha_j \bar{g}_j \right) \right) \right| : \sum_{j=1}^d |\alpha_j|^2 \leq 1 \right\}$$

$$\leq \gamma_1 \|f\|_\infty,$$

so that $\|G\| \leq \gamma_1$.

On the other hand, if $f = (f_1, \ldots, f_d) \in C(K; l_2^d)$, set $R(f)_i = (G(f_i)) \in l_2^d(l_2^d)$, for $1 \leq i \leq d$. Then

$$\|f\|_{C(K;l_2^d)} = \sup_{k \in K} \left(\sum_{i=1}^d |f_i(k)|^2 \right)^{1/2} \quad \text{and} \quad \|R(f)\| = \left(\sum_{i=1}^d \|G(f_i)\|^2 \right)^{1/2},$$

so that

$$\|R(f)\| \leq \pi_2(G) \sup_{k \in K} \Big(\sum_{i=1}^{d} |f_i(k)|^2 \Big)^{1/2} = \pi_2(G) \, \|f\|_{C(K;l_2^d)},$$

and $\pi_2(G) \geq \|R\|$.

We consider R^*. If $e = (e_1, \dots, e_d)$, then $R^*(e) = (\bar{g}_1, \dots, \bar{g}_d)$. Then $\|R^*(e)\| = \mathbf{E}(\chi)$, where $\chi = (\sum_{j=1}^{d} |g_j|^2)^{1/2}$. By Littlewood's inequality, $\sqrt{d} = \|\chi\|_2 \leq \|\chi\|_1^{1/3} \|\chi\|_4^{2/3}$. But

$$\|\chi\|_4^4 = \mathbf{E}\left(\Big(\sum_{j=1}^{d} |g_j|^2 \Big)^2 \right)$$

$$= \sum_{j=1}^{d} \mathbf{E}(|g_j|^4) + \sum_{j \neq k} \mathbf{E}(|g_j|^2 |g_k|^2) = d\gamma_4^4 + d(d-1).$$

Thus

$$\|\chi\|_1^2 \geq d^3 / \|g\|_4^4 = d/(1 + (\gamma_4^4 - 1)/d),$$

so that, since $\|e\| = \sqrt{d}$,

$$\|R\|^2 = \|R^*\|^2 \geq 1/(1 + (\gamma_4^4 - 1)/d).$$

Consequently, $\pi_2(G) \geq \|G\|/(\gamma_1(1 + (\gamma_4^4 - 1)/d)^{1/2})$. Since d is arbitrary, the result follows. $\qquad\square$

18.10 More on cotype

Proposition 18.10.1 *Suppose that $(E, \|.\|_E)$ and $(F, \|.\|_F)$ are Banach spaces and that F has cotype p. If $T \in \Pi_q(E, F)$ for some $1 \leq q < \infty$ then $T \in \Pi_{p,2}$ and $\pi_{p,2}(T) \leq C_p(F)B_q\pi_q(T)$ (where B_q is the constant in Khintchine's inequality).*

Proof Let $j : E \to C(K)$ be an isometric embedding. By Pietsch's domination theorem, there exists a probability measure μ on K such that

$$\|T(x)\|_F \leq \pi_q(T) \left(\int_{C(K)} |j(x)|^q \, d\mu \right)^{1/q} \quad \text{for } x \in E.$$

If $x_1, \ldots, x_N \in E$, then, using Fubini's theorem and Khintchine's inequality,

$$
\left(\sum_{n=1}^{N} \|T(x_n)\|_F^p \right)^{1/p} \leq C_p(F) \left(\mathbf{E} \left(\left\| \sum_{n=1}^{N} \epsilon_n T(x_n) \right\|_F^2 \right) \right)^{1/2}
$$

$$
\leq C_p(F) \left(\mathbf{E} \left(\left\| T \left(\sum_{n=1}^{N} \epsilon_n x_n \right) \right\|_F^q \right) \right)^{1/q}
$$

$$
\leq C_p(F) \pi_q(T) \left(\mathbf{E} \left(\int_K \left| j \left(\sum_{n=1}^{N} \epsilon_n x_n \right) \right|^q d\mu \right) \right)^{1/q}
$$

$$
= C_p(F) \pi_q(T) \left(\int_K \mathbf{E} \left(\left| j \left(\sum_{n=1}^{N} \epsilon_n x_n \right) \right|^q d\mu \right) \right)^{1/q}
$$

$$
\leq C_p(F) B_q \pi_q(T) \left(\int_K \left(\sum_{n=1}^{N} |j(x_n)|^2 \right)^{q/2} d\mu \right)^{1/q}
$$

$$
\leq C_p(F) B_q \pi_q(T) \sup_{\|\phi\|_{E^*} \leq 1} \sum_{n=1}^{N} \left(|\phi(x_n)|^2 \right)^{1/2}.
$$

$\square$

We now have the following generalization of Theorem 16.11.1.

Corollary 18.10.1 *If $(F, \|.\|_F)$ has cotype 2 then $\Pi_q(E, F) = \Pi_2(E, F)$ for $2 \leq q < \infty$.*

We use this to give our final generalization of the little Grothendieck theorem. First we establish a useful result about $C(K)$ spaces.

Proposition 18.10.2 *Suppose that K is a compact Hausdorff space, that F is a finite-dimensional subspace of $C(K)$ and that $\epsilon > 0$. Then there exists a projection P of $C(K)$ onto a finite-dimensional subspace G, with $\|P\| = 1$, such that G is isometrically isomorphic to l_∞^d (where $d = \dim G$) and $\|P(f) - f\| \leq \epsilon \|f\|$ for $f \in F$.*

Proof The unit sphere S_F of F is compact, and so there exists a finite set $f_1, \ldots, f_n \in S_F$ such that if $f \in S_F$ then there exists j such that $\|f - f_j\| \leq \epsilon/3$. If $k \in K$, let $J(k) = (f_1(k), \ldots, f_n(k))$. J is a continuous mapping of K onto a compact subset $J(K)$ of $\mathbf{R}^n$ (or $\mathbf{C}^n$). There is therefore a maximal finite subset S of K such that $\|J(s) - J(t)\| \geq \epsilon/3$ for s, t distinct elements of S. We now set

$$h_s(k) = \max(1 - 3\,\|J(k) - J(s)\|\,/\epsilon, 0)$$

for $s \in S$, $k \in K$. Then $h_s(k) \geq 0$, $h_s(s) = 1$, and $h_s(t) = 0$ for $t \neq s$. Let $h(k) = \sum_{s \in S} h_s(k)$. Then, by the maximality of S, $h(k) > 0$ for each $k \in K$. We now set $g_s = h_s/h$. Then $g_s(k) \geq 0$, $g_s(s) = 1$, $g_s(t) = 0$ for $t \neq s$, and $\sum_{s \in S} g_s(k) = 1$. Let $G = \text{span } g_s$. If $g \in G$ then $\|g\| = \max\{|g(s)|: s \in S\}$, so that G is isometrically isomorphic to l_∞^d, where $d = \dim G$.

If $f \in C(K)$, let $P(f) = \sum_{s \in S} f(s)g_s$. Then P is a projection of $C(K)$ onto G, and $\|P\| = 1$. Further,

$$f_j(k) - P(f_j)(k) = \sum_{s \in S}(f_j(k) - f_j(s))g_s(k)$$

$$= \sum\{(f_j(k) - f_j(s))g_s(k): |f_j(k) - f_j(s)| \leq \epsilon/3\},$$

since $g_s(k) = 0$ if $\|f_j(k) - f_j(s)\| > \epsilon/3$. Thus $\|f_j - P(f_j)\| \leq \epsilon/3$. Finally if $f \in S_F$, there exists j such that $\|f - f_j\| \leq \epsilon/3$. Then

$$\|f - P(f)\| \leq \|f - f_j\| + \|f_j - P(f_j)\| + \|P(f_j) - P(f)\| \leq \epsilon.$$

$$\square$$

Theorem 18.10.1 *If* $(F, \|.\|_F)$ *has cotype 2 and* $T \in L(C(K), F)$ *then* T *is 2-summing, and* $\pi_2(T) \leq \sqrt{3}(C_2(F))^2\,\|T\|$.

Proof First, we consider the case where $K = \{1, \ldots, d\}$, so that $C(K) = l_\infty^d$. Then $T \in \Pi_2(C(K), F)$, and $\pi_4(T) \leq C_2(F)B_4\pi_2(T)$, by Proposition 18.10.1. But $\pi_2(T) \leq (\pi_4(T))^{1/2}\,\|T\|^{1/2}$, by Proposition 16.10.1. Combining these inequalities, we obtain the result.

Next, we consider the general case. Suppose that $f_1, \ldots, f_N \in C(K)$ and that $\epsilon > 0$. Let P and G be as in Proposition 18.10.2. Then

$$\left(\sum_{n=1}^N \|T(f_n)\|_F^2\right)^{1/2} \leq \left(\sum_{n=1}^N \|TP(f_n)\|_F^2\right)^{1/2} + \sqrt{N}\,\|T\|\,\epsilon$$

$$\leq \sqrt{3}(C_2(F))^2\,\|T\|\,\left(\sup_{s \in S}\sum_{n=1}^N |f_n(s)|^2\right)^{1/2} + \sqrt{N}\,\|T\|\,\epsilon,$$

by the finite-dimensional result. Since $\epsilon > 0$ is arbitrary, it follows that

$$\left(\sum_{n=1}^{N} \|T(f_n)\|_F^2 \right)^{1/2} \leq \sqrt{3}(C_2(F))^2 \left(\sup_{k \in K} \sum_{n=1}^{N} |f_n(k)|^2 \right)^{1/2}.$$

□

18.11 Notes and remarks

Littlewood was interested in bilinear forms, rather than linear operators: if B is a bilinear form on $l_\infty^m \times l_\infty^n$ then $B(x, y) = \sum_{i=1}^{m} \sum_{j=1}^{n} x_i b_{ij} y_j$, and $\|B\| = \sup\{B(x, y) : \|x\|_\infty \leq 1, \|y\|_\infty \leq 1\}$. Looking at things this way, it is natural to consider multilinear forms; these (and indeed forms of fractional dimension) are considered in [Ble 01].

Grothendieck's proof depends on the identity

$$\langle x, y \rangle = \cos\left(\frac{\pi}{2} \left(1 - \int_{S^{n-1}} \operatorname{sgn} \langle x, s \rangle \operatorname{sgn} \langle y, s \rangle \, d\lambda(s) \right) \right),$$

where x and y are unit vectors in $l_2^n(\mathbf{R})$ and λ is the rotation-invariant probability measure on the unit sphere S^{n-1}.

In fact, the converse of Proposition 18.7.1 is also true. See [DiJT 95].

Paley's inequality was generalized by Hardy and Littlewood. See [Dur 70] for details.

Kwapień's theorem shows that type and cotype interact to give results that correspond to Hilbert space results. Here is another result in the same direction, which we state without proof.

Theorem 18.11.1 (Maurey's extension theorem) *Suppose that E has type 2 and that F has cotype 2. If $T \in L(G, F)$, where G is a linear subspace of E. There exists $\tilde{T} \in L(E, F)$ which extends T: $\tilde{T}(x) = T(x)$ for $x \in G$.*

Note that, by Kwapień's theorem we may assume that F is a Hilbert space.

In this chapter, we have only scratched the surface of a large and important subject. Very readable accounts of this are given in [Pis 87] and [DiJT 95].

Exercises

18.1 How good a constant can you obtain from the proof of Theorem 18.2.1?

18.2 Suppose that $T \in L(E, F)$ is of cotype p. Show that $T \in \Pi_{p,1}(E, F)$. Compare this with Orlicz' theorem (Exercise 16.6).

18.3   Give an example of an operator T which has no type p for $1 < p \leq 2$, while T^* has cotype 2.

18.4   Suppose that $f(z) = \sum_{k=0}^{\infty} a_k z^k \in H^1$. Let $T(f) = (a_k/\sqrt{k})$. Use Hardy's inequality to show that $T(f) \in l_2$ and that $\|T(f)\|_2 \leq \sqrt{\pi}\,\|f\|_{H^1}$.

   Let $g_k(z) = z^k/\sqrt{k+1}\log(k+2)$. Show that $\sum_{k=0}^{\infty} g_k$ converges unconditionally in H^2, and in H^1. Show that T is not absolutely summing.

   H^1 can be considered as a subspace of $L^1(\mathbf{T})$. Compare this result with Grothendieck's theorem, and deduce that there is no continuous projection of $L^1(\mathbf{T})$ onto H^1.

18.5   Show that γ_1^{-1} is the best constant in Theorem 18.5.2.

References

[Alf 71] E.M. Alfsen (1971). *Compact Convex Sets and Boundary Integrals* (Springer-Verlag).

[Ané 00] C. Ané *et al.* (2000). *Sur les Inégalités de Sobolev Logarithmiques* (Soc. Math. de France, Panoramas et Synthèses, 10).

[App 96] D. Applebaum (1996). *Probability and Information* (Cambridge University Press).

[ArG 80] A. Araujo and E. Giné (1980). *The Central Limit Theorem for Real and Banach Valued Random Variables* (Wiley).

[Bak 94] D. Bakry (1994). L'hypercontractivité et son utilisation en théorie des semigroupes, *Lectures on Probability Theory. École d'Été de Saint Flour 1992* (Springer Lecture Notes in Mathematics, volume 1581).

[Ban 29] S. Banach (1929). Sur les fonctionelles linéaires II, *Studia Math.* **1** 223–239.

[Bar 95] R.G. Bartle (1995). *The Elements of Integration and Lebesgue Measure* (Wiley).

[BaGH 62] L.D. Baumert, S.W. Golomb and M. Hall Jr.(1962) Discovery of a Hadamard matrix of order 92, *Bull. Amer. Math. Soc.* **68** 237–238.

[Bec 75] W. Beckner (1975). Inequalities in Fourier analysis, *Ann. Math.* **102** 159–182.

[BeS 88] C. Bennett and R. Sharpley (1988). *Interpolation of Operators* (Academic Press).

[BeL 76] J. Bergh and J. Löfström (1976). *Interpolation Spaces* (Springer-Verlag).

[Bil 95] P. Billingsley (1995). *Probability and Measure* (Wiley).

[Ble 01] R. Blei (2001). *Analysis in Integer and Fractional Dimensions* (Cambridge University Press).

[BoS 38] H.F. Bohnenblust and A. Sobczyk (1938). Extensions of functionals on complex linear spaces, *Bull. Amer. Math. Soc.* **44** 91–93.

[Bol 90] B. Bollobás (1990). *Linear Analysis* (Cambridge University Press).

[Bon 71] A. Bonami (1971). Étude des coefficients de Fourier des fonctions de $L^p(G)$, *Ann. Inst. Fourier (Grenoble)* **20** 335–402.

[Bre 68] L. Breiman (1968). *Probability* (Addison Wesley).

[BuMV 87] P.S. Bullen, D.S. Mitrinović and P.M. Vasić (1987). *Means and their Inequalities* (Reidel, Boston).

[Cal 63] A.P. Calderón (1963). Intermediate spaces and interpolation, *Studia Math. Special Series* **1** 31–34.

[Cal 64] A.P. Calderón (1964). Intermediate spaces and interpolation, the complex method, *Studia Math.* **24** 113–190.

[Cal 66] A.P. Calderón (1966). Spaces between L^1 and L^∞ and the theorem of Marcinkiewicz, *Studia Math.* **26** 273–299.

[CaZ 52] A.P. Calderón and A. Zygmund (1952). On the existence of certain singular integrals, *Acta Math.* **82** 85–139.

[Car 23] T. Carleman (1923). Sur les fonctions quasi-analytiques, in *Proc. 5th Scand. Math. Cong.* (Helsinki).

[Cau 21] A. Cauchy (1821). *Cours d'Analyse de l'École Royale Polytechnique* (Debures frères, Paris).

[Cla 36] J.A. Clarkson (1936). Uniformly convex spaces, *Trans. Amer. Math. Soc.* **40** 396–414.

[DiJT 95] J. Diestel, H. Jarchow and A. Tonge (1995). *Absolutely Summing Operators* (Cambridge University Press).

[DiU 77] J. Diestel and J.J. Uhl Jr. (1977). *Vector Measures* (American Mathematical Society).

[Doo 40] J.L. Doob (1940). Regularity properties of certain families of chance variables, *Trans. Amer. Math. Soc.* **47** 455–486.

[Dow 78] H.R. Dowson (1978). *Spectral Theory of Linear Operators* (Academic Press).

[Dud 02] R.M. Dudley (2002). *Real Analysis and Probability* (Cambridge University Press).

[DuS 88] N. Dunford and J.T. Schwartz (1988). *Linear Operators Part I: General Theory* (Wiley Classics Library).

[Duo 01] J. Duoandikoetxea (2001). *Fourier Analysis* (Amer. Math. Soc. Graduate Studies in Mathematics 29).

[Dur 70] P.L. Duren (1970). *Theory of H^p Spaces* (Academic Press).

[Enf 73] P. Enflo (1973). On Banach spaces which can be given an equivalent uniformly convex norm, *Israel J. Math.* **13** 281–288.

[Fel 70] W. Feller (1970). *An Introduction to Probability Theory and its Applications, Volume I* (Wiley International Edition).

[Fou 74] J.J.F. Fournier (1974). An interpolation problem for coefficients of H_∞ functions, *Proc. Amer. Math. Soc.* **48** 402–408.

[Gar 70] D.J.H. Garling (1970). Absolutely p-summing operators in Hilbert space, *Studia Math.* **38** 319–331.

[GaG 71] D.J.H. Garling and Y. Gordon (1971). Relations between some constants associated with finite dimensional Banach spaces, *Israel J. Math.* **9** 346–361.

[GiM 91] J.E. Gilbert and M.A.M. Murray (1991). *Clifford Algebras and Dirac Operators in Harmonic Analysis* (Cambridge University Press).

[Gro 75] L. Gross (1975). Logarithmic Sobolev inequalities, *Amer. J. Math.* **97** 1061–1083.

[Gro 93] L. Gross (1993). Logarithmic Sobolev inequalities and contractivity properties of semigroups, *Dirichlet Forms (Varenna, 1992)* 54–88 (Springer Lecture Notes in Mathematics, Volume 1563).

[Grot 53] A. Grothendieck (1953). Résumé de la théorie métrique des produits tensoriels topologiques, *Bol. Soc. Mat. São Paulo* **8** 1–79.

[Had 93] J. Hadamard (1893). Résolution d'une question relative aux déterminantes, *Bull. des sciences Math.(2)* **17** 240–248.

[Hah 27] H. Hahn (1927). Über lineare Gleichungen in linearen Räume, *J. Für Die Reine und Angewandte Math.* **157** 214–229.

[Hal 50] P.R. Halmos (1950). *Measure Theory* (Van Nostrand Reinhold).

[Har 20] G.H. Hardy (1920). Note on a theorem of Hilbert, *Math. Zeitschr.* **6** 314–317.

[HaL 30] G.H. Hardy and J.E. Littlewood (1930). A maximal theorem with function-theoretic applications, *Acta Math.* **54** 81–116.

[HaLP 52] G.H. Hardy, J.E. Littlewood and G. Pólya (1952). *Inequalities*, 2nd edn (Cambridge University Press).

[Hed 72] L. Hedberg (1972). On certain convolution inequalities, *Proc. Amer. Math. Soc.* **36** 505–510.

[HiS 74] M.W. Hirsch and S. Smale (1974). *Differential Equations, Dynamical Systems, and Linear Algebra* (Academic Press).

[Höl 89] O. Hölder (1889) Über ein Mittelwertsatz, *Nachr. Akad. Wiss. Göttingen Math. - Phys. Kl.* 38–47.

[Hor 50] A. Horn (1950). On the singular values of a product of completely continuous operators, *Proc. Nat. Acad. Sci. USA* **36** 374–375.

[Hör 90] L. Hörmander (1990). *The Analysis of Linear Partial Differential Operators I* (Springer-Verlag).

[Hun 64] R.A. Hunt (1964). An extension of the Marcinkiewicz interpolation theorem to Lorentz spaces, *Bull. Amer. Math. Soc.* **70** 803–807.

[Hun 66] R.A. Hunt (1966). On $L(p,q)$ spaces spaces, *L'Enseignement Math. (2)* **12** 249–275.

[Jan 97] S. Janson (1997). *Gaussian Hilbert Spaces* (Cambridge University Press).

[Jen 06] J.L.W.V. Jensen (1906). Sur les fonctions convexes et les inégalités entre les valeurs moyennes, *Acta Math.* **30** 175–193.

[Joh 48] F. John (1948). Extremum problems with inequalities as subsidiary conditions, *Courant Anniversary Volume* 187–204 (Interscience).

[JoL 01,03] W.B. Johnson and J. Lindenstrauss (eds) (2001, 2003). *Handbook of the Geometry of Banach Spaces, Volumes 1 and 2* (Elsevier).

[Kah 85] J.-P. Kahane (1985). *Some Random Series of Functions*, 2nd edn (Cambridge University Press).

[Khi 23] A. Khintchine (1923). Über dyadische Brüche, *Math. Z.* **18** 109–116.

[Kol 25] A.N. Kolmogoroff (1925). Sur les fonctions harmoniques conjuguées et les séries de Fourier, *Fundamenta Math.* **7** 24–29.

[Kön 86] H. König (1986). *Eigenvalue Distribution of Compact Operators* (Birkhäuser).

[Kwa 72] S. Kwapień (1972). Isomorphic characterizations of inner product spaces by orthogonal series with vector valued coefficients, *Studia Math.* **44** 583–595.

[Lac 63] H.E. Lacey (1963). *Generalizations of Compact Operators in Locally Convex Topological Linear Spaces* (Thesis, New Mexico State University).

[La O 94] R. Latała and K. Oleszkiewicz (1994). On the best constant in the Khinchin–Kahane inequality, *Studia Math.* **109** 101–104.

[Lid 59] V.B. Lidskii (1959). Non-self-adjoint operators with a trace (Russian), *Doklady Acad. Nauk SSSR* **125** 485–487.

[LiP 68] J. Lindenstrauss and A. Pełczyński (1968). Absolutely summing operators in $\mathcal{L}_p$ spaces and their applications, *Studia Math.* **29** 275–321.

[LiT 79] J. Lindenstrauss and L. Tzafriri (1979). *Classical Banach Spaces II* (Springer-Verlag).

[Lio 61] J.L. Lions (1961). Sur les espaces d'interpolation: dualité, *Math. Scand.* **9** 147–177.

[Lit 86] J.E. Littlewood (1986). *Littlewood's Miscellany*, edited by Béla Bollobás (Cambridge University Press).

[Lor 50] G.G. Lorentz (1950). Some new functional spaces, *Ann. Math.* **51** 37–55.

[Lux 55] W.A.J. Luxemburg (1955). *Banach Function Spaces*, PhD thesis, Delft Institute of Technology.

[LuZ 63] W.A.J. Luxemburg and A.C. Zaanen (1963). Notes on Banach function spaces I-V, *Indag. Math.* **18** 135–147, 148–153, 239–250, 251–263, 496–504.

[Mar 39] J. Marcinkiewicz (1939). Sur l'interpolation d'opérations, *C. R. Acad. Sci. Paris* **208** 1272–1273.

[Mer 09] J. Mercer (1909). Functions of positive and negative type, and their connection with the theory of integral equations, *Phil. Trans. A* **209** 415–446.

[Min 96] H. Minkowski (1896). *Diophantische Approximationen* (Leipzig).

[Mui 03] R.F. Muirhead (1903). Some methods applicable to identities and inequalities of symmetric algebraic functions of n letters, *Proc. Edinburgh Math. Soc.* **21** 144–157.

[Nel 73] E. Nelson (1973). The free Markov field, *J. Funct. Anal.* **12** 211–227.

[Nev 76] J. Neveu (1976). Sur l'éspérance conditionelle par rapport à un mouvement brownien, *Ann. Inst. Poincaré Sect. B (N.S.)* **12** 105–109.

[Orl 32] W. Orlicz (1932). Über eine gewisse Klasse von Räumen vom Typus B, *Bull. Int. Acad. Polon. Sci. Lett. Cl. Math. Nat.* **A** 207–222.

[Pal 33] R.E.A.C. Paley (1933). On orthogonal matrices, *J. Math. Phys.* **12** 311–320.

[Pee 69] J. Peetre (1969). Sur la transformation de Fourier des fonctions à valeurs vectorielles, *Rend. Sem. Mat. Univ. Padova* **42** 15–26.

[Pel 67] A. Pełczyński (1967). A characterization of Hilbert–Schmidt operators, *Studia Math.* **28** 355–360.

[Pel 77] A. Pełczyński (1977). *Banach Spaces of Analytic Functions and Absolutely Summing Operators* (Amer. Math. Soc. Regional conference series in mathematics, 30).

[Phe 66] R.R. Phelps (1966). *Lectures on Choquet's Theorem* (Van Nostrand).

[Pie 63] A. Pietsch (1963). Zur Fredholmschen Theorie in lokalkonvexe Raüme, *Studia Math.* **22** 161–179.

[Pie 67] A. Pietsch (1967). Absolut p-summierende Abbildungen in Banachräume, *Studia Math.* **28** 333–353.

[Pie 87] A. Pietsch (1987). *Eigenvalues and s-Numbers* (Cambridge University Press).

[Pis 75] G. Pisier (1975). Martingales with values in uniformly convex spaces, *Israel J. Math.* **20** 326–350.

[Pis 87] G. Pisier (1987). *Factorization of Linear Operators and Geometry of Banach Spaces* (Amer. Math. Soc. Regional conference series in mathematics, 60, second printing).

[Pis 89] G. Pisier (1989). *The Volume of Convex Bodies and Banach Space Geometry* (Cambridge University Press).

[Pól 26] G. Pólya (1926). Proof of an inequality, *Proc. London Math. Soc.* **24** 57.

[Pól 50] G. Pólya (1950). Remark on Weyl's note: inequalities between the two kinds of eigenvalues of a linear transformation, *Proc. Nat. Acad. Sci. USA* **36** 49–51.

[Ri(F) 10] F. Riesz (1910). Untersuchungen über Systeme intergrierbarer Funktionen, *Math. Ann.* **69** 449–447.

[Ri(F) 32] F. Riesz (1932). Sur un théorème de MM. Hardy et Littlewood, *J. London M.S.* **7** 10–13.

[Ri(M) 26] M. Riesz (1926). Sur les maxima des formes linéaires et sur les fonctionelles linéaires, *Acta Math.* **49** 465–497.

[Rud 79] W. Rudin (1979). *Real and Complex Analysis* (McGraw-Hill).

[Rud 79] W. Rudin (1990). *Fourier Analysis on Groups* (Wiley Classics Library).

[Ryf 65] J.V. Ryff (1965). Orbits of L^1-functions under doubly stochastic transformations, *Trans. Amer. Math. Soc.* **117** 92–100.

[Scha 50] R. Schatten (1950). *A Theory of Cross-Spaces Ann. Math. Stud.*, **26**.

[Sch 23] I. Schur (1923). Über eine Klasse von Mittelbildungen mit Anwendungen auf die Determinanten theorie, *Sitzungber. d. Berl. Math. Gesellsch.* **22** 9–20.

[Schw 85] H.A. Schwarz (1885). Über ein die Flächen kleinste Flächeneinhalts betreffende problem der variationsrechnung, *Acta Sci. Scient. Fenn.* **15** 315–362.

[Smi 62] F. Smithies (1962). *Integral Equations* (Cambridge University Press).

[Ste 04] J.M. Steele (2004). *The Cauchy–Schwarz Master Class* (Cambridge University Press).

[Stei 70] E.M. Stein(1970). *Singular Integrals and Differentiability of Functions* (Princeton University Press).

[Stei 93] E.M. Stein (1993). *Harmonic Analysis: Real-Variable Methods, Orthogonality and Oscillatory Integrals* (Princeton University Press).

[StW 71] E.M. Stein and G. Weiss (1971). *Introduction to Fourier Analysis on Euclidean Spaces* (Princeton University Press).

[TaL 80] A.E. Taylor and D.C. Lay (1980). *Introduction to Functional Analysis* (Wiley).

[Tho 39] G.O. Thorin (1939). An extension of a convexity theorem due to M. Riesz, *Kung. Fys. Saell. i Lund For.* **8** no. 14.

[Tom 89] N. Tomczak-Jaegermann (1989). *Banach-Mazur Distances and Finite-Dimensional Operator Ideals* (Pitman).

[vLW 92] J.H. van Lint and R.M. Wilson (1992). *A Course in Combinatorics* (Cambridge University Press).

[Vil 39] J. Ville (1939). *Étude Critique de la Notion de Collectif* (Gauthier-Villars).

[Wey 49] H. Weyl (1949). Inequalities between the two kinds of eigenvalues of a linear transformation, *Proc. Nat. Acad. Sci. USA* **35** 408–411.

[Wil 91] D. Williams (1991). *Probability with Martingales* (Cambridge University Press).

[Zyg 56] A. Zygmund (1956).On a theorem of Marcinkiewicz concerning interpolation of operations, *J. Math. Pure Appl.* **35** 223–248.

Index of inequalities

Index

Printed in the United States
By Bookmasters